History of Fisheries and Fishing in South Dakota

**Compiled and Edited
by**

Charles R. Berry, Jr.

Kenneth F. Higgins

David W. Willis

Steven R. Chipps

2007

History of Fisheries and Fishing in South Dakota

Published by South Dakota Department of Game, Fish and Parks
        523 E. Capitol Avenue
        Pierre, SD 57501-3182

Suggested citation formats:

Entire book

Berry, Charles R., K. F. Higgins, D. W. Willis, and S. R. Chipps, editors. 2007. History of fisheries and fishing in South Dakota. South Dakota Department of Game, Fish and Parks, Pierre.

Chapter within the book

Berry, C. 2007. Introduction: Fish, habitat, and anglers. Pages 1–18 *in* C. Berry, K. Higgins, D. Willis, and S. Chipps, editors. History of fisheries and fishing in South Dakota. South Dakota Department of Game, Fish and Parks, Pierre.

Cover photos:

Front — Courtesy of Dr. L. Flake (fly fishing), Charles Schmulz family (historical anglers), Tom Holmlund (ice fishing) and C. Berry (boat on Lake Oahe).

Back — Fish pictures courtesy of U.S. Fish and Wildlife Service; walleye pictures courtesy of South Dakota Office of Tourism.

Book Designer: Tom Holmlund

Printer:

    South Dakota State University
    Printing Laboratory
    Printed on acid-free paper

Library of Congress Control Number 2007935248

ISBN 978-0-9712463-4-8

Department of Game, Fish & Parks
523 East Capitol Avenue, Pierre, SD 57501-3182

Table of Contents

"Enjoy thy stream, O harmless fish;
And when an angler for his dish,
Through gluttony's vile sin,
Attempts, the wretch, to pull thee out,
God give thee strength, O gentle trout,
To pull the rascal in!"

— John Wolcott

Preface

People have harvested fish in what is now South Dakota for several thousand years. Archaeological evidence at campsite digs confirms that Native American Indians living along the Missouri River and its major tributaries were gathering fish to eat and for other uses such as ornamentation and tools. The over-exploitation of fishery resources began with the arrival of the market hunter, the soldier, and the settler. Game fish were taken out by the wagonload until the lakes and streams were depleted. With statehood, South Dakota had the legislative machinery to begin controlling fishing and later, to apply scientific management to fish harvest.

Today, citizens and nonresidents have wonderful opportunities to harvest fish for personal use, recreation, and for sale. The fisheries that developed in the reservoirs on the Missouri River have been one of the great tourist attractions in the state. The sport fishery in South Dakota, however, did not happen on its own—it is the product of research and management efforts primarily directed or funded by the South Dakota Department of Game, Fish and Parks with help from other state, federal and private entities, and from several institutions of higher learning. Our aim was to bring together for the first time the history of each group's conservation efforts.

This book is for the layperson and angler who is interested in South Dakota history and fishing. Most (90%) South Dakota residents believe that healthy fish populations are important to the economy and well being of South Dakota residents (Gigliotti 1999). Professional biologists and students will also find this book useful because fishing activities and fishery management are placed in a historical context, which makes this book unique for the states in the Northern Great Plains. The information is backed up by many years of experience for the 27 authors who were invited to participate—they all volunteered. The information is also backed up by a literature review of some 1,400 articles on South Dakota fishes (Bouchard et al. 2006).

The time was right for this book because of recent surveys that have updated the status of fishes in South Dakota (Chapters 3, 10, 11, 12, and

13), and because some of the authors are nearing the end of their careers. The authors are a mix of university professors, graduate students, and state and federal fisheries biologists, and one Tony Dean—who better to write about modern fishing methods?

We asked each author to include information about three subjects—history, current events, and predictions for the future, but otherwise the chapters are not standardized. We retained the individual writing styles of the authors for variety. Some authors included references in the text whereas others provided a list of readings that were used as references. All references and readings are included at the end of the book. The common names of fishes are used throughout, but a list of common and scientific names can be found in Chapter 3, Table 3.1.

The book addresses three broad themes—the fishes, the aquatic habitats, and the human dimensions of fishing in South Dakota. Chapters 1 through 3 provide basic information on fish and how they populated South Dakota's waters after the glaciers melted. Chapters 4 through 9 cover the types of aquatic habitats found in South Dakota. Chapters 10 through 13 review the kinds of fishes found in these habitats. Chapters 14 through 20 cover fisheries management and the types of anglers that fish South Dakota's waters. Chapters 21 and 22 cover fisheries research and education. Finally, Chapter 23 summarizes the authors' predictions about the future of fishing in South Dakota.

The reader will find quotations at the beginning of each chapter and scattered liberally through the text. We like quotes because they are thought provoking, memorable, and sometimes funny. The sport of fishing has been the subject of many philosophical quotations—some not so complimentary to anglers! For example, the comic asks, "If fish is brain food, why doesn't it show in fishermen?"

Editors: Charles R. Berry, Jr.
Kenneth F. Higgins
David W. Willis
Steven R. Chipps

Acknowledgements

We express sincere thanks to Terri Symens, Secretary, Department of Wildlife and Fisheries Sciences at South Dakota State University (SDSU), for word processing and administrative assistance. Several people volunteered to review and edit parts of the entire book, especially Dr. Charles G. Scalet, Professor and Head of the Department of Wildlife and Fisheries Sciences at SDSU. His comprehensive review, editing, and suggestions were appreciated. Mary Lou Berry, Jessie Y. Sundstrom, Michelle Bouchard, and Dawn M. Gardner helped locate photographs and reference materials. Diane Drake helped with contracting details and Carol Jacobson assisted with many clerical needs during the several years this book was being written, printed, and distributed. Mary Brashier and Eileen Dowd Stukel provided guidance on production, graphic design and printing. The graphic designer was Tom Holmlund. Most fish pictures were from Duane Raver art provided by the U.S. Fish and Wildlife Service.

Robert L. Hanten and Robert P. Hanten were the joint authors of Chapter 15 titled *Commercial Fisheries and Baitfish Industry*. Robert L. Hanten served the people and fisheries resources of South Dakota for more than 35 years, retiring as Chief of Fisheries in 1997. He guided the growth of the fisheries program and mentored most of the current fisheries employees, including his son Robert P. Hanten, and many other authors that contributed to this book. Bob the elder is a member of the Fisheries Management Hall of Excellence of the American Fisheries Society, located at the Ak-Sar-Ben Aquarium, Gretna, Nebraska.

Funding for this project was provided by the South Dakota Department of Game, Fish and Parks and the South Dakota Cooperative Fish and Wildlife Research Unit (U. S. Geological Survey). Doug Hansen and Dennis Unkenholz helped with funding. The Cooperative Fish and Wildlife Research Unit is jointly sponsored by the U. S. Geological Survey, the Wildlife Management Institute, the U. S. Fish and Wildlife Service, South Dakota State University, and the South Dakota Department of Game, Fish and Parks.

Contributing Authors

Douglas C. Backlund
S.D. Game, Fish and Parks
523 E. Capitol, Foss Building
Pierre, SD 57501-3182
605-773-3381
Doug.Backlund@state.sd.us

Michael E. Barnes
S.D. Game, Fish and Parks
McNenny Hatchery
19619 Trout Loop
Spearfish, SD 57783-8905
605-642-6920
Mike.Barnes@state.sd.us

Charles R. Berry, Jr., Ph.D.
U.S.G.S. — S.D. Coop Unit
Box 2140B, SDSU
Brookings, SD 57007-1696
605-688-6121
Charles.Berry@sdstate.edu

Gale A. Bishop, Ph.D. (Retired)
S.D. School of Mines and Technology
501 E. St. Joseph Street
Rapid City, SD 57701-3995
605-394-2467
Gale.Bishop@sdsmt.edu

Brian G. Blackwell, Ph.D.
S.D. Game, Fish and Parks
603 E. 8th Ave.
Webster, SD 57274-1630
605-345-3381
Brian.Blackwell@state.sd.us

Michelle A. Bouchard
Wildlife and Fisheries Sciences
South Dakota State University
Box 2140B
Brookings, SD 57007-1696
605-688-6121
Michelle.Bouchard@sdstate.edu

Michael L. Brown, Ph.D.
Wildlife and Fisheries Sciences
South Dakota State University
Box 2140B
Brookings, SD 57007
605-688-6121
Michael.Brown@sdstate.edu

Robert L. Brown
S.D. Game, Fish and Parks
523 E. Capitol, Foss Building
Pierre, SD 57501-3182
605-773-3381
Bob.Brown@state.sd.us

Steven R. Chipps, Ph.D.
U.S.G.S. — S.D. Coop Unit
Box 2140B, SDSU
Brookings, SD 57007-1696
605-688-6121
Steven.Chipps@sdstate.edu

Rick J. Cordes
S.D. Game, Fish and Parks
McNenny Hatchery
19619 Trout Loop
Spearfish, SD 57783-8905
605-642-6920
Rick.Cordes@state.sd.us

Alan Davis, Ph.D.
Education and Counseling
South Dakota State University
Box 507
Brookings, SD 57007

Tony Dean
Tony Dean Outdoors
Pierre, SD 57501
tony@tonydean.com

Robert P. Hanten
S.D. Game, Fish and Parks
Missouri River Fisheries Center
20641 SD Hwy. 1806
Ft. Pierre, SD 57532-6100
605-223-7700
Robert.Hanten@state.sd.us

Robert L. Hanten (Retired)
S.D. Game, Fish and Parks
29532 SD Hwy. 34
Pierre, SD 57501
605-224-7136

Kenneth F. Higgins, Ph.D. (Retired)
U.S.G.S. — S.D. Coop Unit
Box 2140B, SDSU
Brookings, SD 57007-1696
605-688-6121
Terri.Symens@sdstate.edu

Christopher W. Hoagstrom, Ph.D.
Department of Zoology
Weber State University
2505 University Circle
Ogden, Utah 84408-2505
ChristopherHoagstrom@weber.edu

Kent C. Jensen, Ph.D.
Wildlife and Fisheries Sciences
South Dakota State University
Box 2140B
Brookings, SD 57007-1696
605-688-6121
Kent.Jensen@sdstate.edu

David O. Lucchesi
S.D. Game, Fish and Parks
4500 S. Oxbow Ave.
Sioux Falls, SD 57106-4114
605-362-2700
Dave.Lucchesi@state.sd.us

Kent Keenlyne, Ph.D. (Retired)
U.S. Fish and Wildlife Service
9625 Tyler Ct.
Mobile, AL 36695-7408
251-623-4558
sturgeondoctor@yahoo.com

Ronald M. Koth
S.D. Game, Fish and Parks
3305 W. South St.
Rapid City, SD 57702-8160
605-394-2391
Ron.Koth@state.sd.us

David C. Parris, Ph.D.
New Jersey State Museum
P.O. Box 530
Trenton, NJ 08625
609-292-6330
David.Parris@sos.state.nj.us

Charles G. Scalet, Ph.D.
Wildlife and Fisheries Sciences
South Dakota State University
Box 2140B
Brookings, SD 57007-1696
605-688-6121
Charles.Scalet@sdstate.edu

Jeffrey S. Shearer
S.D. Game, Fish and Parks
523 E. Capitol, Foss Building
Pierre, SD 57501-3182
605-773-3381
Jeff.Shearer@state.sd.us

Sampson M. Stukel
S.D. Game, Fish and Parks
31247 436th Ave.
Yankton, SD 57078
605-668-5464
Sam.Stukel@state.sd.us

Dennis G. Unkenholz
S.D. Game, Fish and Parks
523 E. Capitol, Foss Building
Pierre, SD 57501-3182
605-773-3381
Dennis.Unkenholz@state.sd.us

David W. Willis, Ph.D.
South Dakota State University
Wildlife and Fisheries Sciences
Box 2140B
Brookings, SD 57007-1696
605-688-6121
David.Willis@sdstate.edu

Stephen K. Wilson
National Park Service
Missouri National Recreational River
P.O. Box 666
Yankton, SD 57078
402-667-5524
Stephen_K_Wilson@nps.gov

Chapter 1

Introduction: Fish, Habitat, and Anglers

Charles R. Berry, Jr.

> "There ain't but one time to go
> fishin', and that's whenever you can."
>
> — Diron Talbert (former NFL Player)

The history of fishing in South Dakota is rich and varied. South Dakota has escaped many of the threats to water and fishing that have affected other states. Today, river, lake, and stream fish populations have never been healthier or more abundant. Some say the "good ole days" of fishing are now. Fishing opportunities have been created or restored, and now the challenge is to conserve fish and fish habitat for future generations. How can we manage recreational fishing while promoting tourism that could cause crowded waterways and overfishing? How do we conserve South Dakota's native fishes while promoting watershed developments that could reduce water quality and quantity?

"In the end, we conserve only what we love. We will love only what we understand. We will understand only what we are taught." (B. Dioum, Senegalese poet) This quote prompted the writing of this book. In spite of the importance of fishing as outdoor recreation in South Dakota, there is no comprehensive source describing the history of South Dakota fishing or programs that conserve and enhance our fishery resources.

Fish are invisible to most people, and waterways are often ignored as we hurry across a bridge to our next appointment. This book is intended to help the general public understand that healthy fish populations mean healthy aquatic resources on which we all depend. By seeing South Dakota in terms of its fishes and fishing opportunities, perhaps readers will gain a greater appreciation of the wealth of our <u>water estate</u>. The wealth of our water estate reflects the health of our society (Palmer 1996).

This book covers both historic and contemporary topics about fish, fishing, and fisheries management. Its scope is from the wetlands to the rivers, from the trout streams of the Black Hills to the trophy catfisheries of the James (Jim) and Big Sioux rivers, from the great Missouri River reservoirs to the glacial lakes of the northeastern region. It is written for the general public, but backed up by a comprehensive literature review. Some 1,400 references have been found on fishes of South Dakota (Bouchard et al. 2006).

Many agencies and advocacy groups have participated in conserving and developing South Dakota's fisheries. Scientists, teachers, Conservation Officers, aquaculturists, fishery managers, conservation organizations, and service clubs have been involved with the three main aspects of any fishery—the fish, the aquatic habitat, and the angler. This book is roughly divided into sections that address these three themes.

The Fish: A Primitive Creature?

A fish is defined as a cold-blooded animal with a backbone and fins that uses gills to obtain oxygen from water throughout its life (Figure 1.1). Fish have most of the organ systems and physiological processes of mammals; some organ systems are simplified whereas others are greatly enhanced with fascinating specializations. Most walleye anglers know about the unique vision advantage that walleyes have over their prey. The crystalline mirror behind the eye enables a walleye to feed in the dim light of morning or evening or in the "mud line" on a windy shore during the "walleye chop."

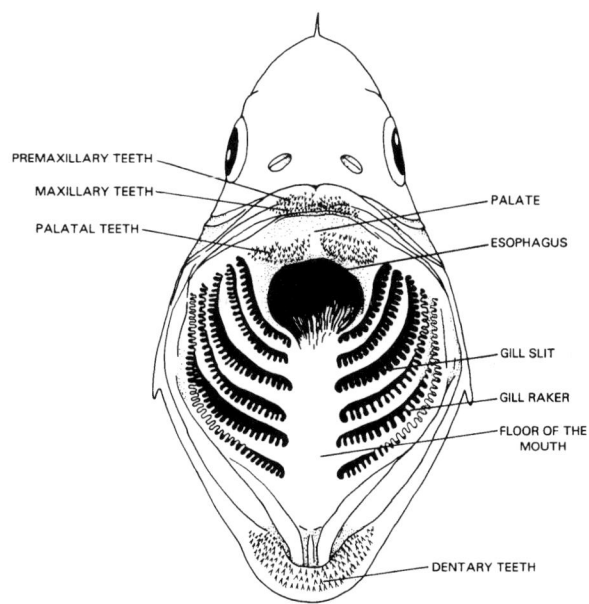

Figure 1.1   Oral cavity and pharynx of the perch; knowing the anatomy helps unhook the fish with a minimum of stress (From Chiasson and Radke 1991).

Information in the following chapters on fish management, fish hatcheries, water pollution, and fish research is based on the needs of the fish as an organism. There is an organizational hierarchy of biology from the cellular level to the organismal level to the population level and beyond. The ***explanation*** for change at any level can be found in the levels below and the ***consequences*** can be found in the levels above. In fishery terms, this means a change in a fish population can usually be explained by changes in the structure and function (e.g., respiration, reproduction, growth) of individual fish, and the consequence of the population change can be found in the responses of the aquatic community (e.g., diversity, productivity).

Fish obtain oxygen at the gills where water and blood are separated by only one cell layer. Four gill bars on each side of the head support the delicate respiratory gill filaments and supply blood to the gills (Figure 1.1). The gills have great surface area for exchanging oxygen for carbon dioxide, and balancing salt and water concentrations in the blood. The respiratory surface area of the gills is 3-7 times that of the fish's skin surface area. The muscles and bones in the fish's head are complex because of the actions needed to pump the thick, heavy water (compared to air). The fish's respiratory processes use about 20% of the fish's daily energy expenditures. Anglers who appreciate the importance of the gills are especially careful when unhooking a hooked fish, and can improve survival of released fish.

Fishing tackle companies design lures that appeal to fish senses and stimulate feeding responses. It has been said that lures are designed to "catch more anglers than fish." Fish do have much greater sensory abilities, however, than humans because they are bombarded with a much greater variety of sensory signals than we experience. Imagine living in a weightless state in a thick, cold fog, and having the ability to see around corners with polarized and infrared vision. Imagine smelling the odors of other animals and tasting the waters that pass you in the veil. Imagine hearing every animal near you and even hearing echoes of your own noises reflecting back to you. Imagine sensing the earth's magnetism and electrical

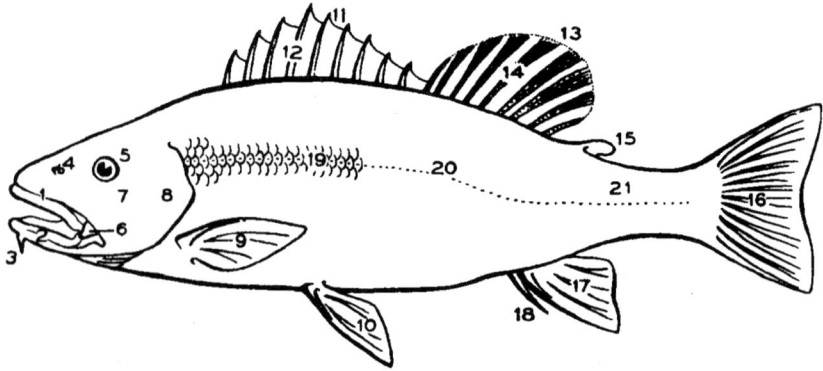

Figure 1.2 Generalized fish showing characters used in identifying fish. 1 = upper jaw (premaxillary and maxillary bone), 2 = lower jaw (dentary bone), 3 = barbel, 4 = nostril, 5 = eye, 6 = maxillary barbel, 7 = cheek, 8 = gill cover or operculum, 9 = pectoral fin, 10 = pelvic fin, 11 = spiny dorsal fin, 12 = fin spine, 13 = soft dorsal fin, 14 = fin ray, 15 = adipose fin, 16 = caudal fin, 17 = anal fin, 18 = anal spine, 19 = lateral line scales, 20 = lateral line, 21 = caudal peduncle.

fields. This fantasy gives a remote concept of a fish in the water in terms of human senses (Royce 1972). Fish interpret and integrate light, water pressure and flow, chemical odors and tastes, gravity, and even electromagnetic fields to give them a sense of their environment.

Fish also take the sense of touch to extremes. They have free nerve endings in the skin to sense touch and temperature, but they also have a sense of *distant touch*. Distant touch refers to the lateral line (Figure 1.2) that can sense pressure waves moving through the water. The lateral line is used to "feel" the presence of other fish by sensing the pressure waves sent out as they swim. The lateral line sense is used by blind fish to maintain their place in the school, and by predators and prey to locate each other. Catfish go to extremes with their sense of taste. They have so many taste buds in the mouth, on the barbells, and at certain locations on the skin that they might be called "swimming tongues."

Fish can talk! Not as we do, of course, but they make noises that resemble clicks, knocks, snaps, croaks, pops, moans, booms, grunts, purrs, and whistles by vibrating muscles, rubbing bones, and expelling air from the bladder (Reebs 2001). Common carp can even discriminate between blues music and classical music (Chase 2001). Will musical crank baits be next?

Fish do not have ear openings, but do have inner ear bones called otoliths (Figure 1.3). Sound travels through the fish and makes the otolith and nerve endings vibrate in a way that sends signals to the brain. Sound (and light) can be used to guide fish movements, and may be useful for scaring them away from dangerous places, like the intakes of Oahe Dam, as current research is investigating.

Fish biologists use otoliths to determine the

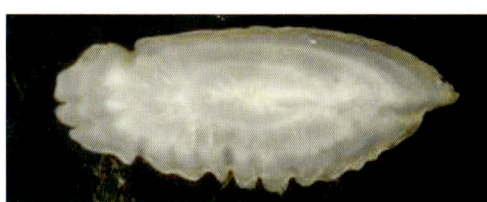

Figure 1.3 This fish ear stone or otolith from a yellow perch is only 1/4 inch long but annual growth rings are visible when the otolith is prepared and viewed with a microscope (photo by Q. Phelps).

growth and age of fish (Soupir et al. 1997). Also, biologists can mark the otoliths of young fish by adding chemicals to the diet or water so that the fate of stocked fish can be determined. Walleye and yellow perch fingerlings have been marked in this way for study in South Dakota waters (Unkenholz et al. 1997).

When a fishing lure has been presented properly, its action or color or size or smell will trigger a fish to bite—what species is it? Most anglers identify fish by the body shape, skin coloration, location of the fins, and other external anatomy. Most game fish are pictured in the annual fishing handbook (South Dakota Game, Fish and Parks [SDGFP] 2006), and the wonderful paintings of Joseph Tomelleri (Tomelleri and Eberle 1990; *http://www.sdgfp.info/Publications/Index.htm*, accessed 9/21/06) are on the GFP web page. Pictures of the most common fishes can also be found in Neumann and Willis (1994). These pictures are usually sufficient for the angler to identify a fish; however, fisheries scientists use taxonomic keys to accurately determine fish identity. A taxonomic key uses more stable features of the fish than color patterns. The features are called meristic characteristics (e.g., counts of fin rays, scales, etc.) and morphometric characteristics (e.g., body shape measurements).

Anglers should be aware of some of the meristic differences among South Dakota fishes because fish colors are not always reliable. For example, black crappies have 7-8 fin spines supporting the spiny dorsal fin (Figure 1.4) whereas white crappies have 5-6 fin spines. A fish is a northern pike if the cheeks and gill cover (Figure 1.2) are fully scaled; a muskellunge if only the upper half of the cheeks are scaled. Many fishing reports list catches of blue catfish, but blue catfish are actually fairly rare in South Dakota. The reports are probably about catches of the more common channel catfish. Distinguish between the two species by counting rays in the anal fin (Figure 1.4). Blue catfish have a long anal fin with a straight edge and more than 30 rays, whereas channel catfish have a shorter anal fin with a rounded edge and less than 30 rays.

The angler assumes a great responsibility after catching a fish. Knowing a little about the fish's anatomy and physiology will help an angler to reduce fish stress and ethically handle the fish before cooking or releasing. Many fish are recycled by catch-and-release fishing, so anglers can improve survival of released fish by following certain recommendations (Box 1.1).

Hooking location strongly affects post-release survival (Persons and Hirsch 1994, DuBois et al. 1994). Fish hooked in the lips or mouth area have much greater survival than those hooked in the gills or gullet (esophagus, Figure 1.1).

It's important to play and land a fish as fast as possible, and release as quickly as possible, particularly in summer.

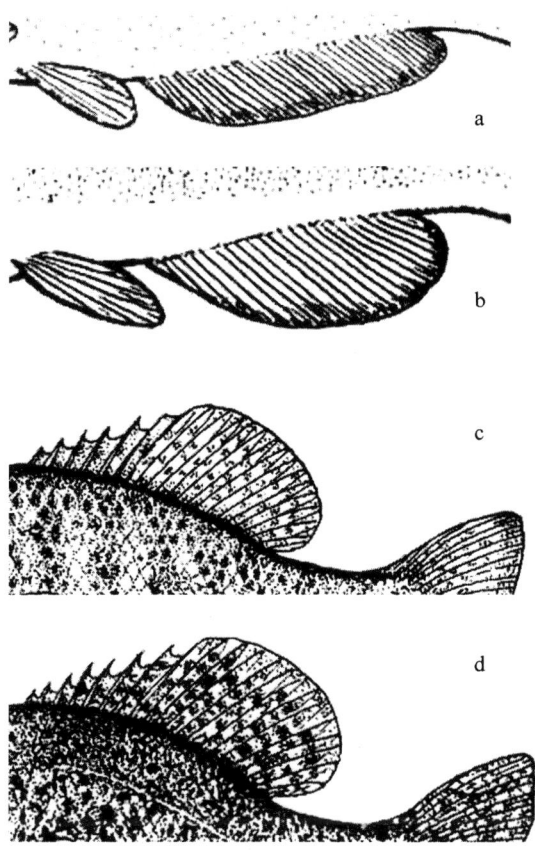

Figure 1.4 Meristic differences in common South Dakota fishes. Blue catfish anal fin with straight edge (a) and channel catfish anal fin with rounded edge (b); white crappie dorsal fin with six spines (c) and black crappie dorsal fin with seven spines (d). (Adapted from Eddy 1969).

Studies at Lake Francis Case fishing tournaments showed that most released walleyes survive when the reservoir is 45°F, but most die when it is 67°F (Graeb et al. 2005). The key to success is to keep the fish wet and release it quickly.

Bringing an exhausted fish out of the water is like placing a plastic bag over the head of a marathon runner. Fish get "winded" just like human runners. Both the fish and the

1.1
Guidelines for Catch-and-Release Fishing

- Land and unhook the fish as quickly as possible.
- Fish with barbless hooks or crimped barbs.
- Protect the mucous covering by placing the fish on a wet towel, using wet hands, or unhooking the fish in the water.
- Avoid hard polypropylene nets that remove mucus.
- Cut the line when releasing fish hooked in the gullet.
- Use long-nosed pliers.
- To remove a hook from the gullet, reach through the gill cover, turn the hook so that the shank sticks through the gills, reach into the mouth, pull the hook out of the mouth.
- Keep the fish if a gill arch is broken and bleeding.
- Revive exhausted fish by placing one hand under the tail and hold the bottom lip with the other to cause the jaw to gape and gently move the fish back and forth to cause water to flow over the gills.
- Return fish caught in deep water by: 1) "fizzing," which means sticking a needle through the body into the bladder, then release the fish, or 2) forcing the fish back down in an open-bottom "release basket" – when the fish revives, it swims out, or by using a "release sinker," which has a barbless hook and the sinker eye on the same end – the fish is hooked in the jaw going down and released when the sinker is lifted.

runner need oxygen and both have sore muscles from the build up of lactic acid. About 55% of any fish is muscle, and most muscle functions anaerobically resulting in lactic acid accumulation that reduces the ability of the blood to take up oxygen. Stress weakens the fish's immune system resulting in *increased* susceptibility to disease and likely mortality after being released (Killen et al. 2003).

Preserving the protective mucous layer and scales of the skin is also important. The skin is a compound organ that contains blood vessels, sensory cells, nerve fibers, several layers of tissue (hypodermis, dermis, epidermis), pigment cells, chemical communication cells, scales, and mucus. The skin makes up about 8% of any fish's weight. The thick dermis wrapping the body functions like an external tendon because it is linked to the muscles to enhance swimming. The dermis is made of thick tissue called collagen, which is so tough that the skin can be tanned to produce "fish leather." The mucus and scales prevent water and bacteria from entering the fish, so preventing damage to this coating is important when handling caught fish.

A fish caught at more than the 30-foot depth may have a distended air bladder and cannot easily dive after being released. Several methods have been suggested for reducing air volume in the bladder (Box 1.1). Fish become stressed when kept in boat live wells so keep water temperature within 5°F of the natural water temperature, and add a little salt. Yes, salt! When fish are stressed, their blood chemistry is altered and the salt-water balance is upset because fish are leaky at the gills and thin oral membranes. Fish are 90% water and maintain salt balance by absorbing salt, so supply non-iodized salt to make a 0.5% solution in the holding tank (2 pounds of salt in 50 gallons of water).

When a fish is caught, the angler unknowingly joins a debate about recreational fishing—do fish feel pain? We know that fish have a nervous system that transmits nervous impulses, and they have other features and behaviors that point to the conclusion that they probably do feel pain. Fish learn to avoid being caught (Young and Hayes 2004). The crux of the debate is whether the cerebral cortex is developed enough to have the capacity to perceive and be aware of sensory stimuli, rather than just react to stimuli. Anglers can find a simple discussion of the pain issue in Rose (1999-2000), whereas more technical details can be found in scientific literature (Rose 2002, Sneddon et al. 2003). Regardless of whether fish feel pain or not, it is the ethical responsibility of each angler to minimize stress and to have respect for all caught fish.

When the fish is a keeper, kill it with a sharp blow to the head and put it on ice to preserve the eating quality. Fish should be gilled, gutted, washed, and iced as soon as possible. Most muscle is white muscle but there is a red muscle mass below the lateral line (Figure 1.2) that is visible on each side of the fish after it is skinned. Some anglers refer to the red muscle as the "mud stripe." The term mud stripe is a good one. The red muscle contains oils that hold contaminants from pollution, and the oils hold a muddy, fishy flavored substance called geosmin, which is produced by algae and subsequently ingested by fish (Lovell and Sackey 1973). Filleting a fish to remove the red muscle requires practice, but it usually improves the flavor of the fillet.

The orange or reddish muscle of wild trout and salmon comes from the orange pigments in the diet. Fish culturists can artificially color fish by feeding a substance called canthanaxon. Historically, fish culturists fed fish paprika to get the red-colored meat in hatchery fish. Recently fish farmers have been forced to reveal any artificial coloring practice to customers.

Eating fish is good for your health because fish fat is really oil, and fish are a lean, low-calorie source of protein. It makes sense that a cold-blooded fish would store energy as oil rather than fat because oils require less heat and energy to be converted to an energy source for swimming. Fish oils are unsaturated (carbon chains incompletely linked) and have Omega-3 factors (e.g., EPA, eicosapentaenoic acid) that help the human cardio-vascular system. These same polyunsaturated fats, however, represent a problem when freezing fish.

The unsaturated fats are more prone to oxidation and becoming rancid than are saturated fats found in other meats. Most of the oxidation problem can be overcome by thorough wrapping with plastic wraps or freezer paper to exclude air, and by rapidly freezing small packages. Freezing in water adds a mushy texture to a fillet. When thawing fish, avoid oxidation and off flavors by thawing in cold, running water as fast as possible, not in a refrigerator, in hot water, or at room temperature.

Contaminants are only a minor problem in South Dakota fishes. Just as fish absorb off-flavors from water, they also take up impurities, especially through the food web. The ingestion of contaminants is of concern to humans because small, harmless quantities of contaminants are bioaccumulated. As big fish eat little fish, the toxins accumulate at higher concentrations at every level of the food web and humans are at the top of the food web.

At present, 45 states have fish consumption advisories, including South Dakota. South Dakota fish are tested for metals, pesticides, and other contaminants. Of the 73 waters tested, only four lakes have consumption advisories because of mercury contamination (SDGFP 2006). Consumption advisories are intended to guide people in selecting sizes and species of fish that are low in a particular contaminant; they are not restrictions or bans or limits on which fish to eat or not eat.

Aquatic Habitat: The Scale Of It

Habitat for fish includes all of the physical, chemical, and biological features of the environment needed to sustain fish life. Examples include water quality, migration routes, spawning grounds, feeding and resting sites, and shelter from predators and adverse environmental conditions. Habitat quality influences the numbers, sizes, and species of fish in a particular area. Habitat change can change a fish community for better or worse (McMahon et al. 1996).

South Dakota has a variety of aquatic habitats. About 12% of the fishing in South Dakota is done in the Black Hills, 30%

on the Missouri River system of reservoirs and tailraces, and 58% on all the other wetlands in the state (Giglotti 2005). The lakes, streams, rivers, reservoirs, and ponds are the subject of many of the following chapters. Anglers have access to hundreds of public fishing waters and can obtain contour maps of about 100 lakes (SDGFP 2006).

Contour maps show anglers important kinds of fish habitat—water depth, flats, shoreline shape, points, bays, and narrows. By knowing the habitat preferences of fish, anglers can concentrate their fishing efforts in the hot spots. For example, walleyes holding in deep river pools during the day move to shallower reefs, humps, and rock piles during late evening and early morning. In lakes, walleyes move from the deep lake regions to the shoreline during late evening and early morning. Channel catfish and flathead catfish hide among snags during the day and move upstream to riffle areas or low head dams to feed at night. These habitats are parts of a lake or river segment, but consider the larger habitat picture. Where is the lake or river segment in the watershed?

Aquatic habitat is a multi-scale issue. The term <u>scale</u>, which has recently made its way into the ecological literature (Thompson 2001), is easy to understand. The scale of agriculture has grown from small farms to large farms, from local markets to international ones. Water pollution problems were once thought to impact one creek or lake, but we now understand that a local problem anywhere in the Mississippi River Basin may be linked to the dead zone in the Gulf of Mexico. Wildlife biologists have found that grassland birds select a local habitat patch because of larger landscape types around the patch (Bakker et al. 2002). Advanced technology (e.g., satellite images) and advanced research approaches (e.g., geographical information systems) have enabled fish biologists to study aquatic habitat at several scales.

There is a hierarchical organization of physical fish habitat (Frissell et al. 1986). Beginning with "microhabitat" (e.g., one stone on the stream bottom), the next habitat in the hierarchy is the "pool" or "riffle," the next is a complex of pools and riffles called a "reach," the next is a stream "segment" between the upstream and downstream entrance of two tributaries, and the last is the stream "system" of all tributaries and the main stem. The system is located in a watershed, so changes in the watershed cascade down hill, down stream, and down wind to affect habitat at smaller and smaller scales. Fish need clean water with a natural hydrograph. A hydrograph is the annual flow pattern, which in South Dakota streams usually appears as peak flows in spring and low flows in winter. Water quality and quantity are covered in several of the following chapters.

The "new idea" in water quality management is to work at the watershed scale because habitat improvement at the microhabitat scale really depends on conditions at higher hierarchical levels (Williams et al. 1997). In simple terms, we have rediscovered the rule that "land health and water health are not two issues but one" (Leopold 1968).

Anglers: The Human Dimension

Although Mother Nature's influences are considerable, anglers are the key to the future of fishing. Angler behaviors can shape fish populations (yes, we can over fish a big lake), angler opinions can shape fish management programs (angler opinions are welcomed at SDGFP Commission meetings), and angler attitudes can influence the non-fishing public (most non-anglers support recreational fishing). We have become more aware of the importance of the human dimension in fisheries management—so much so that the SDGFP has a full-time human dimension specialist who conducts routine surveys to help state fishery biologists plan regulations and management. For example, about 7,300 anglers returned questionnaires for a response rate of 66% in 2003 (Gigliotti 2005). The high response rate shows the interest and concern of South Dakota anglers for their sport. Several chapters in this book are about anglers, and in the chapters about fish management, much of the discussion deals with human dimension effects.

Anglers in South Dakota are a mix of ethic backgrounds, of residents and nonresidents, men and women, young and old, trophy and family fishers, fly anglers and worm dunkers, and boat and shore anglers (Figure 1.5). What a mix of constituents! About 130,000 residents and 70,000 nonresidents usually fish each year in South Dakota. For our state's population of 760,000, we have a higher percentage of anglers than most states. About 9% of our anglers are youth, ages 12 to 18 years. This proportion has been stable over the last few years. Increasing recently is the number of anglers over 100 years old. More women are buying fishing licenses these days, thanks in part to educational programs such as Becoming an Outdoors Woman (BOW) and Becoming an Outdoors Family (BOF). The key feature of these programs is hands-on workshops. For example, the BOW program introduces women to 20 activities equally balanced between hunting and shooting, fishing, and non-harvest sports like canoeing and camping.

The number of anglers has been declining nationally and in South Dakota for the last few years. Fishing, however, touches most people's lives; only about 12% of citizens over 18 years of age have ***never*** fished at some time in their lives. This

Figure 1.5   Anglers at Lake Sinai (a), kids fishing pond (b), Blue Dog Lake (c), and Lake Oahe (d). (photos by C. Berry)

fact may explain why 85% of the people support fishing as a type of outdoor recreation. Anglers give up fishing, but usually for social reasons such as lack of time, lost interest, or loss of former fishing friends, not because of pollution, crowding, or lack of success.

It is no secret that fishing is important to the economy of South Dakota. Fishing creates about 4,500 jobs and it generates $90 million in salaries and wages, and $12 million in taxes. Fishing is a "user pays" sport in South Dakota. Anglers pay about $5 million for licenses. The figure is split about equally among residents and nonresidents, even though there are twice as many residents buying licenses. The state's fishing program is supported only by license fees and excise taxes that anglers pay on certain purchases (e.g., motor boat fuel, fishing tackle). The excise taxes are collected under the Federal Aid in Sport Fish Restoration Act and reapportioned to states. The act is commonly referred to as the Dingell-Johnson Act or D-J; initials of the Senators that created it in 1950. South Dakota uses its D-J funds for research, dam construction, operating fish hatcheries, and other management activities. The D-J reports since 1951 are available on the SDGFP home page or from the State Library.

Why do people fish? Non-anglers would be surprised to learn that the highest motivation is not to catch a lot of fish or a big fish. Only 3% of anglers fish for trophies and only 7% fish in tournaments (Kellert 1999). Angler opinion surveys usually show that people fish to get outdoors, to relax, and to be with family. These reasons are being given more often these days and fishing for sport less often. It is understandable then that the main reasons listed by dissatisfied anglers who still fish are usually pollution, litter, and interferences from other people (Giglotti 2005, Kellert 1999).

The future of fishing depends on the knowledge, attitudes, and ethics of anglers today. The increased use of technology has put tremendous pressure on fish populations. More anglers can concentrate more effective gear on a lake in a shorter period of time than ever before. Yes, we can quickly over fish a population! A change in thinking is needed to conserve fishing quality. Fish managers suggest following the principle of *selective harvest*. An angler practicing selective harvest doesn't always try to *limit out*. Rather, the angler practices catch-and-release fishing, or harvests just enough for a meal, or fishes for under-used species.

Fishing ethics becomes more and more important with the increased use of our waterways and increased pressure on fish populations. There are many ways to be ethical while fishing and they are all under the control of the angler. "A particular virtue of wildlife ethics is that the hunter ordinarily has no gallery to applaud or disapprove of his conduct. Whatever his acts, they are dictated by his own conscience, rather than by a mob of onlookers" (Leopold 1968). For anglers, those thoughts can be manifested in many types of fishing behavior. Don't high grade! High grading is keeping small fish alive to be released later if a larger fish is caught. Anglers generate a lot of refuse, from used bait containers to fish guts. Recycle fishing line and of course, do not litter. Avoid use of lead sinkers because of the lead toxicity problem.

Do you continue to harvest fish even with a possession limit in the freezer? This is not only illegal; it is unethical and definitely hurts the fish population. Take time to rinse boats, boots, and gear to remove exotic plants and animals. Have respect for the fish by killing or releasing it quickly, and care for the carcass to preserve table quality.

Respect other anglers and comply with fishing and boating regulations. Take a kid fishing. Studies show that kids with hunting, fishing and bird watching experiences have a much greater knowledge of the environment and a much deeper ecological understanding than kids who do not participate in outdoor activities (Kellert 1999).

Anglers are on the water more than any other group. It follows then that they should be effective proponents of environmental conservation and sound fisheries management. Anglers need to learn more about rivers, fishing, and fisheries management. They need to be involved in government issues that affect their pastime and South Dakota's water estate.

If they are informed, anglers can be an important force in keeping our nation's waters clean and fish populations healthy. The "Conservation President" Theodore Roosevelt stated, *"The Nation behaves well if it treats the natural resources as assets, which it must turn over to the next generation increased, and not impaired in value."*

We conserve only what we love. We will love only what we understand. We will understand only what we are taught. Enjoy reading about South Dakota fisheries and fishing in the chapters that follow.

Acknowledgements

The SDGFP funded many of the studies used in the writing of this chapter. The South Dakota Cooperative Fish and Wildlife Research Unit is jointly sponsored by SDGFP, the U.S. Geological Survey, The Wildlife Management Institute, the U.S. Fish and Wildlife Service, and South Dakota State University. ❖

Chapter 2

Prehistoric Fishes and Fishing in South Dakota

David C. Parris, Gale A. Bishop and Kenneth F. Higgins

" Below this on a hill on the L.S. we found the backbone of a fish, 45 feet long, tapering to the tail. Some teeth, &c. Those joints were separated, and all petrified. "

— Lewis and Clark Journals, September 11, 1804 near the Cheyenne River

If you enjoy fishing, you belong to one of the greatest traditions in the world. You have thousands of years of human history behind you, and there are men and women almost everywhere who share your enthusiasm. It should also please you to know that thousands more years of human prehistory and more than 500 million years of fossils are part of the lore of fishing. Few places on earth have contributed as much to our knowledge of fishes and fishing as has South Dakota. Although it isn't possible to give a complete story of fish and fishing prehistory in just a few pages, it is certainly worthwhile to know how much scientists have learned in South Dakota. In a very real sense, the paleontologists and archaeologists, whether amateur or professional, are "anglers" too (Figure 2.1).

South Dakota's Fossil Fishes

Fossil evidence from all over South Dakota documents the presence of both freshwater and marine fishes of the past. This record extends from the Paleozoic Era, more than 500 million years ago, to the present. Most of the evidence comes from durable parts of the fishes, the skeletons, bones, teeth, scales, and otoliths (ear stones). However, trace fossils, such as nesting traces or hibernation areas, are sometimes found. More specimens have been found in the western part of the state, primarily because rocks are more commonly exposed west of the Missouri River, notably in the Black Hills and Badlands. Also, Ice Age (Pleistocene) deposits of the last continental glaciation (16,000-10,000 years ago) cover much of eastern South Dakota, and such deposits seldom have fossil fish.

A summary of much of South Dakota's fossil fish record is given in Figure 2.2. It is only

Figure 2.1 Young amateur paleontologist after a successful "fishing" trip to the Ree Heights fossil site, South Dakota.

Figure 2.2 Geologic time table with an overview of the record of fossil fish from South Dakota.

GEOLOGIC TIME SCALE

ERA	PERIOD	MILLIONS OF YEARS AGO	CHARACTERISTIC LIFE FORMS and IMPORTANT EVENTS	FOSSIL FISH
CENOZOIC	QUATERNARY	0–1.8	*HUMANS*	Freshwater fishes in lake and sinkhole sediments
	TERTIARY	1.8–65	*MAMMALS*	Fossil catfish in bad-land clay formations
MESOZOIC	CRETACEOUS	65–144	*DINOSAURS* First flowering plants	Freshwater fish in Hell Creek Formation Many kinds of fish in chalk and shale formations
	JURASSIC	144–200	*DINOSAURS* First birds	Bony fish in Sundance Formation
	TRIASSIC	200–250	*ADVANCED REPTILES* First Mammals	
PALEOZOIC	PERMIAN	250–290	*EARLY REPTILES*	Bony fish in Minnekahta Formation
	CARBONIFEROUS	290–354	*AMPHIBIANS* "Coal" Forests	
	DEVONIAN	354–417	*FISHES* First Amphibians	Earliest sharks in Englewood Formation
	SILURIAN	417–443	*INVERTEBRATES* First land plants First land invertebrates	
	ORDOVICIAN	443–505	*INVERTEBRATES* First jawless fishes	Jawless fish in Whitewood Formation
	CAMBRIAN	505–543	*INVERTEBRATES*	Earliest known fish in Deadwood Formation

a basic introduction that demonstrates an exceptional spectrum of resources. No doubt many discoveries remain to be made, and knowledge will be enhanced by further research. Many specimens are in study collections or on exhibition at the Museum of Geology at the South Dakota School of Mines and Technology in Rapid City or at the Black Hills Institute of Geological Research in Hill City. Many important discoveries have been made by amateur collectors, all of which supplement the findings of professional paleontologists.

If you are interested in collecting fossil fish, however, you should be aware that all fossils of vertebrates (backboned animals, including all fishes) are basically rarities. All may yield important information about the geologic past. If you find (or are given) a fish fossil, you should carefully retain a record of all information about it; get more data if you possibly can. Any facts about where and when it was collected, and by whom, will be significant. All natural and cultural history of the fossil should be recorded. We recommend that you contact a specialist at a museum or university to share the information with, and carefully consider whether you should retain it or not.

Fossil vertebrates are protected by laws and regulations wherever they are found on state, federal or Native American tribal trust lands; all of these tracts require special collecting permits. Fossils on privately owned lands in the United States belong to the landowners, and you should seek written permission to search for fossils on private property. Many South Dakota landowners have generously allowed scientists to collect fossils on their property, and these include some of the most famous of fossil sites.

The Earliest Fossil Fish

South Dakota's amazing fossil resources have brought international fame to our state. The oldest fish fossils found here are practically as old as any in the world. They are thus among the oldest vertebrates as well. Many are from marine environments, in contrast with our basic impression of the Great Plains being remote from bodies of saltwater. The story begins in the Paleozoic Era, meaning the time of ancient life.

The oldest fish in the world (*Anatolepis*) comes from the Black Hills region, in the Bear Lodge Mountains of eastern Wyoming and adjacent to the border of South Dakota. The rock formation in which it was discovered is named after

Deadwood, South Dakota, and is more than 500 million years old. It would appear likely that this, the oldest known fossil vertebrate, will eventually be found in South Dakota as well. Other ancient fish are preserved as fragments in the Whitewood Formation in the northern Black Hills of South Dakota, and are about 400 million years old. Barely recognizable as fish, the fragments are pieces of bony armor from primitive jawless fishes called agnathans. Elsewhere in the world, where more complete specimens have been found, their body shapes are seen to be vaguely fish-like, but appear unlike anything now alive. Strange to say, these bizarre creatures do have a few living relatives, one of which is still alive in South Dakota waters. The lamprey, which also lacks a true jaw, belongs to this most ancient and primitive class of vertebrates.

Another very ancient group of fishes, of which there are many living relatives, is the sharks. Modern sharks are essentially marine fishes (as are the vast majority of their extinct relatives) and thus they are absent from present-day living fishes in South Dakota. The fossil record of sharks (and the related rays and skates) is very extensive in the Great Plains, however, extending from hundreds of millions of years ago until about 50 million years ago when marine waters receded from the mid-continent for the last time. Sharks belong to a class of vertebrates called Chondrichthyes, meaning that they have skeletons mostly of cartilage. Because cartilage is not easily preserved as a fossil, the vast majority of fossil sharks are known primarily from teeth. Fossils of one particularly early group of shark forms, known as bradyodonts, have been found in 300 million year old rocks in the Black Hills. There are a number of other shark forms known in later rocks of the Paleozoic Era, which extends to about 250 million years ago.

The most numerous of fish are the bony fish, the Osteichthyes, and their fossils are also found in the Paleozoic rocks of the Black Hills. One notable occurrence is in a rock formation called the Minnekahta Limestone. Entire fossilized skeletons of a fish called *Platysomus* have been found on the rock surfaces, readily recognizable by any observer.

The "Middle Ages" of Fish

The record of fish fossils from South Dakota is even more extensive in the next era, the Mesozoic, or age of "Middle Life." Inland seas covered what is now South Dakota for much of that time, and fish fossils occur in many of the resulting sediments. The era is divided into three periods; from oldest to youngest they are named the Triassic, Jurassic, and Cretaceous. Of these three, the Cretaceous, meaning "chalk-bearing," has the most outstanding record of fish fossils. During that time, 150 million to 60 million years ago, a seaway covered much of the western interior of North America, forming a waterway that connected the Gulf of Mexico to the Arctic Ocean (Figure 2.3). What is now South Dakota was located near the center of this seaway, and great thicknesses of chalk, mud, and sand were deposited as sediment, all of which were capable of preserving fossils of fish.

Among the most interesting of Cretaceous rocks is one called the Carlile Formation, which is found in the Black Hills region. One unit within it is called the Turner Sandy Member, which contains such vast numbers of shark teeth that even the most casual observer would easily see them. At least 17 different kinds of sharks have been found in it, including a shell-crushing shark with

Figure 2.3   North America during the Late Cretaceous period, showing the Interior Seaway that brought ancient fishes to South Dakota (After Gallagher 1990).

button-like teeth called *Ptychodus*. Fossil fish skulls have been found along with fossil lobsters in the Carlile Formation. In Grant County in northeastern South Dakota, rock quarries have exposed a rock unit equal to the Carlile, and it too has fossils of sharks and other fishes.

In September, 2003, a field crew from the South Dakota School of Mines and Technology and the U.S. Bureau of Reclamation recovered a very large fossil fish from the Carlile Formation near the Belle Fourche Reservoir. It was a good example of classic field paleontology: collecting using plaster jacket methods. The sequence of techniques includes delineation of the large fossil, trenching around it, protecting the exposed fossil with tissue paper, encasing it in plaster bandages, flipping it over (the scariest step), and finally completely enclosing it with more plaster bandages. Complicated, but successful, it also was a good example of why the recovery of such precious vertebrate fossils is no casual undertaking. It may be a lengthy, laborious, and expensive operation, and possibly even hazardous.

The classic chalk formation of the Great Plains is called the Niobrara Formation. It is famous throughout the world for its exquisite preservation of marine fossils, including many fish. It is widespread in South Dakota, especially in the southeast and southwestern parts of the state, but is often poorly exposed. As with all chalks, it is composed largely of microscopic fossils, but has large ones as well. One of the most famous is the huge bulldog fish, *Xiphactinus audax*, of which a spectacular specimen is on exhibition at the Museum of Geology of the South Dakota School of Mines and Technology in Rapid City (Figure 2.4). Its massive jaws are armed with sizeable teeth firmly implanted in sockets. (Most fish have teeth that are fused along the edges of the jawbones.)

Another fossil fish of the Niobrara chalks is called *Enchodus*, a fierce-looking creature with well-developed stabbing teeth. It is represented by hundreds of South Dakota specimens, ranging in length from panfish size to more than a yard long. Paleontologists jokingly call it the saber-toothed herring!

Figure 2.4   Skeleton of a 10-ft long bulldog fish (*Xiphactinus audax*) of the Cretaceous Period displayed at the Museum of Geology, South Dakota School of Mines and Technology, Rapid City.

Many of the fossil fishes of the Niobrara chalks are also found in the rock that overlies it, the Pierre Shale. It is casually, but appropriately, called the South Dakota State Rock, not only because it was named for Pierre/Fort Pierre, but because it truly characterizes much of the state. It is widespread both east and west of the Missouri River, and it gives a gray to black color to rock exposures that are familiar to travelers all across the state. The name is widespread in the Great Plains and eastern Rocky Mountains wherever the rock occurs. The clays in Pierre Shale can become "gumbo" when wet, creating difficulties for anyone who ventures off of paved roads in any weather wetter than dew. Like the Niobrara Formation, the Pierre Shale was deposited in marine waters and contains wonderful treasures of fossil fishes.

The Museum of Geology staff at the South Dakota School of Mines and Technology have collected fossils from the Pierre Shale throughout its history, notably in the Black Hills and within the Missouri River basin. Many of the fish fossils are of particular interest to scientists for the information they have yielded about prehistoric life. For example, the fossilized stomach contents of a huge sea lizard (mosasaur) on exhibition at the Museum of Geology demonstrated that fish were among the prey animals of a predator that also ate birds and smaller mosasaurs! Another Pierre Shale specimen on exhibition is a fish called *Apsopelix*. It was found in a concretion (a particularly hard nodule of rock), preserved uncompressed, and thus three-dimensional as in life. Found near Oacoma by Lauren Hill, and donated by Samual Bice, it seemed to be unique, but another very similar specimen of the same fish species has recently been found and donated by Lee Azure of the Crow Creek Tribe.

An important discovery occurred in 1995 when members of a South Dakota School of Mines and Technology field party found a fossil fish spine in the Pierre Shale in Hyde County. Fossil spines of this type are well known from formations along the Eastern Seaboard and in Europe and Africa and are attributed to a fossil fish called *Cylindracanthus*, meaning a cylindrical spine. It was categorized to be a fish spine, even though no other skeletal parts were found, and no one is totally certain as to where it belongs on a fish. The South Dakota specimen, however, is better than any other ever found, and it has well-preserved tiny teeth, which confirms that *Cylindracanthus* is related to sturgeons and paddlefishes of present-day South Dakota. Young paddlefish have similar teeth, and their skeletons are largely cartilage, unlikely to be preserved.

The living paddlefish are freshwater species, of course, but they too are known from Mesozoic rocks of South Dakota. The last rock formation to be deposited during that era is the Hell Creek Formation of the northwestern part of the state. It represents primarily freshwater environments, and two kinds of paddlefish are among the 19 fossil fish species known from it. An imprint of a fossil paddlefish, collected from the essentially equivalent Lance Formation of Wyoming, is in the collections of the Black Hills Institute of Geological Research.

Life in "Modern" Times

The Cenozoic Era, meaning the Age of New Life, began about 65 million years ago. With the extinction of the dinosaurs and many other life forms at the end of the Mesozoic, it was indeed a new phase of earth history. Early in the Cenozoic, the seas receded from the interior of the continent for the last time, and the geologic record of South Dakota has since then consisted only of non-marine rocks. Fish fossils are not so likely to be found in such rocks, and the famed fossil beds of the Big Badlands are not good fossil fish territory. Although a few remains of fossil fishes have been recovered from South Dakota rock formations laid down during this time in a time slot called the Tertiary Period, they have not added a great deal to our knowledge of fish prehistory. Notable exceptions are the records of early catfishes, a fish family which was common in the Tertiary, and that has remained of prime importance in South Dakota ever since.

The Ice Age (Pleistocene Epoch) occurred during the last one million years. It is represented by many fossil localities in South Dakota, one of which is the Ree Heights fish fossil site in Hand County. Twelve species of fossil fishes were found there from sediments of an ancient lake environment. They include five families: minnows (*Cyprinidae*), catfishes (*Ictaluridae*), killifishes (*Fundulidae*), sunfishes (*Centrarchidae*), and perches (*Percidae*). They are from the last stage of the Ice Age, called the Wisconsinan, but most of the species still live in South Dakota today. They provide a suitable introduction to the archaeological record of fish and fishing which was soon to follow. They are among the fish species most used by the first human inhabitants (Native American Indians) here, and are among those most appreciated by South Dakotans and others today.

Prehistoric Indian Fisheries

Archaeological evidence indicates that native peoples were catching fish for food at least 10,000 years ago during the Paleo-Indian Period in what is now South Dakota. Minnows were found along with mammoth bones at the Lange-Ferguson Site in Shannon County. Most evidence from zooarchaeology (the study of animal remains from archaeological sites), however, relates to times much closer to the present.

The prime time of the cultural complex known as the Initial Middle Missouri has been particularly well studied in South Dakota. About 30 sites have been excavated, mostly along the Big Sioux, James, and Missouri rivers. Bones from these sites reveal important facts about food resources of the prehistoric occupants. For example, bones found in cache pits at the Heath Site along the Big Sioux River in Lincoln County indicate that fish, turtles, and frogs were used there during prehistoric times from about A.D. 500 to A.D. 1500. Detailed studies enable determination of sizes and weights of the meat resources, and may even suggest the methods by which they were obtained.

The zooarchaeologist who specializes in fish remains, however, faces a daunting task dealing with skeletons that are often tiny and fragile, fall apart easily, and may have only a few identifiable bones. That is not to say that the work is impossible, but often the effort is very laborious and extensive detailed studies may seem unjustified. Furthermore, the bone assemblages in South Dakota archaeological sites are often heavily dominated by large mammals, and research is often concentrated on the resources that seem to have been most important to the cultures of the past.

Because of the fragility of fish bones, the fish resources are likely to be under-represented in archaeological sites. This is amply demonstrated by sites that have substantial numbers of fishing hooks, net sinkers, and other items associated with fishing, but which have few fish bones. Small fish can be eaten whole, leaving essentially no archaeological record at all. Some species, while present in a site, can be difficult to enumerate, as with sturgeons, whose few preservable parts (armor scutes) all look exactly alike except for size.

The motivations of indigenous peoples for their subsistence presumably were much different from those of today's sport fishermen. For example, a catch of minnows (which we see as mere baitfish) could have been just as important to their food supply as any large fish would be to us today. Of South Dakota's native fish species, nearly half are minnows, potentially important resources in a hunter/gatherer society, but likely to be poorly represented in an archaeological site.

Besides being an important source of protein, fish were certainly used in other ways. Perforated vertebrae (backbones) found at some sites indicate that some fish bones were used as ornaments. Modified fin spines may have been used as awls and fasteners. It is important to answer the questions of how fish were caught and how they were used, but sometimes, in the absence of good evidence, we can only make educated guesses.

Simple Fishing Methods and Artifacts

The most commonly recognized fishing artifacts are hooks. Occasionally, hooks were made from bone by cutting a flat piece into a small rectangle, removing the center area, then splitting the remainder down the long side, thus producing two hooks. Small grooves were usually made in the longer end in order to attach a cord. Short ends were sharpened to a point; often there was no barb. Bone fishhooks found in South Dakota are usually less than 1½ inches long by ¾ inch wide (Figure 2.5) and were probably only used for catching small to medium sized fish, most probably those that were abundant and easy to catch near shore.

Larger catfish, or scaled fish such as northern pike, walleye, and buffalo fish, may have been harvested using gorge hooks, or by spears, arrows, or even by hand. Even today, large flathead catfish of 50 pounds or more may be caught by hand using a

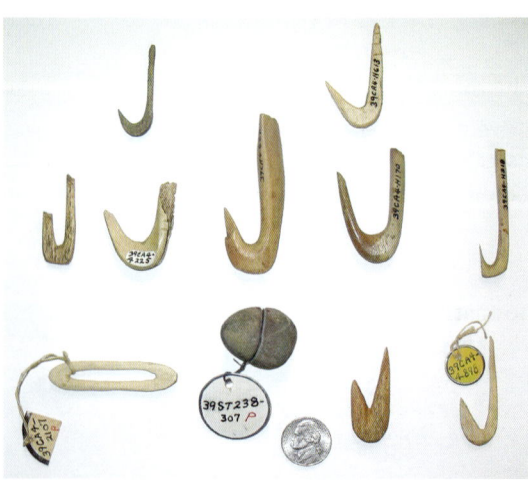

Figure 2.5   Fish hooks, a net sinker, and a possible pre-formed pair of hooks from the Rygh archaeological site in Campbell County and the Durkin Village site in Stanley County (courtesy of the South Dakota Archaeological Research Center).

technique known as "noodling." Still a popular fishing activity in the lower Midwest, noodling consists of wading into shallow waters and probing by hand for fish that are resting in bank holes and pockets. (Sometimes a sizeable catfish spine or an occasional snapping turtle can make noodling a bit hazardous!)

Smaller fish and minnows were probably taken with basketry, either in the form of a pocket drag or a catchment-holding pen. It is also conceivable that woven baskets or framed animal hides could be used to form dippers or even crude types of fish nets. Cordage imprints on pottery shards show that native peoples were knowledgeable about making cord and twine and how to attach it to things. It is also possible that they were able to make simple gill or other webbed nets in combination with pliable woody stems such as willow. On August 15, 1804, a Lewis and Clark expedition journal (written by Captain Clark) related how 318 fish were taken in one day and nearly 800 another, using a drag of willows and bark in a creek. The catch was said to include pike, bass, perch, redhorse, catfish, and other species. Perhaps they were mimicking gear used by Indians along the Missouri River. In any case, their crude equipment is indirect evidence that native peoples were likely capable of taking many fish in the smaller streams and creeks, and probably in the larger rivers as well. Certainly the members of the expedition were observant of Indian fishing methods and made use of such techniques, for Captain Lewis had noted in his journal of September 4 of the previous year how he and his men had tried gigging for fish on the Ohio River "after the Indian method."

Of course, dead or dying fish can be salvaged when trapped in small pools or river oxbows during droughts. Fish are also susceptible to winterkill (oxygen depletion), injury, old age, and disease. Often dead fish end up on beaches and river banks due to wind and wave action, making them easily accessible for the taking. Doubtless there were many other ways of taking fish unbeknown to us that were formerly used by indigenous peoples in what was to become South Dakota.

Fish From Archaeological Sites

While biologists have noted more than 90 species of fishes that are native to South Dakota, they are rather unevenly represented in the archaeological record. These 90 species are classified into about 24 families, only half of which have any significant occurrences to report. There are various reasons for this; for example, the afore-mentioned lampreys (Family *Petromyzontidae*) have no preservable bones. Sturgeons (Family *Acipenseridae*) have skeletons mostly of cartilage, but the bony armor scutes are readily preserved. Atlantic and Pacific coastal sites often have sturgeon scutes, but there are few records from inland sites. This may be due to actual differences in procurement and use, a problem of interpretation for the zooarchaeologist.

Fish families with a significant archaeological record in South Dakota are:

Gars (*Lepisosteidae*): With a substantial skeleton and a body armor of matted enamel scales, the gars are readily preserved and easily identified by archaeologists. A number of occurrences are known.

Bowfins (*Amiidae*): Although the one living species, *Amia calva*, is of uncertain status in South Dakota, it was found in a historical archaeology dig at Fort Randall. This is one situation where the archaeological record supplements current biological surveys. Most of the skeleton of this ancient fish is highly distinctive and readily identified, and it has a fossil record extending back many millions of years.

Pikes (*Esocidae*): The northern pike (*Esox lucius*) is native to South Dakota and has many durable and distinctive bones, especially in its skull. It is known from several sites.

Mooneyes (*Hiodontidae*): The goldeye (*Hiodon alosoides*) is a common South Dakota fish today, but it has only a sparse archaeological record. It has been found in nearby Iowa sites.

Minnows (*Cyprinidae*): This is the most complicated family for archaeological potential with about 35 species in South Dakota, far more than any other family of fishes. Most are small and their bones could be easily missed by even the most sophisticated means of recovery; however, they have very distinctive bones in the throat/gill region (the pharyngeals). These bones have been found in many sites and are highly identifiable, often to the species level. Minnows are so common that one would expect them to occur in almost any site near water, and their remains might be considered to be more environmental than cultural. Incidentally, the common carp (*Cyprinus carpio*) is a member of the minnow family, although it attains considerable size. It is not native to North America and should not be expected in prehistoric sites.

Suckers (*Catostomidae*): This family of very common fishes has been recorded from many South Dakota archaeological sites. Most records are of the white sucker (*Catostomus commersonii*), which is found throughout the state. Native American peoples considered them to be good quality food. Suckers frequently can be identified from the comb-shaped pharyngeal bones of their throat/gill region.

Catfishes (*Ictaluridae*): The freshwater catfishes have excellent archaeological records wherever they occur, and South Dakota is no exception. Not only are they abundant and widespread, but some species attain great size. Most of the bones of the skeleton are very distinctive and durable, notably the fin spines that are such a hazard to handlers. Catfishes are almost universally regarded as excellent food resources and may be caught by a variety of methods. The fabled association of catfishes with the Missouri River is supported by the archaeological record, with many records from prehistoric sites along that drainage.

It would appear that the archaeological records of catfishes in South Dakota should be reviewed and upgraded. There are eight native species and all are now known to be readily identifiable from the pectoral fin spines alone. Many published records do not identify them to species, but a review study would likely refine the identifications.

Temperate basses (*Moronidae*): The white bass (*Morone chrysops*) is the only species of this family native to South Dakota. There are sparse records of white bass from sites in nearby Iowa.

2.1

Walleye *Sander vitreus:* The South Dakota State Fish

The State Fish: The walleye was adopted as the State Fish in 1982. The scientific name has changed from *Stizostedion vitreum* to *Sander vitreus.*

Is walleye a native species? No one doubts that the walleye is native to the Mississippi River basin, but scanty historical records have caused debate about its native status in the upper Missouri River and in South Dakota. Genetic evidence suggests that walleye survived glaciation in the Missouri River basin.

Post-glacial dispersal: Walleye could have invaded the upper Missouri River by two routes. Walleye invaded the Hudson Bay drainages from the Mississippi River drainage during deglaciation about 14,000 years ago by using the Ancient River Warren, which carved the valley presently filled by Big Stone Lake and Lake Traverse. So, they could have moved into the Missouri Basin via headwater connections between River Warren or Minnesota River and the Big Sioux River. A second possibility is that the very-mobile walleye migrated upstream in the Missouri River during a time of low turbidity. The post-glacial climate of the Upper Missouri Basin was sometimes warm (increases flow, erosion and river turbidity) and sometimes cold (decreases river turbidity) for centuries.

Conclusion: Walleye probably used two routes to immigrate to South Dakota after glaciation. They would have been free to move throughout state waters depending on changing environmental conditions over 14,000 years. The walleye was in some and likely many waters of the state before settlement. Walleye were found in Rock Creek (Mitchell) and Choteau Creek (Springfield) in 1894.

Sunfishes (*Centrarchidae*): This family includes the well-known smallmouth (*Micropterus dolomieu*) and largemouth (*Micropterus salmoides*) basses, as well as a number of small panfishes, such as bluegills (*Lepomis macrochirus*) and crappies (*Pomoxis* spp.). Although greatly favored by anglers and considered good food fishes, there are few archaeological records. Most of the skeleton is not very identifiable nor diagnostic, but the scales are rather easily recognized as belonging to the family. Careful archaeological analysis might assist biologists in establishing the original ranges of centrarchid species now greatly affected by deliberate introductions, although they have been recorded from only a few sites.

Perches (*Percidae*): Among South Dakota's most favored sport fishes, the eight species now found here include the sauger (*Sander canadensis*), walleye (*Sander vitreus*), and yellow perch (*Perca flavescens*). Although readily taken by modern anglers, they might have been much more difficult for Native American inhabitants to catch in prehistoric times. There are some archaeological records, as would be expected, because the perch family has provided food to many cultures.

Drums (*Sciaenidae*): The freshwater drum (*Aplodinotus grunniens*) has readily identifiable otoliths (ear stones) and crushing teeth on the pharyngeal bones of its throat/gill region. These distinctive elements have been identified at a number of sites.

Evidence from paleontology confirms that fish have been present in South Dakota for hundreds of millions of years. Archaeological evidence shows that fish have been used as food for as long as humans have lived here. Future studies will help to fill in the knowledge gaps about the kinds and amounts of fish used by native peoples and about the kinds of gear and methods used to catch them. A regional comparative reference collection with bones, spines, scutes, and otoliths from all possible fish species past and present would greatly facilitate future archaeological studies. Much remains to be learned about fish from South Dakota's past.

Acknowledgements

We are grateful to the South Dakota State Historical Society Archaeological Research Center for assistance and permission to photograph artifacts. Renee Boen and Candy Taft were particularly helpful in our visits there. The Fawcett family (Leonard, Dorothy, Dennis, and Connie) of Ree Heights, Hand County, graciously permitted us to visit the fossil locality on their ranch, where we collected specimens and photographed the site. Archaeologist Lauren Milideo assisted with photography of artifacts. James E. Martin was very helpful with discussions of the fossil record. See the References section for a list of sources that were used to write this chapter, particularly Evetts 1979; Haug et. al. 1994; Kivett and Jensen 1976; Martin et. al. 1998; Ossian 1973; Parris et. al. 2001; Sundstrom 1996; and Zimmerman 1985.   ❖

Chapter 3

Recent Zoogeography of South Dakota Fishes

Christopher W. Hoagstrom, Steven S. Wall, Jason G. Kral and Brian G. Blackwell

> " One fish, two fish, red fish, blue fish. Black fish, blue fish, old fish, new fish... Say! What a lot of fish there are. "
>
> — Dr. Seuss

Fish zoogeography is the study of fish distribution patterns. The patterns reveal the history of aquatic habitats because waters that share the same fish species were connected at some previous time, even if they are unconnected today. The patterns also indicate the habitat requirements of fishes because the range of every fish species is limited to where habitat conditions are suitable for it. When different fish species share a similar range it suggests that they have similar distributional histories or similar habitat requirements. Thus, the study of fish zoogeography provides general insight into the origin of fish populations and the habitat requirements of fish species. In this chapter we summarize the distribution of fish species among major river drainages of South Dakota.

South Dakota Geography

The geography of South Dakota is diverse. It includes: (1) glaciated regions with abundant natural lakes and wetlands, (2) unglaciated regions with deep valleys, canyons, and badlands, and (3) mountains that are covered with forest. The climate of South Dakota varies from relatively moist and warm in the southeast to relatively dry and cold in the northwest. South Dakota is divided into two physiographic provinces that have different geology, elevation, and glacial histories. The Central Lowlands Province, which has a relatively low elevation and relatively old geology, includes roughly the northeastern 30% of the state. The remaining 70% is in the Great Plains Province, which is higher in elevation and has relatively young geology. The Missouri Coteau, an ancient plateau that was covered by glacial deposits, is the divide between the two provinces. Several uplifts in the Great Plains of South Dakota rise like islands from the surrounding plain (Kornfeld and Osborn 2003). These islands have unique ecological conditions including relatively high precipitation. The largest island-like uplift is the Black Hills. Smaller uplifts in South Dakota include the Pine Ridge, Slim Buttes, Short Pine Hills, and Cave Hills. Humans have altered the geography of the state by creating large reservoirs in many river basins. In addition, humans have drained natural wetlands (e.g., tiling agricultural fields), created artificial wetlands (e.g., dugouts and stock dams), and altered stream courses (e.g., channelization, ditching).

South Dakota contains portions of two major watersheds: the Hudson Bay and the Gulf of Mexico. The South Dakota portion of the Hudson Bay Watershed is in the northeast corner of the state and it comprises less than 1% of the state. All other South Dakota waters are within the Mississippi

River drainage that is part of the Gulf of Mexico Watershed. Two major sub-drainages of the Mississippi River drainage are within South Dakota: the Upper Mississippi and the Missouri. The Upper Mississippi River sub-drainage includes roughly 2% of the state. The Missouri River sub-drainage makes up the remaining 97% of South Dakota. The main-stem Missouri River traverses the state and roughly divides it in half. It receives inflows from a series of tributary drainages that lie both to the west and to the east.

How Fishes Colonize New Waters

Fishes can only expand their range and colonize new waters if a suitable aquatic route is available to them. An aquatic route is only usable if it is navigable and inhabitable enough so that a sufficient number of fishes from existing populations can use the route to colonize other suitable habitats and establish new populations. There are at least four types of aquatic routes that are used by fishes to colonize new waters: (1) direct stream connections, (2) headwater trans-divide connections, (3) stream captures, and (4) human made connections. Fishes of South Dakota have used each type of aquatic route.

Direct stream connections are the most obvious. Fishes can swim downstream or upstream or some combination of the two (downstream in one stream and then upstream in an adjoining stream) to reach new suitable habitats. Many fishes used the Missouri River to expand their range into South Dakota.

Headwater connections are less common, but occur in regions that have low relief, such as glaciated portions of South Dakota. In these regions the divides between stream drainages may be low and indistinct and may even include lakes or wetlands that connect different stream drainages, enabling fishes to go back and forth. These connections may be seasonal or may only occur infrequently during very wet conditions. A good example of headwater connectedness is Lake Hendricks, which spans the divide between the Big Sioux and Minnesota river drainages.

Stream captures result from one stream eroding head ward until it cuts across the path of another stream and incorporates upstream portions into its drainage. When this happens, fishes of the captured stream become part of the fauna of the capturing stream. If the capture is not instantaneous, fishes from the capturing stream are also able to invade other portions of the captured stream. Stream captures have been hypothesized to be important in western South Dakota (Bailey and Allum 1962, Mayden 1987).

Human made connections include canals between river drainages, hauling tanks that are used to transport fishes from one river drainage (or from a hatchery) to another, and bait buckets. Fish managers have been active in introducing species, especially when new habitats are created (i.e., Missouri River reservoirs), but the practice has been curtailed because the negative impacts of introduced and exotic species are now understood.

Fishes of South Dakota

The distribution and abundance of fishes in South Dakota is always changing because some species expand their ranges into new areas while others suffer range declines. Based on recent (post-1990) inventories, at least 111 species of fish are present in South Dakota today. At least 122 fish species, however, have been present in some portion of South Dakota over the last 150 years since Anglo-Americans settled the state. Some of the 122 species were present at the beginning of settlement (1856) and others are present today. This chapter discusses the distributions of all 122 species known from South Dakota.

The number of species known from South Dakota (122) is low compared to most other states because the cold and dry climate is relatively harsh for fishes. South Dakota is in the central part of a continuum of fish species diversity. States to the east and south have more species because they are warmer and wetter, while states to the north and west have fewer species because they are colder and dryer. This trend is also present within the state. Southeastern South Dakota has the most fish species while northwestern South Dakota has the fewest.

Most people are familiar with only a few of the fish species present in South Dakota such as black bullhead, channel catfish, rainbow trout, largemouth bass, and walleye. There are several other species that are similar to these well-known ones. For example, there are three species of "bullheads" in the state. There are also other groups of fish species that are much different from the well-known fishes. The terms "minnow" and "sucker" refer to particular groups of fish species. Many minnows and suckers, as well as members of other such groups, are present in South Dakota.

The 122 fishes that are known from South Dakota can be loosely grouped into three categories: creek fishes (71), river fishes (36), and cold-water fishes (15) (Table 3.1). Creek fishes, such as common shiners, fathead minnows (Figure 3.1), creek chubs, and white suckers, are typical of small streams and wetlands and are associated with relatively coarse substrate (e.g., cobble and gravel). When creek species inhabit wetlands or lakes they normally use shallow or near-shore habitats.

Figure 3.1 The fathead minnow (*Pimephales promelas*) is an example of a creek fish.

When they inhabit larger streams they are normally restricted to shoreline habitats such as backwaters or creek-like habitats such as riffles. These species are relatively sensitive to siltation because it clouds the water and smothers and clogs spaces within coarse substrates that are important for nesting and nursery habitat, for invertebrate prey, for nutrient cycling, and for moderating stream temperatures (water that flows through spaces between gravel does not warm or cool as rapidly as surface water; thus, streams that have abundant inter-gravel flow have more stable temperatures).

Table 3.1   List of fish species that have been documented in South Dakota within the last 150 years (1856 to 2005). A total of 122 fish species has been found of which 71 are typical of creek habitats, 36 are typical of river habitats, and 15 are typical of cold-water habitats.

COMMON NAME	SCIENTIFIC NAME	CATEGORY[1]
Silver lamprey	*Ichthyomyzon unicuspis*	Creek
Lake sturgeon	*Ascipenser fulvescens*	River
Pallid sturgeon	*Scaphirhynchus albus*	River
Shovelnose sturgeon	*Scaphirhynchus platorynchus*	River
Paddlefish	*Polyodon spathula*	River
Longnose gar	*Lepisosteus osseus*	Creek
Shortnose gar	*Lepisosteus platostomus*	Creek
Bowfin	*Amia calva*	Creek
Goldeye	*Hiodon alosoides*	River
Mooneye	*Hiodon tergisus*	Creek
American eel	*Anguilla rostrata*	River
Skipjack herring	*Alosa chrysochloris*	Creek
Alewife	*Alosa pseudoharengus*	Creek
Gizzard shad	*Dorosoma cepedianum*	Creek
Central stoneroller	*Campostoma anomalum pullum*	Creek
Goldfish	*Carassius auratus*	Creek
Lake chub	*Couesius plumbeus*	Cold-water
Grass carp	*Ctenopharyngodon idella*	River
Red shiner	*Cyprinella lutrensis lutrensis*	Creek
Spotfin shiner	*Cyprinella spiloptera*	Creek
Common carp	*Cyprinus carpio*	Creek
Western silvery minnow	*Hybognathus argyritis*	River
Brassy minnow	*Hybognathus hankinsoni*	Creek
Plains minnow	*Hybognathus placitus*	River
Silver carp	*Hypophthalmichthys molitrix*	River
Bighead carp	*Hypophthalmichthys nobilis*	River
Common shiner	*Luxilus cornutus*	Creek
Sturgeon chub	*Macrhybopsis gelida*	River
Shoal chub	*Macrhybopsis hyostoma*	River
Sicklefin chub	*Macrhybopsis meeki*	River
Silver chub	*Macrhybopsis storeriana*	River
Pearl dace	*Margariscus margarita nachtriebi*	Creek
Hornyhead chub	*Nocomis biguttatus*	Creek
Golden shiner	*Notemigonus crysoleucas*	Creek
Emerald shiner	*Notropis atherinoides*	River
River shiner	*Notropis blennius*	River
Bigmouth shiner	*Notropis dorsalis*	Creek
Blackchin shiner	*Notropis heterodon*	Creek
Blacknose shiner	*Notropis heterolepis*	Creek
Spottail shiner	*Notropis hudsonius*	River
Carmine shiner	*Notropis percobromus*	Creek
Silverband shiner	*Notropis shumardi*	River
Plains sand shiner	*Notropis stramineus missuriensis*	Creek
Eastern sand shiner	*Notropis stramineus stramineus*	Creek
Topeka shiner	*Notropis topeka*	Creek
Mimic shiner	*Notropis volucellus*	Creek
Suckermouth minnow	*Phenacobius mirabilis*	Creek
Northern redbelly dace	*Phoxinus eos*	Creek
Southern redbelly dace	*Phoxinus erythrogaster*	Creek
Finescale dace	*Phoxinus neogaeus*	Creek
Bluntnose minnow	*Pimephales notatus*	Creek
Fathead minnow	*Pimephales promelas*	Creek
Flathead chub	*Platygobio gracilis*	River
Longnose dace	*Rhinichthys cataractae cataractae*	Creek
Western blacknose dace	*Rhinichthys obtusus*	Creek
Rudd	*Scardinus erythrophthalmus*	Creek
Creek chub	*Semotilus atromaculatus*	Creek
Northern river carpsucker	*Carpiodes carpio carpio*	River
Central quillback carpsucker	*Carpiodes cyprinus hinei*	River
Highfin carpsucker	*Carpiodes velifer*	River
Longnose sucker	*Catostomus catostomus*	Cold-water
White sucker	*Catostomus commersonii*	Creek
Mountain sucker	*Catostomus platyrhynchus*	Cold-water
Blue sucker	*Cycleptus elongatus*	River

Table 3.1 continued.

COMMON NAME	SCIENTIFIC NAME	CATEGORY[1]
Northern hog sucker	*Hypentelium nigricans*	Creek
Smallmouth buffalo	*Ictiobus bubalus*	River
Bigmouth buffalo	*Ictiobus cyprinellus*	River
Black buffalo	*Ictiobus niger*	River
Golden redhorse	*Moxostoma erythrurum*	Creek
Shorthead redhorse	*Moxostoma macrolepidotum*	Creek
Black bullhead	*Ameiurus melas*	Creek
Yellow bullhead	*Ameiurus natalis*	Creek
Brown bullhead	*Ameiurus nebulosus*	Creek
Blue catfish	*Ictalurus furcatus*	River
Channel catfish	*Ictalurus punctatus*	River
Stonecat	*Noturus flavus*	Creek
Tadpole madtom	*Noturus gyrinus*	Creek
Flathead catfish	*Pylodictis olivaris*	River
Grass pickerel	*Esox americanus vermiculatus*	Creek
Northern pike	*Esox lucius*	Creek
Muskellunge	*Esox masquinongy*	Creek
Central mudminnow	*Umbra limi*	Creek
Rainbow smelt	*Osmerus mordax*	Cold-water
Cisco	*Coregonus artedi*	Cold-water
Lake whitefish	*Coregonus clupeaformis*	Cold-water
Cutthroat trout	*Oncorhynchus clarkii*	Cold-water
Coho salmon	*Oncorhynchus kisutch*	Cold-water
Rainbow trout	*Oncorhynchus mykiss*	Cold-water
Kokanee (sockeye) salmon	*Oncorhynchus nerka*	Cold-water
Chinook salmon	*Oncorhynchus tschawytscha*	Cold-water
Bonneville cisco	*Prosopium gemmifer*	Cold-water
Brown trout	*Salmo trutta*	Cold-water
Brook trout	*Salvelinus fontinalis*	Cold-water
Lake trout	*Salvelinus namaycush*	Cold-water
Trout-perch	*Percopsis omiscomaycus*	Creek
Burbot	*Lota lota maculosa*	River
Western banded killifish	*Fundulus diaphanus menona*	Creek
Northern plains killifish	*Fundulus kansae*	Creek
Plains topminnow	*Fundulus sciadicus*	Creek
Brook stickleback	*Culaea inconstans*	Creek
Mottled sculpin	*Cottus bairdii*	Creek
White bass	*Morone chrysops*	River
Striped bass	*Morone saxatilis*	River
Rock bass	*Ambloplites rupestris*	Creek
Green sunfish	*Lepomis cyanellus*	Creek
Pumpkinseed	*Lepomis gibbosus*	Creek
Orangespotted sunfish	*Lepomis humilis*	Creek
Bluegill	*Lepomis macrochirus macrochirus*	Creek
Redear sunfish	*Lepomis microlophus*	Creek
Smallmouth bass	*Micropterus dolomieu*	Creek
Largemouth bass	*Micropterus salmoides salmoides*	Creek
White crappie	*Pomoxis annularis*	Creek
Black crappie	*Pomoxis nigromaculatus*	Creek
Iowa darter	*Etheostoma exile*	Creek
Johnny darter	*Etheostoma nigrum*	Creek
Yellow perch	*Perca flavescens*	Creek
Logperch	*Percina caprodes*	Creek
Blackside darter	*Percina maculata*	Creek
Slenderhead darter	*Percina phoxocephala*	Creek
Sauger	*Sander canadensis*	River
Walleye	*Sander vitreus vitreus*	River
Freshwater drum	*Aplodinotus grunniens*	River

[1]Fishes are divided into three categories based on their general habitat associations and based on our observations in South Dakota and literature. River fishes are most abundant in larger streams, commonly associate with fine substrates, and are relatively tolerant of turbidity. Creek fishes are normally associated with smaller streams, coarser substrates, and have relatively low tolerance of turbidity. Cold-water fishes are restricted to high elevation streams, cold springs, deep lakes and impoundments, and reservoir tailwaters.

In contrast, river fishes are associated with fine substrates (e.g., sand). Many of these species (e.g., pallid sturgeon, shovelnose sturgeon) have streamlined bodies that make them efficient swimmers and embedded scales that resist abrasion from sand (Figure 3.2). Some (e.g., paddlefish, sturgeon chub, sicklefin chub) have highly developed taste buds that are found on their barbels, fins, heads, and bodies because they rely on their senses of smell and taste to find food in low light environments such as deep or turbid waters.

Figure 3.2. The river carpsucker (*Carpiodes carpio*) is an example of a river fish.

Others (e.g., goldeye, sauger) have highly developed eyesight. River fishes may also feed in shallow, clear water at night to avoid high daytime temperatures, predation from terrestrial animals (e.g., fish eating birds), and to take advantage of nocturnal prey species (e.g., dragonfly larvae). These physical and behavioral adaptations enable river fishes to live in swift water over unstable substrate. When river fishes occupy lakes and impoundments they often use open-water and deep-water habitats. When they use lake shorelines they often select wave swept areas. Some river fishes may occupy small streams if there are large habitats such as extensive pools. Others may use small streams during the spawning season, in which case they make spawning "runs" upstream to spawning areas. River fishes are relatively sensitive to changes in the flow regime, sediment supply, and channel shape of a stream, partly because the combination of high flows and a wide river channel with unstable substrate creates a wide variety of habitat types. One of the most important habitat types is shallow, low-velocity areas such as backwaters and embayments that are nursery habitats for most species and thus necessary for species to maintain their populations.

Cold-water fishes are typical of northern, high elevation, and deep lake environments (Figure 3.3). In South Dakota, cold-water habitats are restricted to mountain streams of the Black Hills, cold springs, deep impoundments, and reservoir tailwaters. Cold-water habitats normally have coarse substrates and clear waters. Cold-water

Figure 3.3. The rainbow trout (*Oncorhynchus mykiss*) is an example of a cold-water fish.

fishes that are native to South Dakota primarily use streams but many introduced cold-water fishes also use lakes.

Fish biologists distinguish between ***native*** fish and ***introduced*** fish species. Native species were present in South Dakota at the time when Anglo-American settlement began. These species colonized the waters of South Dakota without human aid. They were tolerant of the prevailing environmental conditions and were able to coexist with other native species that were present. In contrast, humans brought in introduced species either intentionally or accidentally. Intentional introductions began with the earliest pioneers and thus predated fish inventories (Bailey and Allum 1962). Introduced species only became established where they could tolerate environmental conditions and coexist with other species that were already present. Because of this, the number of species that could be fruitfully introduced to South Dakota was limited, but in special cases, species were introduced on a regular basis where they could not become established to provide recreational put-and-take fisheries.

Most native fish species in South Dakota colonized the state within the last 18,000 years. Prior to that the eastern portion of the state was covered by glacial ice and fish could not survive there. During glaciation, the western portion of the state may have supported some fish species but conditions were tundra-like and many species that are present today probably could not have survived at that time. As the glaciers receded, the climate warmed and by 9,500 years ago active glacial ice was completely gone from South Dakota. Since then dramatic climate changes have occurred including repeated extreme dry periods. The frequency and duration of extreme dry periods has declined through the last 4,000 years and it is likely that most native South Dakota fishes, particularly those that depend upon permanent aquatic habitats, colonized waters of the state during that period.

In this chapter we summarize the fish faunas of 14 distinct river drainages and of the Missouri River Valley (Figure 3.4, Table 3.2). Factors that determine the composition of each fish fauna include the climate, the variety of aquatic habitats, the proximity of drainages to others that have potential fish colonists, the amount of human activity that affects aquatic

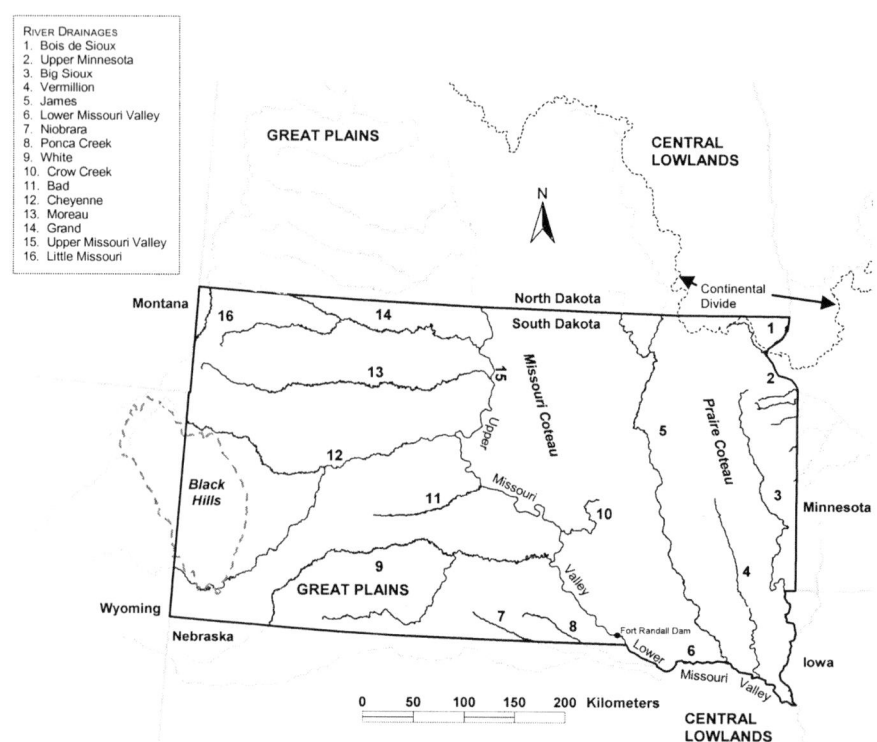

Figure 3.4 Map of South Dakota showing river drainages, geomorphic provinces, and other important regions of the state.

Table 3.2 **Geographical characterization of South Dakota river drainages.**

DRAINAGE	DRAINAGE AREA (km²)	POLITICAL REGION[1]	PHYSIOGRAPHIC PROVINCE[2]
Bois de Sioux River	965	East River	Central Lowlands
Minnesota River	2,529	East River	Central Lowlands
Big Sioux River	10,056	East River	Central Lowlands
Vermillion River	4,267	East River	Central Lowlands
James River	19,308	East River	Central Lowlands
Niobrara River[3] Ponca Creek[3]	3,218	West River	Great Plains
White River	13,274	West River	Great Plains
Crow Creek[4]	?	East River	Great Plains
Bad River	5,069	West River	Great Plains
Cheyenne River	38,455	West River	Great Plains
Moreau River	8,105	West River	Great Plains
Grand River	8,320	West River	Great Plains
Little Missouri River	973	West River	Great Plains
Missouri River Valley[4,5]	26,725	Both	Both

[1]South Dakotans refer to the portion of the state east of the Missouri River as "East River" and the portion west of the Missouri River as "West River."

[2]The two physiographic provinces have different elevations and geology. See text for more description.

[3]Drainage area is reported for the Niobrara River and Ponca Creek drainages combined.

[4]Drainage area for Crow Creek is unknown and is a portion of the drainage area reported for the Missouri River Valley.

[5]Drainage area reported for the Missouri River Valley is contributing drainages within South Dakota excluding the major drainages. The actual drainage area for the Missouri River Valley is much greater and increases as the Missouri River traverses South Dakota. For example, drainage area at Bismarck, North Dakota, is 299,918 km2 and drainage area at Sioux City, Iowa, is 506,191 km².

habitats, and the number of nonnative species that humans introduce. The river drainages vary in size (area). Larger river drainages have more fish species because they have more aquatic habitats and also support larger fish populations that are more resistant to extinction (Hoagstrom and Berry 2006). In short, river drainage geography has a lot to do with what fish species are present.

Hudson Bay Watershed, Red River of the North Sub-Drainage

Bois de Sioux River drainage

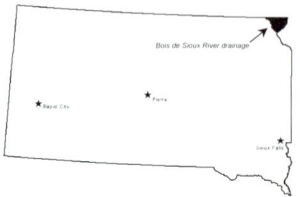

The Bois de Sioux River begins at the outlet of Lake Traverse on the South Dakota-Minnesota border. Jim Creek of South Dakota and the Mustinka River of Minnesota contribute runoff to Lake Traverse. The largest South Dakota cities that lie within the Bois de Sioux River drainage are New Effington and Rosholt.

A total of 33 fish species, of which 25 were native and eight were introduced, was documented from the Bois de Sioux River drainage of South Dakota during the last 150 years (Table 3.3). At least 20 of these (14 native and six nonnative) are currently present, including the rare golden redhorse, but it is likely that more species are present because the recent (post-1990) composition of the Bois de Sioux fish fauna is known only from surveys of Lake Traverse. Stream and wetland fishes in the Bois de Sioux River drainage have not been extensively sampled and their recent distribution is poorly known. Black bullhead, white bass, yellow perch, and walleye are important game species in Lake Traverse (MDNR 2005b).

The Bois de Sioux River drainage was entirely glaciated. As the glaciers melted, the Bois de Sioux River drainage was inundated by melt water that formed Lake Agassiz. The primary route native fishes used to reach the Bois de Sioux River drainage as the glaciers receded was the River Warren, which carried glacial melt water from Lake Agassiz to the Mississippi River. Today, the ancient valley of the River Warren is partly filled by Lake Traverse. Some native fishes may have also reached the Bois de Sioux River drainage from the Great Lakes-Saint Lawrence Watershed when it became the new outlet for glacial melt water, replacing the River Warren.

Table 3.3 **A total of 33 fish species (25 native, 8 introduced) is known from the Bois de Sioux River drainage in South Dakota. Native species were present at the time of settlement by Anglo-Americans. Humans transported introduced species to the Bois de Sioux River drainage. Recent speciesa (20 total of which 14 are native and 6 are introduced) were present in Bois de Sioux River drainage fish surveys since 1990. The symbol + indicates yes and the symbol – indicates no.**

COMMON NAME	SCIENTIFIC NAME	NATIVE	INTRODUCED	RECENT[a]
Common carp	Cyprinus carpio	-	+	+
Brassy minnow	Hybognathus hankinsoni	+	-	?
Common shiner	Luxilus cornutus	+	-	*
Hornyhead chub	Nocomis biguttatus	+	-	?
Emerald shiner	Notropis atherinoides	+	-	+
Bigmouth shiner	Notropis dorsalis	+	-	*
Blacknose shiner	Notropis heterolepis	+	-	?
Bluntnose minnow	Pimephales notatus	+	-	*
Fathead minnow	Pimephales promelas	+	-	*
Longnose dace	Rhinichthys cataractae cataractae	+	-	*
Creek chub	Semotilus atromaculatus	+	-	*
Central quillback carpsucker	Carpiodes cyprinus hinei	+	-	+
White sucker	Catostomus commersonii	+	-	+
Bigmouth buffalo	Ictiobus cyprinellus	+	-	+
Golden redhorse	Moxostoma erythrurum	+	-	+
Shorthead redhorse	Moxostoma macrolepidotum	+	-	+
Black bullhead	Ameiurus melas	+	-	+
Yellow bullhead	Ameiurus natalis	-	+	+
Brown bullhead	Ameiurus nebulosus	+	-	+
Channel catfish	Ictalurus punctatus	+	-	+
Northern pike	Esox lucius	+	-	+
Brook stickleback	Culaea inconstans	+	-	*
White bass	Morone chrysops	-	+	+
Orangespotted sunfish	Lepomis humilis	-	+	*
Bluegill	Lepomis macrochirus macrochurus	-	+	*
Largemouth bass	Micropterus salmoides salmoides	-	+	+
White crappie	Pomoxis annularis	-	+	+
Black crappie	Pomoxis nigromaculatus	-	+	+
Iowa darter	Etheostoma exile	+	-	*
Johnny darter	Etheostoma nigrum	+	-	+
Yellow perch	Perca flavescens	+	-	+
Walleye	Sander vitreus vitreus	+	-	+
Freshwater drum	Aplodinotus grunniens	+	-	+

[a]Post-1990 fish collections were limited to Lake Traverse so it is likely many stream species not collected in those surveys are still present in the state. Species likely to be present are noted with an * if they are a tolerant species and their persistence is likely or a ? if they are a sensitive species and their persistence may be questionable.

Gulf of Mexico Watershed, Upper Mississippi River Sub-Drainage

Minnesota River tributary drainage

The Minnesota River drainage begins on the eastern edge of the Prairie Coteau in northeastern South Dakota. The Minnesota River itself begins at the outlet of Big Stone Lake on the South Dakota-Minnesota border. The Little Minnesota River and the Whetstone River of South Dakota and Fish Creek of Minnesota flow into Big Stone Lake. The Yellow Bank River and Lac qui Parle River are tributaries to the Minnesota River that begin in South Dakota. Their headwaters are in the Prairie Couteau and they flow eastward into the state of Minnesota before joining the Minnesota River. Sisseton and Milbank are the largest South Dakota cities that lie within the Minnesota River drainage.

A total of 60 fish species, of which 56 were native and four were introduced, was documented from the Minnesota River drainage of South Dakota during the last 150 years (Table 3.4). At least 47 of these (45 native, two introduced) are currently present. The relatively high number of fish species in the Minnesota River drainage of South Dakota, a relatively small drainage, is attributed to relatively high precipitation and moderate winter temperatures that provide a relatively favorable climate for fishes. Streams of the Minnesota River drainage maintain the last populations of rare hornyhead chubs and rare carmine shiners (formerly known as rosyface shiners) within the state. The state endangered blacknose shiner, state endangered central mudminnow, and state endangered western banded killifish, state threatened northern redbelly dace, rare golden redhorse, and rare blackside darter are also present. Ten species of native fishes including longnose gar, bowfin, American eel, skipjack herring, blackchin shiner, northern hog sucker, trout-perch, smallmouth bass, white crappie, and slenderhead darter, were missing from recent (post-1990) fish surveys. White bass, yellow perch,

Table 3.4   A total of 60 fish species (56 native, 4 introduced) is known from the Minnesota River drainage in South Dakota. Native species were present at the time of settlement by Anglo-Americans. Humans transported introduced species to the Minnesota River drainage. Recent species (47 total of which 45 are native and 2 are introduced) were present in Minnesota River drainage fish surveys since 1990. The symbol + indicates yes and the symbol − indicates no.

COMMON NAME	SCIENTIFIC NAME	NATIVE	INTRODUCED	RECENT
Longnose gar	*Lepisosteus osseus*	+	-	-
Shortnose gar	*Lepisosteus platostomus*	+	-	+
Bowfin	*Amia calva*	+	-	-
American eel	*Anguilla rostrata*	+	-	-
Skipjack herring	*Alosa chrysochloris*	+	-	-
Central stoneroller	*Campostoma anomalum anomalum*	+	-	+
Common carp	*Cyprinus carpio*	-	+	+
Brassy minnow	*Hybognathus hankinsoni*	+	-	+
Common shiner	*Luxilus cornutus*	+	-	+
Hornyhead chub	*Nocomis biguttatus*	+	-	+
Golden shiner	*Notemigonus crysoleucas*	+	-	+
Emerald shiner	*Notropis atherinoides*	+	-	+
Bigmouth shiner	*Notropis dorsalis*	+	-	+
Blackchin shiner	*Notropis heterodon*	+	-	-
Blacknose shiner	*Notropis heterolepis*	+	-	+
Spottail shiner	*Notropis hudsonius*	+	-	+
Carmine shiner	*Notropis percobromus*	+	-	+
Eastern sand shiner	*Notropis stramineus stramineus*	+	-	+
Northern redbelly dace	*Phoxinus eos*	+	-	+
Bluntnose minnow	*Pimephales notatus*	+	-	+
Fathead minnow	*Pimephales promelas*	+	-	+
Western blacknose dace	*Rhinichthys obtusus*	+	-	+
Creek chub	*Semotilus atromaculatus*	+	-	+
Central quillback carpsucker	*Carpiodes cyprinus hinei*	+	-	+
White sucker	*Catostomus commersonii*	+	-	+
Northern hog sucker	*Hypentelium nigricans*	+	-	-
Bigmouth buffalo	*Ictiobus cyprinellus*	+	-	+
Golden redhorse	*Moxostoma erythrurum*	+	-	+
Shorthead redhorse	*Moxostoma macrolepidotum*	+	-	+
Black bullhead	*Ameiurus melas*	+	-	+
Yellow bullhead	*Ameiurus natalis*	+	-	+
Brown bullhead	*Ameiurus nebulosus*	+	-	+
Channel catfish	*Ictalurus punctatus*	+	-	+
Stonecat	*Noturus flavus*	+	-	-
Tadpole madtom	*Noturus gyrinus*	+	-	+
Northern pike	*Esox lucius*	+	-	+
Central mudminnow	*Umbra limi*	+	-	+
Rainbow trout	*Oncorhynchus mykiss*	-	+	+
Brown trout	*Salmo trutta*	-	+	+

Table 3.4 continued.

COMMON NAME	SCIENTIFIC NAME	NATIVE	INTRODUCED	RECENT
Brook trout	*Salvelinus fontinalis*	-	+	-
Trout-perch	*Percopsis omiscomaycus*	+	-	-
Western banded killifish	*Fundulus diaphanus menona*	+	-	+
Brook stickleback	*Culaea inconstans*	+	-	+
White bass	*Morone chrysops*	+	-	+
Rock bass	*Ambloplites rupestris*	+	-	+
Green sunfish	*Lepomis cyanellus*	+	-	+
Pumpkinseed	*Lepomis gibbosus*	+	-	+
Orangespotted sunfish	*Lepomis humilis*	+	-	+
Bluegill	*Lepomis macrochirus macrochirus*	+	-	+
Smallmouth bass	*Micropterus dolomieu*	+	-	-
Largemouth bass	*Micropterus salmoides salmoides*	+	-	+
White crappie	*Pomoxis annularis*	+	-	-
Black crappie	*Pomoxis nigromaculatus*	+	-	+
Iowa darter	*Etheostoma exile*	+	-	+
Johnny darter	*Etheostoma nigrum*	+	-	+
Yellow perch	*Perca flavescens*	+	-	+
Blackside darter	*Percina maculata*	+	-	+
Slenderhead darter	*Percina phoxocephala*	+	-	-
Walleye	*Sander vitreus vitreus*	+	-	+
Freshwater drum	*Aplodinotus grunniens*	+	-	+

and walleye are the most important game fish in Big Stone Lake (MDNR 2005a). One of the few brown trout fisheries of eastern South Dakota is in Gary Creek, a tributary to the Lac qui Parle River (Milewski and Willis 1989, Pope and Willis 1994).

The Minnesota River drainage in South Dakota was entirely glaciated. After the glaciers receded, fishes could have reached the Minnesota River drainage from the Mississippi River or from the Big Sioux River drainage. If fishes colonized from the Big Sioux River drainage they would have used connections where headwater streams of these two river basins adjoin in wetlands of the Prairie Coteau. In prehistoric time there were periods with a wetter climate than today and it is likely that many of these streams were connected then. At least one such connection still exists at Lake Hendricks. When Lake Hendricks has a high water level it overflows both into the Lac qui Parle River of the Minnesota River drainage and into Deer Creek of the Big Sioux River drainage. At these times, fishes in the lake can disperse in either direction and fishes from either river drainage can enter the lake.

Gulf of Mexico Watershed, Missouri River Sub-drainage

Big Sioux River tributary drainage

The Big Sioux River begins in the Prairie Coteau of northeastern South Dakota and includes a large number of Prairie Coteau lakes. It flows south through eastern South Dakota and joins the Missouri River on the South Dakota-Nebraska border. The lower Big Sioux River forms the South Dakota-Iowa border. The Rock River of Minnesota and Iowa is the largest tributary of the Big Sioux River. Stray Horse Creek, Hidewood Creek, Medary Creek, Battle Creek, Bachelor Creek, Skunk Creek, Split Rock Creek, and Brule Creek are major tributaries of this river in South Dakota. Watertown, Brookings, Flandreau, Madison, Sioux Falls, Canton, and North Sioux City are the largest South Dakota cities that lie within the Big Sioux River drainage.

A total of 82 fish species, of which 71 were native and 11 were introduced, was documented from the Big Sioux River drainage of South Dakota during the last 150 years (Table 3.5). At least 68 of these (58 native, 10 introduced) are currently present. This is the largest number of species for any tributary river drainage in South Dakota (but the Missouri River Valley contains more species) and is explained by a relatively wet and warm climate that is favorable for fishes. The Big Sioux River drainage contains the last known populations of the rare logperch in South Dakota. Many populations of the federally endangered Topeka shiner are also present. The state endangered blacknose shiner, state endangered western banded killifish, state threatened northern redbelly dace, rare central quillback carpsucker, rare blue

Table 3.5 A total of 82 fish species (71 native, 11 introduced) is known from the Big Sioux River drainage in South Dakota. Native species were present at the time of settlement by Anglo-Americans. Humans transported introduced species to the Big Sioux River drainage. Recent species (68 total of which 58 are native and 10 are introduced) were present in Big Sioux River drainage fish surveys since 1990. The symbol + indicates yes and the symbol – indicates no.

COMMON NAME	SCIENTIFIC NAME	NATIVE	INTRODUCED	RECENT
Shovelnose sturgeon	*Scaphirhynchus platorynchus*	+	-	+
Paddlefish	*Polyodon spathula*	+	-	-
Longnose gar	*Lepisosteus osseus*	+	-	-
Shortnose gar	*Lepisosteus platostomus*	+	-	+
Goldeye	*Hiodon alosoides*	+	-	+
Mooneye	*Hiodon tergisus*	+	-	-
American eel	*Anguilla rostrata*	+	-	-
Gizzard shad	*Dorosoma cepedianum*	+	-	+
Central stoneroller	*Campostoma anomalum anomalum*	+	-	+
Goldfish	*Carassius auratus*	-	+	+
Grass carp	*Ctenopharyngodon idella*	-	+	+
Red shiner	*Cyprinella lutrensis lutrensis*	+	-	+
Common carp	*Cyprinus carpio*	-	+	+
Western silvery minnow	*Hybognathus argyritis*	+	-	-
Brassy minnow	*Hybognathus hankinsoni*	+	-	+
Silver carp	*Hypophthalmichthys molitrix*	-	+	+
Bighead carp	*Hypophthalmichthys nobilis*	-	+	+
Common shiner	*Luxilus cornutus*	+	-	+
Silver chub	*Macrhybopsis storeriana*	+	-	+
Hornyhead chub	*Nocomis biguttatus*	+	-	-
Golden shiner	*Notemigonus crysoleucas*	+	-	-
Emerald shiner	*Notropis atherinoides*	+	-	+
Bigmouth shiner	*Notropis dorsalis*	+	-	+
Blacknose shiner	*Notropis heterolepis*	+	-	+
Spottail shiner	*Notropis hudsonius*	+	-	+
Carmine shiner	*Notropis percobromus*	+	-	-
Eastern sand shiner	*Notropis stramineus stramineus*	+	-	+
Topeka shiner	*Notropis topeka*	+	-	+
Suckermouth minnow	*Phenacobius mirabilis*	+	-	-
Northern redbelly dace	*Phoxinus eos*	+	-	+
Southern redbelly dace	*Phoxinus erythrogaster*	+	-	+
Bluntnose minnow	*Pimephales notatus*	+	-	+
Fathead minnow	*Pimephales promelas*	+	-	+
Flathead chub	*Platygobio gracilis*	+	-	-
Western blacknose dace	*Rhinichthys obtusus*	+	-	+
Rudd	*Scardinius erythrophthalmus*	-	+	+
Creek chub	*Semotilus atromaculatus*	+	-	+
Northern river carpsucker	*Carpiodes carpio carpio*	+	-	+
Central quillback carpsucker	*Carpiodes cyprinus hinei*	+	-	+

Table 3.5 continued.

COMMON NAME	SCIENTIFIC NAME	NATIVE	INTRODUCED	RECENT
Highfin carpsucker	*Carpiodes velifer*	+	-	-
White sucker	*Catostomus commersonii*	+	-	+
Blue sucker	*Cycleptus elongatus*	+	-	+
Northern hogsucker	*Hypentelium nigricans*	+	-	-
Smallmouth buffalo	*Ictiobus bubalus*	+	-	+
Bigmouth buffalo	*Ictiobus cyprinellus*	+	-	+
Black buffalo	*Ictiobus niger*	+	-	-
Shorthead redhorse	*Moxostoma macrolepidotum*	+	-	+
Black bullhead	*Ameiurus melas*	+	-	+
Yellow bullhead	*Ameiurus natalis*	+	-	+
Blue catfish	*Ictalurus furcatus*	+	-	+
Channel catfish	*Ictalurus punctatus*	+	-	+
Stonecat	*Noturus flavus*	+	-	+
Tadpole madtom	*Noturus gyrinus*	+	-	+
Flathead catfish	*Pylodictis olivaris*	+	-	+
Northern pike	*Esox lucius*	+	-	+
Muskellunge	*Esox masquinongy*	-	+	+
Central mudminnow	*Umbra limi*	+	-	+
Cisco	*Coregonus artedi*	-	+	+
Rainbow trout	*Oncorhynchus mykiss*	-	+	+
Trout-perch	*Percopsis omiscomaycus*	+	-	+
Western banded killifish	*Fundulus diaphanus menona*	+	-	+
Plains topminnow	*Fundulus sciadicus*	+	-	+
Brook stickleback	*Culaea inconstans*	+	-	+
White bass	*Morone chrysops*	+	-	+
Rock bass	*Ambloplites rupestris*	+	-	+
Green sunfish	*Lepomis cyanellus*	+	-	+
Pumpkinseed	*Lepomis gibbosus*	+	-	+
Orangespotted sunfish	*Lepomis humilis*	+	-	+
Bluegill	*Lepomis macrochirus macrochirus*	+	-	+
Redear sunfish	*Lepomis microlophus*	-	+	-
Smallmouth bass	*Micropterus dolomieu*	-	+	+
Largemouth bass	*Micropterus salmoides salmoides*	+	-	+
White crappie	*Pomoxis annularis*	+	-	+
Black crappie	*Pomoxis nigromaculatus*	+	-	+
Iowa darter	*Etheostoma exile*	+	-	+
Johnny darter	*Etheostoma nigrum*	+	-	+
Yellow perch	*Perca flavescens*	+	-	+
Logperch	*Percina caprodes*	+	-	+
Blackside darter	*Percina maculata*	+	-	+
Sauger	*Sander canadensis*	+	-	+
Walleye	*Sander vitreus vitreus*	+	-	+
Freshwater drum	*Aplodinotus grunniens*	+	-	+

sucker, rare plains topminnow, and rare blackside darter are present as well. A comparison of fish distributions in 1967 and 1994 suggested that water quality had improved through that period (Dieterman and Berry 1998). However, thirteen native fish species including paddlefish, longnose gar, mooneye, American eel, western silvery minnow, hornyhead chub, golden shiner, carmine shiner, suckermouth minnow, flathead chub, highfin carpsucker, northern hogsucker, and black buffalo were missing from recent (post-1990) fish surveys. Black bullhead, bluegill, black crappie, largemouth bass, yellow perch, and walleye are important game fish in lakes and impoundments of the Big Sioux River drainage (SDGFP 2005b, c). Channel catfish, flathead catfish, and walleye are important game fish in the Big Sioux River.

The Big Sioux River drainage was entirely glaciated. The river formed during glacial melting. Meltwater from ice lobes that lay to the east and west drained into the Big Sioux River and excavated the river valley (Hogan 1995). After glaciation, fishes could have reached the Big Sioux River drainage from several different directions. The mainstem Missouri River is the most obvious colonization route. Sixty fish species, however, are distributed throughout headwater streams of the Big Sioux, Minnesota, Des Moines, and Boyer river drainages (Hoagstrom and Berry 2006). These species might have colonized the Big Sioux River drainage through headwater connections with those eastern streams during prehistoric wet periods. A headwater connection still exists between the Big Sioux River and Minnesota River drainages at Lake Hendricks on the South Dakota-Minnesota state boundary.

Vermillion River tributary drainage

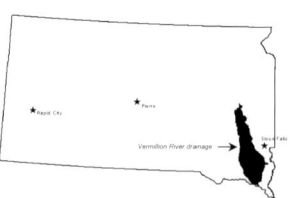

The headwaters of the Vermillion River drainage include Cherry Lake, Spirit Lake, Lake Preston, Lake Whitewood, and Lake Thompson. The East and West Forks of the Vermillion River begin south and west of Lake Thompson. Both forks flow south and eventually join near Parker, South Dakota, to form the Vermillion River. Turkey Ridge Creek and Clay Creek are major tributaries to the Vermillion River. The Vermillion River joins the Missouri River along the South Dakota-Nebraska border near Vermillion, South Dakota. Parker, Lennox, Centerville, and Vermillion are the largest South Dakota cities that lie within the Vermillion River drainage.

A total of 58 fish species, of which 47 were native and 11 were introduced, was documented from the Vermillion River drainage during the last 150 years (Table 3.6). At least 46 of these (36 native, 10 introduced) are currently present. The Vermillion River drainage contains many populations of the federally endangered Topeka shiner. It is an important habitat for some large river fishes that have been impacted by impoundments on the main stem Missouri River, such as the rare blue sucker. Twelve native fishes including silver lamprey, paddlefish, American eel, western silvery minnow, river shiner, silverband shiner, flathead chub, central quillback carpsucker,

Table 3.6   A total of 58 fish species (47 native, 11 introduced) is known from the Vermillion River drainage in South Dakota. Native species were present at the time of settlement by Anglo-Americans. Humans transported introduced species to the Vermillion River drainage. Recent species (46 total of which 36 are native and 10 are introduced) were present in Vermillion River drainage fish surveys since 1990. The symbol + indicates yes and the symbol – indicates no.

COMMON NAME	SCIENTIFIC NAME	NATIVE	INTRODUCED	RECENT
Silver lamprey	*Ichthyomyzon unicuspis*	+	-	-
Shovelnose sturgeon	*Scaphirhynchus platorynchus*	+	-	+
Paddlefish	*Polyodon spathula*	+	-	-
Longnose gar	*Lepisosteus osseus*	+	-	+
Shortnose gar	*Lepisosteus platostomus*	+	-	+
Goldeye	*Hiodon alosoides*	+	-	+
American eel	*Anguilla rostrata*	+	-	-
Gizzard shad	*Dorosoma cepedianum*	+	-	+
Central stoneroller	*Campostoma anomalum anomalum*	+	-	+
Red shiner	*Cyprinella lutrensis lutrensis*	+	-	+
Spotfin shiner	*Cyprinella spiloptera*	-	+	+
Common carp	*Cyprinus carpio*	-	+	+
Western silvery minnow	*Hybognathus argyritis*	+	-	-
Brassy minnow	*Hybognathus hankinsoni*	+	-	+
Common shiner	*Luxilus cornutus*	+	-	+

Table 3.6 continued.

COMMON NAME	SCIENTIFIC NAME	NATIVE	INTRODUCED	RECENT
Silver carp	*Hypophthalmichthys molitrix*	-	+	+
Golden shiner	*Notemigonus crysoleucas*	+	-	+
Emerald shiner	*Notropis atherinoides*	+	-	+
River shiner	*Notropis blennius*	+	-	-
Bigmouth shiner	*Notropis dorsalis*	+	-	+
Silverband shiner	*Notropis shumardi*	+	-	-
Eastern sand shiner	*Notropis stramineus stramineus*	+	-	+
Topeka shiner	*Notropis topeka*	+	-	+
Fathead minnow	*Pimephales promelas*	+	-	+
Flathead chub	*Platygobio gracilis*	+	-	-
Western blacknose dace	*Rhinichthys obtusus*	+	-	+
Creek chub	*Semotilus atromaculatus*	+	-	+
Northern river carpsucker	*Carpiodes carpio carpio*	+	-	+
Central quillback carpsucker	*Carpiodes cyprinus hinei*	+	-	-
White sucker	*Catostomus commersonii*	+	-	+
Blue sucker	*Cycleptus elongatus*	+	-	+
Smallmouth buffalo	*Ictiobus bubalus*	+	-	+
Bigmouth buffalo	*Ictiobus cyprinellus*	+	-	+
Shorthead redhorse	*Moxostoma macrolepidotum*	+	-	+
Black bullhead	*Ameiurus melas*	+	-	+
Channel catfish	*Ictalurus punctatus*	+	-	+
Stonecat	*Noturus flavus*	+	-	+
Tadpole madtom	*Noturus gyrinus*	+	-	+
Flathead catfish	*Pylodictis olivaris*	+	-	+
Northern pike	*Esox lucius*	-	+	+
Rainbow trout	*Oncorhynchus mykiss*	-	+	+
Burbot	*Lota lota maculosa*	+	-	-
Plains topminnow	*Fundulus sciadicus*	+	-	-
Brook stickleback	*Culaea inconstans*	+	-	+
White bass	*Morone chrysops*	+	-	+
Green sunfish	*Lepomis cyanellus*	+	-	+
Orangespotted sunfish	*Lepomis humilis*	+	-	+
Bluegill	*Lepomis macrochirus macrochirus*	-	+	+
Smallmouth bass	*Micropterus dolomieu*	-	+	+
Largemouth bass	*Micropterus salmoides salmoides*	-	+	+
White crappie	*Pomoxis annularis*	-	+	+
Black crappie	*Pomoxis nigromaculatus*	-	+	+
Iowa darter	*Etheostoma exile*	+	-	+
Johnny darter	*Etheostoma nigrum*	+	-	+
Yellow perch	*Perca flavescens*	+	-	+
Sauger	*Sander canadensis*	+	-	+
Walleye	*Sander vitreus vitreus*	+	-	+
Freshwater drum	*Aplodinotus grunniens*	+	-	+

bigmouth buffalo, burbot, plains topminnow, and white bass were missing from recent (post-1990) fish surveys. Bluegill, largemouth bass, black crappie, yellow perch, and walleye are important game fish in the Vermillion River drainage (SDGFP 2005c).

The Vermillion River drainage was entirely glaciated and the river valley was created roughly 12,000 years ago by glacial meltwater (Christensen and Stephens 1967). Fishes colonized the Vermillion River drainage from the Missouri River. They also possibly used headwater connections with the Big Sioux River or James River drainage that were presumably present during prehistoric wet periods. Headwater connections between the Vermillion River and neighboring Big Sioux River and James River drainages are most probable in the region between Madison and Mitchell, South Dakota.

James River tributary drainage

The James River begins in the Missouri Coteau of central North Dakota. It flows south and enters South Dakota near Hecla, South Dakota. The Elm River, Moccasin Creek, Mud Creek, Snake Creek, Turtle Creek, Shue Creek, and Firesteel Creek are major James River tributaries in South Dakota. The James River joins the Missouri River along the South Dakota-Nebraska border near Yankton, South Dakota. Aberdeen, Redfield, Miller, Huron, and Mitchell are the largest South Dakota cities that lie within the James River drainage.

A total of 64 fish species, of which 49 were native and 15 were introduced, was documented from the James River drainage within South Dakota during the last 150 years (Table 3.7). At least 57 of these (45 native, 12 introduced) are currently present. Several populations of the federally endangered Topeka shiner inhabit the James River drainage of South Dakota. The James River drainage provides important habitat for some large river fishes that have been impacted by impoundments on the main stem Missouri River such as

Table 3.7   **A total of 64 fish species (49 native, 15 introduced) is known from the James River drainage in South Dakota. Native species were present at the time of settlement by Anglo-Americans. Humans transported introduced species to the James River drainage. Recent species (57 total of which 45 are native and 12 are introduced) were present in James River drainage fish surveys since 1990. The symbol + indicates yes and the symbol – indicates no.**

COMMON NAME	SCIENTIFIC NAME	NATIVE	INTRODUCED	RECENT
Shovelnose sturgeon	*Scaphirhynchus platorynchus*	+	-	+
Paddlefish	*Polyodon spathula*	+	-	+
Longnose gar	*Lepisosteus osseus*	+	-	+
Shortnose gar	*Lepisosteus platostomus*	+	-	+
Goldeye	*Hiodon alosoides*	+	-	+
American eel	*Anguilla rostrata*	+	-	+
Gizzard shad	*Dorosoma cepedianum*	+	-	+
Central stoneroller	*Campostoma anomalum anomalum*	+	-	+
Grass carp	*Ctenopharyngodon idella*	-	+	+
Red shiner	*Cyprinella lutrensis lutrensis*	+	-	+
Common carp	*Cyprinus carpio*	-	+	+
Brassy minnow	*Hybognathus hankinsoni*	+	-	+
Silver carp	*Hypophthalmichthys molitrix*	-	+	+
Bighead carp	*Hypophthalmichthys nobilis*	-	+	+
Common shiner	*Luxilus cornutus*	+	-	+
Golden shiner	*Notemigonus crysoleucas*	+	-	-
Emerald shiner	*Notropis atherinoides*	+	-	+
Bigmouth shiner	*Notropis dorsalis*	+	-	+
Blacknose shiner	*Notropis heterolepis*	+	-	-
Spottail shiner	*Notropis hudsonius*	+	-	+
Eastern sand shiner	*Notropis stramineus stramineus*	+	-	+
Topeka shiner	*Notropis topeka*	+	-	+
Bluntnose minnow	*Pimephales notatus*	+	-	+
Fathead minnow	*Pimephales promelas*	+	-	+
Western blacknose dace	*Rhinichthys obtusus*	+	-	+
Creek chub	*Semotilus atromaculatus*	+	-	+
Northern river carpsucker	*Carpiodes carpio carpio*	+	-	+
Central quillback carpsucker	*Carpiodes cyprinus hinei*	+	-	+
White sucker	*Catostomus commersonii*	+	-	+
Blue sucker	*Cycleptus elongatus*	+	-	+
Smallmouth buffalo	*Ictiobus bubalus*	+	-	+
Bigmouth buffalo	*Ictiobus cyprinellus*	+	-	+
Black buffalo	*Ictiobus niger*	+	-	-
Golden redhorse	*Moxostoma erythrurum*	+	-	-
Shorthead redhorse	*Moxostoma macrolepidotum*	+	-	+
Black bullhead	*Ameiurus melas*	+	-	+
Yellow bullhead	*Ameiurus natalis*	+	-	+
Blue catfish	*Ictalurus furcatus*	+	-	-
Channel catfish	*Ictalurus punctatus*	+	-	+

Table 3.7 continued.

COMMON NAME	SCIENTIFIC NAME	NATIVE	INTRODUCED	RECENT
Stonecat	*Noturus flavus*	+	-	+
Tadpole madtom	*Noturus gyrinus*	+	-	+
Flathead catfish	*Pylodictis olivaris*	+	-	+
Northern pike	*Esox lucius*	+	-	+
Muskellunge	*Esox masquinongy*	-	+	+
Rainbow trout	*Oncorhynchus mykiss*	-	+	+
Brown trout	*Salmo trutta*	-	+	+
Plains topminnow	*Fundulus sciadicus*	+	-	+
Brook stickleback	*Culaea inconstans*	+	-	+
White bass	*Morone chrysops*	+	-	+
Rock bass	*Ambloplites rupestris*	-	+	-
Green sunfish	*Lepomis cyanellus*	+	-	+
Pumpkinseed	*Lepomis gibbosus*	-	+	-
Orangespotted sunfish	*Lepomis humilis*	+	-	+
Bluegill	*Lepomis macrochirus macrochirus*	-	+	+
Smallmouth bass	*Micropterus dolomieu*	-	+	+
Largemouth bass	*Micropterus salmoides salmoides*	-	+	+
White crappie	*Pomoxis annularis*	-	+	+
Black crappie	*Pomoxis nigromaculatus*	-	+	+
Iowa darter	*Etheostoma exile*	+	-	+
Johnny darter	*Etheostoma nigrum*	+	-	+
Yellow perch	*Perca flavescens*	+	-	+
Sauger	*Sander canadensis*	+	-	+
Walleye	*Sander vitreus vitreus*	+	-	+
Freshwater drum	*Aplodinotus grunniens*	+	-	+

the rare central quillback carpsucker and the rare blue sucker. Five native fishes including golden shiner, blacknose shiner, black buffalo, golden redhorse, and blue catfish were missing from recent (post-1990) fish surveys. Bluegill, black crappie, smallmouth bass, largemouth bass, yellow perch, and walleye are important game fish in lakes and impoundments of the James River drainage (SDGFP 2005c). Channel catfish and flathead catfish are important game fish in the James River.

The James River drainage was entirely glaciated. Meltwater runoff from the James River Lobe of the most recent glacier formed the lower James River Valley (downstream from Redfield, South Dakota). As the James River Lobe melted, it left a deep valley that filled to form Lake Dakota. The Lake Dakota outlet was continually eroded by glacial meltwater and eventually the lake was emptied and continued erosion formed the upper James River Valley. After glaciation, native fishes presumably reached the James River drainage from the Missouri River. They also may have used headwater connections with the Big Sioux River, Vermillion River, Crow Creek, or Sheyenne River (North Dakota, Hudson Bay Watershed) drainages during prehistoric wet periods. This seems most probable in the region west of Webster and Watertown, South Dakota, where the headwaters of James River tributaries intermingle with waters of the Big Sioux River drainage and along the western edge of the Prairie Coteau, between Miller and Wessington Springs, South Dakota where the headwaters of Turtle Creek and Firesteel Creek (James River tributaries) intermingle with those of Crow Creek and Smith Creek (Crow Creek tributaries) on the Missouri Coteau.

Niobrara River tributary drainage

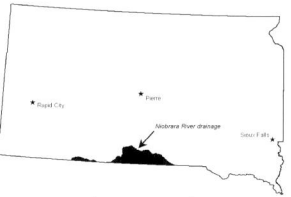

Only a small portion of the Niobrara River drainage, the upper Keya Paha River drainage, lies within South Dakota. The Keya Paha River begins at the junction of Antelope Creek and Rock Creek near Hidden Timber, South Dakota. Antelope Creek and Rock Creek flow northeastward from the Nebraska Sandhills. Once formed, the Keya Paha River turns southeast. Sand Creek, Lost Creek, and Cottonwood Creek are major tributaries to the Keya Paha River in South Dakota. The Keya Paha River flows into Nebraska east of Wewela, South Dakota, and eventually joins the Niobrara River. The Niobrara River joins the Missouri River along the South Dakota-Nebraska border near Running Water, South Dakota. Mission is the largest South Dakota city that lies within the Niobrara River drainage.

A total of 40 fish species, of which 33 were native and seven were introduced, was documented from the Niobrara River drainage of South Dakota within the last 150 years (Table 3.8). At least 35 of these (28 native, seven intro-

Table 3.8 A total of 40 fish species (33 native, 7 introduced) is known from the Niobrara River drainage in South Dakota. Native species were present at the time of settlement by Anglo-Americans. Humans transported introduced species to the Niobrara River drainage. Recent species (35 total of which 28 are native and 7 are introduced) were present in Niobrara River drainage fish surveys since 1990. The symbol + indicates yes and the symbol – indicates no.

COMMON NAME	SCIENTIFIC NAME	NATIVE	INTRODUCED	RECENT
Central stoneroller	*Campostoma anomalum anomalum*	+	-	+
Goldfish	*Carassius auratus*	-	+	+
Red shiner	*Cyprinella lutrensis lutrensis*	+	-	+
Common carp	*Cyprinus carpio*	-	+	+
Western silvery minnow	*Hybognathus argyritis*	+	-	+
Brassy minnow	*Hybognathus hankinsoni*	+	-	+
Plains minnow	*Hybognathus placitus*	+	-	-
Common shiner	*Luxilus cornutus*	+	-	+
Silver chub	*Macrhybopsis storeriana*	+	-	+
Pearl dace	*Margariscus margarita nachtriebi*	+	-	+
Golden shiner	*Notemigonus crysoleucas*	+	-	+
Emerald shiner	*Notropis atherinoides*	+	-	-
Bigmouth shiner	*Notropis dorsalis*	+	-	+
Blacknose shiner	*Notropis heterolepis*	+	-	+
Plains sand shiner	*Notropis stramineus missuriensis*	+	-	+
Northern redbelly dace	*Phoxinus eos*	+	-	+
Finescale dace	*Phoxinus neogaeus*	+	-	+
Fathead minnow	*Pimephales promelas*	+	-	+
Flathead chub	*Platygobio gracilis*	+	-	+
Longnose dace	*Rhinichthys cataractae cataractae*	+	-	+
Western blacknose dace	*Rhinichthys obtusus*	+	-	+
Creek chub	*Semotilus atromaculatus*	+	-	+
Northern river carpsucker	*Carpiodes carpio carpio*	+	-	+
White sucker	*Catostomus commersonii*	+	-	+
Shorthead redhorse	*Moxostoma macrolepidotum*	+	-	+
Black bullhead	*Ameiurus melas*	+	-	+
Channel catfish	*Ictalurus punctatus*	+	-	+
Stonecat	*Noturus flavus*	+	-	+
Northern pike	*Esox lucius*	-	+	+
Plains topminnow	*Fundulus sciadicus*	+	-	+
Green sunfish	*Lepomis cyanellus*	+	-	+
Bluegill	*Lepomis macrochirus macrochirus*	-	+	+
Largemouth bass	*Micropterus salmoides salmoides*	-	+	+
Black crappie	*Pomoxis nigromaculatus*	-	+	+
Iowa darter	*Etheostoma exile*	+	-	+
Johnny darter	*Etheostoma nigrum*	+	-	+
Yellow perch	*Perca flavescens*	-	+	+

duced) including the state endangered blacknose shiner, state endangered finescale dace, state threatened pearl dace, state threatened northern redbelly dace, rare silver chub, and rare plains topminnow, are still present. The native plains minnow and emerald shiner were missing from recent (post-1990) fish surveys.

The Niobrara River drainage of South Dakota lies near the southwestern edge of glaciation and thus if any fishes survived there during glacial periods they were presumably cold-water species. After glaciation there were periods of extreme drought that must have reduced stream flows and thus only drought resistant fish would have been present. Most of the modern day fishes presumably colonized the Niobrara River drainage of South Dakota from the Missouri River during the last 4,000 years as the climate became wetter. It is possible some species reached the Niobrara River drainage from the White River drainage via headwater connections that were presumably present in the northern Nebraska Sandhills during prehistoric wet periods (Mayden 1987). Headwater connections might also have been present between Keya Paha River and Ponca Creek tributaries northwest of Paxton, South Dakota. Fishes that occupy relatively cool waters (e.g., brassy minnow, common shiner, pearl dace, bigmouth shiner, northern redbelly dace, finescale dace, western blacknose dace) most likely colonized the Niobrara River drainage during cold periods and have declined during the warming of the last 200 years.

Ponca Creek
tributary drainage

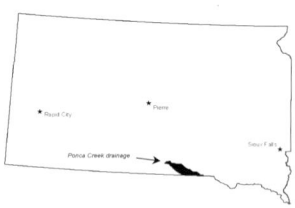

Ponca Creek begins near Colome, South Dakota. It flows southeast and enters Nebraska near Naper, Nebraska. Ponca Creek joins the Missouri River along the South Dakota-Nebraska border near Verdel, Nebraska. Whiskey Creek of Nebraska is the largest Ponca Creek tributary. Ponca Creek does not have any major tributaries within South Dakota.

Gregory is the largest South Dakota city that lies within the Ponca Creek drainage.

A total of 17 fish species, of which 11 were native and six were introduced, was documented from the Ponca Creek drainage during the last 150 years (Table 3.9). All 17 of these, including the state threatened pearl dace, are currently present. There were no pre-1990 surveys so it is unknown whether any additional species were present prior to human impacts.

The Ponca Creek drainage lies very near to the southwestern edge of glaciation and thus presumably supported only cold-water fish at that time, if any fish were present. Post-glacial drought periods must have severely impacted Ponca Creek fishes because even today, when conditions are relatively wet, the waters of the drainage are limited and support relatively few fish species. Modern-day fishes most likely colonized Ponca Creek from the main stem Missouri River during the

Table 3.9 A total of 17 fish species (11 native, 6 introduced) is known from the Ponca Creek drainage in South Dakota. Native species were present at the time of settlement by Anglo-Americans. Humans transported introduced species to the Ponca Creek River drainage. Recent species (17 total of which 11 are native and 6 are introduced) were present in Ponca Creek River drainage fish surveys since 1990. The symbol + indicates yes and the symbol – indicates no.

COMMON NAME	SCIENTIFIC NAME	NATIVE	INTRODUCED	RECENT
Red shiner	*Cyprinella lutrensis lutrensis*	+	-	+
Common carp	*Cyprinus carpio*	-	+	+
Pearl dace	*Margariscus margarita nachtriebi*	+	-	+
Bigmouth shiner	*Notropis dorsalis*	+	-	+
Plains sand shiner	*Notropis stramineus missuriensis*	+	-	+
Fathead minnow	*Pimephales promelas*	+	-	+
Longnose dace	*Rhinichthys cataractae cataractae*	+	-	+
Creek chub	*Semotilus atromaculatus*	+	-	+
White sucker	*Catostomus commersonii*	+	-	+
Black bullhead	*Ameiurus melas*	+	-	+
Northern pike	*Esox lucius*	-	+	+
Green sunfish	*Lepomis cyanellus*	+	-	+
Bluegill	*Lepomis macrochirus macrochirus*	-	+	+
Largemouth bass	*Micropterus salmoides salmoides*	-	+	+
Black crappie	*Pomoxis nigromaculatus*	-	+	+
Johnny darter	*Etheostoma nigrum*	+	-	+
Yellow perch	*Perca flavescens*	-	+	+

last 4,000 years as droughts became less frequent and severe. All of the fishes in Ponca Creek are typical of creeks, but they presumably dispersed up the Missouri River either via flood plain habitats or during multi-decade droughts or very cold periods when Missouri River flows were presumably low. Fishes might also have moved between the Ponca Creek and Niobrara River drainages via headwater connections where the drainage divide includes small basins (e.g., Lambert Lake, northwest of Paxton, South Dakota) that probably filled with water during prehistoric wet periods. Sharp bends in Ponca Creek tributaries near Naper, Nebraska, suggest stream captures of Keya Paha tributaries by Ponca Creek tributaries.

White River
tributary drainage

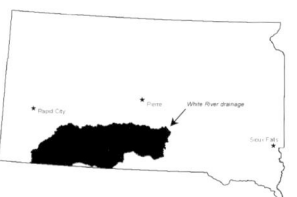

The White River begins on the Pine Ridge of northwestern Nebraska. It flows north and enters South Dakota where it eventually turns east. The Little White River is the primary tributary of the White River. White Clay Creek, Wounded Knee Creek, Porcupine Creek, Medicine Root Creek, Potato Creek, Bear-in-the-lodge Creek, Blackpipe Creek, and Cottonwood Creek are major tributaries to the Upper White River (upstream from the Little White River) in South Dakota, which flows through the White River Badlands. The Little White River flows from the northern Nebraska Sandhills and receives Rosebud Creek, Cut Meat Creek, and Pine Creek as major tributaries. Pine Ridge, Martin, White River, and Winner are the largest South Dakota cities that lie within the White River drainage.

A total of 46 fish species, of which 34 were native and 12 were introduced, was documented from the White River drainage of South Dakota during the last 150 years (Table 3.10). At least 40 of these (30 native, 10 introduced) are currently present including the state endangered finescale dace, state threatened sturgeon chub, state threatened pearl dace, state threatened northern redbelly dace, and rare plains

Table 3.10 A total of 46 fish species (34 native, 12 introduced) is known from the White River drainage in South Dakota. Native species were present at the time of settlement by Anglo-Americans. Humans transported introduced species to the White River drainage. Recent species (40 total of which 30 are native and 10 are introduced) were present in White River drainage fish surveys since 1990. The symbol + indicates yes and the symbol − indicates no.

COMMON NAME	SCIENTIFIC NAME	NATIVE	INTRODUCED	RECENT
Goldeye	*Hiodon alosoides*	+	−	+
Gizzard shad	*Dorosoma cepedianum*	+	−	−
Central stoneroller	*Campostoma anomalum anomalum*	+	−	+
Goldfish	*Carassius auratus*	−	+	+
Red shiner	*Cyprinella lutrensis lutrensis*	+	−	+
Common carp	*Cyprinus carpio*	−	+	+
Western silvery minnow	*Hybognathus argyritis*	+	−	+
Brassy minnow	*Hybognathus hankinsoni*	+	−	+
Plains minnow	*Hybognathus placitus*	+	−	+
Sturgeon chub	*Macrhybopsis gelida*	+	−	+
Pearl dace	*Margariscus margarita nachtriebi*	+	−	+
Golden shiner	*Notemigonus crysoleucas*	+	−	+
Emerald shiner	*Notropis atherinoides*	+	−	+
Bigmouth shiner	*Notropis dorsalis*	+	−	+
Blacknose shiner	*Notropis heterolepis*	+	−	−
Plains sand shiner	*Notropis stramineus missuriensis*	+	−	+
Northern redbelly dace	*Phoxinus eos*	+	−	+
Finescale dace	*Phoxinus neogaeus*	+	−	+
Fathead minnow	*Pimephales promelas*	+	−	+
Flathead chub	*Platygobio gracilis*	+	−	+
Longnose dace	*Rhinichthys cataractae cataractae*	+	−	+
Western blacknose dace	*Rhinichthys obtusus*	+	−	−
Creek chub	*Semotilus atromaculatus*	+	−	+
Northern river carpsucker	*Carpiodes carpio carpio*	+	−	+
White sucker	*Catostomus commersonii*	+	−	+
Shorthead redhorse	*Moxostoma macrolepidotum*	+	−	+
Black bullhead	*Ameiurus melas*	+	−	+
Channel catfish	*Ictalurus punctatus*	+	−	+
Stonecat	*Noturus flavus*	+	−	+
Flathead catfish	*Pylodictis olivaris*	+	−	−
Northern pike	*Esox lucius*	−	+	+
Rainbow trout	*Oncorhynchus mykiss*	−	+	+
Brow trout	*Salmo trutta*	−	+	−
Brook trout	*Salvelinus fontinalis*	−	+	−
Plains topminnow	*Fundulus sciadicus*	+	−	+
Brook stickleback	*Culaea inconstans*	+	−	+
Green sunfish	*Lepomis cyanellus*	+	−	+
Pumpkinseed	*Lepomis gibbosus*	−	+	+
Bluegill	*Lepomis macrochirus macrochirus*	−	+	+
Largemouth bass	*Micropterus salmoides salmoides*	−	+	+
White crappie	*Pomoxis annularis*	−	+	+
Black crappie	*Pomoxis nigromaculatus*	−	+	+
Iowa darter	*Etheostoma exile*	+	−	+
Yellow perch	*Perca flavescens*	−	+	+
Sauger	*Sander canadensis*	+	−	+
Walleye	*Sander vitreus vitreus*	+	−	+

topminnow. Four species including gizzard shad, blacknose shiner, western blacknose dace, and flathead catfish, were missing from recent (post-1990) fish surveys. Channel catfish are an abundant sport fish in the White River and there is a put-and-take rainbow trout fishery on the Lacreek National Wildlife Refuge.

The White River drainage was not glaciated except where it joins the Missouri River. Immediately after glaciation, roughly 11,000 years ago, the environment of the White River drainage was similar to that found in northeastern South Dakota today (Martin 1987). The rough and arid badland topography developed during dry periods that occurred periodically since that time. Modern fishes of the White River drainage, particularly species that are sensitive to drought, probably invaded during the last 4,000 years via the main stem Missouri River. Some species may have reached the White River drainage via tributary headwater connections with the Niobrara River drainage in the northern Nebraska Sandhills (Mayden 1987). Stream captures by the Cheyenne River drainage (Harksen 1969, Wayne et al. 1991) also may have given fish access to the White River drainage. It is possible some fishes survived post-glacial droughts in upper portions of the White River drainage because uplifts, such as the Pine Ridge, might have mediated climate changes (Kornfeld 2003).

Crow Creek tributary drainage

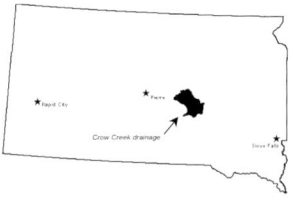

The Crow Creek drainage lies between the Missouri Coteau and the Missouri River. Smith Creek and Elm Creek are major tributaries of Crow Creek. Smith Creek begins on the eastern edge of the Missouri Coteau northwest of Wessington Springs, South Dakota, and Elm Creek begins south of Ree Heights, South Dakota. Gann Valley is the largest South Dakota city that lies within the Crow Creek drainage.

A total of 26 fish species, of which 23 were native and three were introduced, was documented from the Crow Creek drainage during the last 150 years (Table 3.11). The number of species presently found in the Crow Creek drainage is unknown because no recent (post-1990) fish surveys have been conducted.

The Crow Creek drainage was entirely glaciated. Fish fossils indicate that early post-glacial inhabitants of the Crow Creek drainage (14,000 to 12,000 years ago) included black

Table 3.11 A total of 26 fish species (23 native, 3 introduced) is known from the Crow Creek drainage in South Dakota. Native species were present at the time of settlement by Anglo-Americans. Humans transported introduced species to the Crow Creek River drainage. Recent species present in Crow Creek River drainage are unknown due to a lack of fish surveys since 1990. The symbol + indicates yes and the symbol – indicates no.

COMMON NAME	SCIENTIFIC NAME	NATIVE	INTRODUCED	RECENT
Silver lamprey	Ichthyomyzon unicuspis	+	-	?
Longnose gar	Lepisosteus osseus	+	-	?
Goldeye	Hiodon alosoides	+	-	?
Central stoneroller	Campostoma anomalum anomalum	+	-	?
Lake chub	Couesius plumbeus	+	-	?
Red shiner	Cyprinella lutrensis lutrensis	+	-	?
Common carp	Cyprinus carpio	-	+	?
Western silvery minnow	Hybognathus argyritis	+	-	?
Brassy minnow	Hybognathus hankinsoni	+	-	?
Plains sand shiner	Notropis stramineus missuriensis	+	-	?
Suckermouth minnow	Phenacobius mirabilis	+	-	?
Northern redbelly dace	Phoxinus eos	+	-	?
Fathead minnow	Pimephales promelas	+	-	?
Western blacknose dace	Rhinichthys obtusus	+	-	?
Creek chub	Semotilus atromaculatus	+	-	?
Northern river carpsucker	Carpiodes carpio carpio	+	-	?
White sucker	Catostomus commersonii	+	-	?
Shorthead redhorse	Moxostoma macrolepidotum	+	-	?
Black bullhead	Ameiurus melas	+	-	?
Northern pike	Esox lucius	-	+	?
Brook stickleback	Culaea inconstans	+	-	?
Green sunfish	Lepomis cyanellus	+	-	?
Orangespotted sunfish	Lepomis humilis	+	-	?
Iowa darter	Etheostoma exile	+	-	?
Yellow perch	Perca flavescens	-	+	?
Walleye	Sander vitreus vitreus	+	-	?

head, least madtom, western banded killifish, orangespotted sunfish, pumpkinseed, bluegill, largemouth bass, Iowa darter, yellow perch, and channel darter (Ossian 1973). The native fish fauna included only two of these fossil species (black bullhead, orangespotted madtom). This suggests that the fish species composition of the Crow Creek drainage changed dramatically throughout the post-glacial period. In fact, least madtom (*Noturus hildebrandi*) and channel darter (*Percina copelandi*) have never been found alive in South Dakota and their modern day distributions are far to the south. Their presence in prehistoric South Dakota suggests that prehistoric fish faunas could have included many species that have been absent from the state during the last 150 years.

The native Crow Creek drainage fish fauna included three fish species that were otherwise restricted to eastern South Dakota (silver lamprey, longnose gar, suckermouth minnow) and one that was otherwise restricted to western South Dakota (lake chub). During prehistoric wet periods, fishes could have colonized the drainage from the Missouri River or via headwater connections with the James River drainage. Such connections seem likely to have occurred between Turtle Creek (James River drainage) and Crow Creek in the vicinity of Collins Slough, south of Miller, South Dakota, and between Firesteel Creek (James River drainage) and Smith Creek (Crow Creek drainage) along the Missouri Coteau west of Wessington Springs, South Dakota. Colonization from both east and west might explain the presence of both eastern and western South Dakota fishes in this central South Dakota river drainage.

Bad River
tributary drainage

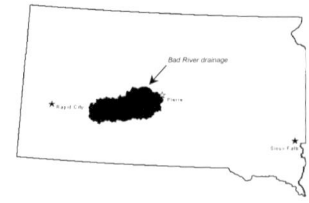

The South Fork of the Bad River is formed by the confluence of Whitewater Creek and Big Buffalo Creek, which flow from the White River Badlands. The North Fork of the Bad River begins near Horse Tooth Peak. The two forks join near

Philip, South Dakota. The Bad River flows eastward and eventually joins the Missouri River near Fort Pierre, South Dakota. White Willow Creek, Dry Creek, and Plum Creek are major tributaries to the Bad River. Wall, Kadoka, Murdo, and Fort Pierre are the largest South Dakota cities that lie within the Bad River drainage.

A total of 27 fish species, of which 18 were native and nine were introduced, was documented from the Bad River drainage during the last 150 years (Table 3.12). At least 24

Table 3.12 A total of 27 fish species (18 native, 9 introduced) is known from the Bad River drainage in South Dakota. Native species were present at the time of settlement by Anglo-Americans. Humans transported introduced species to the Bad River drainage. Recent species (24 total of which 16 are native and 8 are introduced) were present in Bad River drainage fish surveys since 1990. The symbol + indicates yes and the symbol – indicates no.

COMMON NAME	SCIENTIFIC NAME	NATIVE	INTRODUCED	RECENT
Goldeye	*Hiodon alosoides*	+	-	+
Red shiner	*Cyprinella lutrensis lutrensis*	+	-	+
Common carp	*Cyprinus carpio*	-	+	+
Western silvery minnow	*Hybognathus argyritis*	+	-	-
Plains minnow	*Hybognathus placitus*	+	-	+
Golden shiner	*Notemigonus crysoleucas*	+	-	+
Emerald shiner	*Notropis atherinoides*	+	-	+
Plains sand shiner	*Notropis stramineus missuriensis*	+	-	+
Fathead minnow	*Pimephales promelas*	+	-	+
Flathead chub	*Platygobio gracilis*	+	-	+
Longnose dace	*Rhinichthys cataractae cataractae*	+	-	-
Northern river carpsucker	*Carpiodes carpio carpio*	+	-	+
White sucker	*Catostomus commersonii*	+	-	+
Shorthead redhorse	*Moxostoma macrolepidotum*	+	-	+
Black bullhead	*Ameiurus melas*	+	-	+
Channel catfish	*Ictalurus punctatus*	+	-	+
Northern pike	*Esox lucius*	-	+	+
Green sunfish	*Lepomis cyanellus*	-	+	+
Pumpkinseed	*Lepomis gibbosus*	-	+	-
Orangespotted sunfish	*Lepomis humilis*	+	-	+
Bluegill	*Lepomis macrochirus macrochirus*	-	+	+
Smallmouth bass	*Micropterus dolomieu*	-	+	+
Largemouth bass	*Micropterus salmoides salmoides*	-	+	+
Black crappie	*Pomoxis nigromaculatus*	-	+	+
Yellow perch	*Perca flavescens*	-	+	+
Sauger	*Sander canadensis*	+	-	+
Walleye	*Sander vitreus vitreus*	+	-	+

of these (16 native, eight introduced) are currently present. The native western silvery minnow and longnose dace were missing from recent (post-1990) fish surveys.

The Bad River drainage was not glaciated except where it joins the Missouri River (Duchossois 1993). The entire drainage, however, lies relatively close to the glacial front and presumably only cold-water fishes could have survived glaciation there. Post-glacial droughts would have also impacted fishes because the Bad River drainage is relatively small and even today supports relatively few fish species. Most of the modern day native fishes probably colonized the Bad River drainage during the last 4,000 years via the main stem Missouri River. The dominance of creek fishes within the Bad River drainage suggests that, although atypical of large rivers, creek fishes were able to reach the Bad River drainage via the main stem Missouri River.

Cheyenne River tributary drainage

The Cheyenne River begins in the Powder River Basin of northeastern Wyoming. The north and south forks of the Cheyenne River (i.e., the Belle Fourche and Upper Cheyenne rivers) flow to the north and south of the Black Hills. Mountain streams flow from the Black Hills and contribute to the flow of both forks. The Redwater River, Owl Creek, Whitewood Creek, Horse Creek, Willow Creek, Bear Butte Creek, Alkali Creek, and Elm Creek are major tributaries to the Belle Fourche River in South Dakota. Cottonwood Creek, Hat Creek, Horsehead Creek, the Fall River, Beaver Creek, French Creek, Battle Creek, Spring Creek, Rapid Creek, Boxelder Creek, and Elk Creek are major tributaries to the Upper Cheyenne River in South Dakota. The Lower Cheyenne River is formed where the Belle Fourche and Upper Cheyenne rivers join. Cherry Creek and Plum Creek are major tributaries to the Lower Cheyenne River. Spearfish, Belle Fourche, Deadwood, Lead, Sturgis,

Hot Springs, Custer, and Rapid City are the largest South Dakota cities that lie within the Cheyenne River drainage.

A total of 55 fish species, of which 32 were native and 23 were introduced, was documented from the Cheyenne River drainage of South Dakota during the last 150 years (Table 3.13). Of these, at least 54 species (31 native, 23 introduced) are currently present. The Cheyenne River drainage contains the only known populations of the rare lake chub, rare longnose sucker, rare mountain sucker, and northern plains killifish in the state. The state endangered finescale dace, state threatened sturgeon chub, and rare plains topminnow are also present. One native species, burbot, was missing from recent (post-1990) fish surveys. Impoundments within the Cheyenne River drainage are of two types: (1) cold-water impoundments of the Black Hills and (2) warm-water impoundments of the Cheyenne River Valley. Rainbow trout, brown trout, splake trout (brook trout-lake trout hybrids), and lake trout are important game fishes of cold-water impoundments (SDGFP 2005d). Channel catfish, black crappie, smallmouth bass, largemouth bass, and walleye are important game fishes

Table 3.13 A total of 55 fish species (32 native, 23 introduced) is known from the Cheyenne River drainage in South Dakota. Native species were present at the time of settlement by Anglo-Americans. Humans transported introduced species to the Cheyenne River drainage. Recent species (54 total of which 31 are native and 23 are introduced) were present in Cheyenne River drainage fish surveys since 1990. The symbol + indicates yes and the symbol – indicates no.

COMMON NAME	SCIENTIFIC NAME	NATIVE	INTRODUCED	RECENT
Goldeye	*Hiodon alosoides*	+	-	+
Gizzard shad	*Dorosoma cepedianum*	-	+	+
Goldfish	*Carassius auratus*	-	+	+
Lake chub	*Couesius plumbeus*	+	-	+
Red shiner	*Cyprinella lutrensis lutrensis*	+	-	+
Common carp	*Cyprinus carpio*	-	+	+
Western silvery minnow	*Hybognathus argyritis*	+	-	+
Brassy minnow	*Hybognathus hankinsoni*	+	-	+
Plains minnow	*Hybognathus placitus*	+	-	+
Sturgeon chub	*Macrhybopsis gelida*	+	-	+
Golden shiner	*Notemigonus crysoleucas*	+	-	+
Emerald shiner	*Notropis atherinoides*	+	-	+

Table 3.13 continued.

COMMON NAME	SCIENTIFIC NAME	NATIVE	INTRODUCED	RECENT
Spottail shiner	*Notropis hudsonius*	-	+	+
Plains sand shiner	*Notropis stramineus missuriensis*	+	-	+
Finescale dace	*Phoxinus neogaeus*	+	-	+
Fathead minnow	*Pimephales promelas*	+	-	+
Flathead chub	*Platygobio gracilis*	+	-	+
Longnose dace	*Rhinichthys cataractae cataractae*	+	-	+
Rudd	*Scardinius erythrophthalmus*	-	+	+
Creek chub	*Semotilus atromaculatus*	+	-	+
Northern river carpsucker	*Carpiodes carpio carpio*	+	-	+
Longnose sucker	*Catostomus catostomus*	+	-	+
White sucker	*Catostomus commersonii*	+	-	+
Mountain sucker	*Catostomus platyrhynchus*	+	-	+
Shorthead redhorse	*Moxostoma macrolepidotum*	+	-	+
Black bullhead	*Ameiurus melas*	+	-	+
Yellow bullhead	*Ameiurus natalis*	-	+	+
Channel catfish	*Ictalurus punctatus*	+	-	+
Stonecat	*Noturus flavus*	+	-	+
Northern pike	*Esox lucius*	-	+	+
Rainbow smelt	*Osmerus mordax*	-	+	+
Cutthroat trout	*Oncorhynchus clarkii*	-	+	+
Rainbow trout	*Oncorhynchus mykiss*	-	+	+
Kokanee salmon	*Oncorhynchus nerka*	-	+	+
Brown trout	*Salmo trutta*	-	+	+
Brook trout	*Salvelinus fontinalis*	-	+	+
Lake trout	*Salvelinus namaycush*	-	+	+
Burbot	*Lota lota maculosa*	+	-	-
Northern plains killifish	*Fundulus kansae*	+	-	+
Plains topminnow	*Fundulus sciadicus*	+	-	+
Brook stickleback	*Culaea inconstans*	+	-	+
White bass	*Morone chrysops*	-	+	+
Rock bass	*Ambloplites rupestris*	-	+	+
Green sunfish	*Lepomis cyanellus*	-	+	+
Orangespotted sunfish	*Lepomis humilis*	+	-	+
Bluegill	*Lepomis macrochirus macrochirus*	-	+	+
Smallmouth bass	*Micropterus dolomieu*	-	+	+
Largemouth bass	*Micropterus salmoides salmoides*	-	+	+
White crappie	*Pomoxis annularis*	-	+	+
Black crappie	*Pomoxis nigromaculatus*	-	+	+
Iowa darter	*Etheostoma exile*	+	-	+
Yellow perch	*Perca flavescens*	-	+	+
Sauger	*Sander canadensis*	+	-	+
Walleye	*Sander vitreus vitreus*	+	-	+
Freshwater drum	*Aplodinotus grunniens*	+	-	+

of warm-water impoundments (SDGFP 2005d). Rainbow trout, brown trout, and brook trout are important game fishes in cold-water, Black Hills streams (SDGFP 2005d). Channel catfish are an important game fish in the Cheyenne River and Belle Fourche River. Smallmouth bass are sometimes common in the Upper Cheyenne River below Angostura Dam and saugers are seasonally present in the Lower Cheyenne River and Belle Fourche River.

The Cheyenne River drainage was not glaciated except in its lowest reaches. Fishes may have survived post-glacial droughts within streams of the Black Hills because the effects of climate change were moderated there (Kornfeld 2003). Fossil evidence from Beaver Creek on Wind Cave National Park suggests that creek chubs and white suckers did survive such droughts (Martin et al. 1993). However, most fishes probably colonized the Cheyenne River during the last 4,000 years via the main stem Missouri River. It is also possible that some species reached the Cheyenne River drainage via stream captures between the Cheyenne River drainage and the Yellowstone River, Little Missouri River, White River, or Platte River drainages. The Cheyenne River drainage captured portions of the Little Missouri and White River drainages on the northern and southern edges of the Black Hills. Headwater streams of the Yellowstone River and Platte River drainages possibly captured headwater streams of the Cheyenne River drainage in the Powder River Basin of Wyoming.

Moreau River
tributary drainage

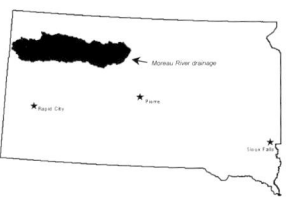

The North and South Forks of the Moreau River begin in the Short Pine Hills of northwestern South Dakota. Battle Creek and Sand Creek are major tributaries of the South Fork. The two forks join to form the Moreau River near Zeona, South Dakota. Deep Creek, Rabbit Creek, Thunder Butte Creek, Red Earth Creek, and the Little Moreau River are major tributaries of the Moreau River. Bison, Timber Lake, and Eagle Butte are the largest South Dakota cities that lie within the Moreau River drainage.

A total of 33 fish species, of which 21 were native and 12 were introduced, was documented from the Moreau River drainage during the last 150 years (Table 3.14). All fish spe-

Table 3.14   A total of 33 fish species (21 native, 12 introduced) is known from the Moreau River drainage in South Dakota. Native species were present at the time of settlement by Anglo-Americans. Humans transported introduced species to the Moreau River drainage. Recent species (33 total of which 21 are native and 12 are introduced) were present in Moreau River drainage fish surveys since 1990. The symbol + indicates yes and the symbol – indicates no.

COMMON NAME	SCIENTIFIC NAME	NATIVE	INTRODUCED	RECENT
Goldeye	*Hiodon alosoides*	+	-	+
Common carp	*Cyprinus carpio*	-	+	+
Western silvery minnow	*Hybognathus argyritis*	+	-	+
Brassy minnow	*Hybognathus hankinsoni*	+	-	+
Plains minnow	*Hybognathus placitus*	+	-	+
Golden shiner	*Notemigonus crysoleucas*	+	-	+
Emerald shiner	*Notropis atherinoides*	+	-	+
Plains sand shiner	*Notropis stramineus missuriensis*	+	-	+
Fathead minnow	*Pimephales promelas*	+	-	+
Flathead chub	*Platygobio gracilis*	+	-	+
Longnose dace	*Rhinichthys cataractae cataractae*	+	-	+
Creek chub	*Semotilus atromaculatus*	+	-	+
Northern river carpsucker	*Carpiodes carpio carpio*	+	-	+
White sucker	*Catostomus commersonii*	+	-	+
Shorthead redhorse	*Moxostoma macrolepidotum*	+	-	+
Black bullhead	*Ameiurus melas*	+	-	+
Channel catfish	*Ictalurus punctatus*	+	-	+
Stonecat	*Noturus flavus*	+	-	+
Northern pike	*Esox lucius*	-	+	+
Rainbow trout	*Oncorhynchus mykiss*	-	+	+
Brown trout	*Salmo trutta*	-	+	+
Brook stickleback	*Culaea inconstans*	+	-	+
White bass	*Morone chrysops*	-	+	+
Green sunfish	*Lepomis cyanellus*	-	+	+
Bluegill	*Lepomis macrochirus macrochirus*	-	+	+
Smallmouth bass	*Micropterus dolomieu*	-	+	+
Largemouth bass	*Micropterus salmoides salmoides*	-	+	+
Black crappie	*Pomoxis nigromaculatus*	-	+	+
Iowa darter	*Etheostoma exile*	+	-	+
Johnny darter	*Etheostoma nigrum*	-	+	+
Yellow perch	*Perca flavescens*	-	+	+
Sauger	*Sander canadensis*	+	-	+
Walleye	*Sander vitreus vitreus*	+	-	+

cies that have ever been recorded from the drainage are still present, but this drainage holds no rare, threatened, or endangered species. Channel catfish are the dominant game fish of the main stem Moreau River.

The Moreau River drainage was not glaciated except in its lowest reaches. The entire drainage, however, lies relatively close to the glacial front and presumably only cold-water fishes could have survived during glaciation. Extreme postglacial droughts also presumably limited fish survival within the Moreau River drainage unless uplifts such as the Slim Buttes or Short Pine Hills maintained climatic refugia. Most of the modern day native fishes probably reached the Moreau River from the main stem Missouri River. The dominance of creek fishes within the Moreau River drainage suggests that the main stem Missouri River was suitable for their dispersal at least some of the time.

Grand River tributary drainage

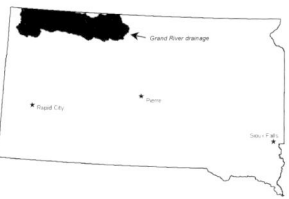

The North and South Forks of the Grand River begin in the Cave Hills of northwestern South Dakota and southwestern North Dakota. Clarks Fork Creek and Big Nasty Creek are major tributaries to the South Fork. The two forks join to form the Grand River near Shadehill, South Dakota. Flat Creek, Black Horse Butte Creek, Firesteel Creek, and High Bank Creek are major tributaries to the Grand River. Buffalo, Lemmon, and McIntosh are the largest South Dakota cities that lie within the Grand River drainage.

A total of 39 fish species, of which 23 were native and 16 were introduced, was documented from the Grand River drainage of South Dakota during the last 150 years (Table 3.15). Of these, at least 36 fish species (22 native, 14 introduced) are currently present, including the state threatened northern redbelly dace. One native species, sturgeon chub, was absent from recent (post-1990) collections. Channel catfish, black crappie, smallmouth bass, yellow perch, and walleye

Table 3.15   A total of 39 fish species (23 native, 16 introduced) is known from the Grand River drainage in South Dakota. Native species were present at the time of settlement by Anglo-Americans. Humans transported introduced species to the Grand River drainage. Recent species (36 total of which 22 are native and 14 are introduced) were present in Grand River drainage fish surveys since 1990. The symbol + indicates yes and the symbol – indicates no.

COMMON NAME	SCIENTIFIC NAME	NATIVE	INTRODUCED	RECENT
Goldeye	*Hiodon alosoides*	+	-	+
Gizzard shad	*Dorosoma cepedianum*	-	+	+
Common carp	*Cyprinus carpio*	-	+	+
Western silvery minnow	*Hybognathus argyritis*	+	-	+
Brassy minnow	*Hybognathus hankinsoni*	+	-	+
Plains minnow	*Hybognathus placitus*	+	-	+
Sturgeon chub	*Macrhybopsis gelida*	+	-	-
Golden shiner	*Notemigonus crysoleucas*	+	-	+
Emerald shiner	*Notropis atherinoides*	+	-	+
Spottail shiner	*Notropis hudsonius*	-	+	+
Plains sand shiner	*Notropis stramineus missuriensis*	+	-	+
Northern redbelly dace	*Phoxinus eos*	+	-	+
Bluntnose minnow	*Pimephales notatus*	-	+	-
Fathead minnow	*Pimephales promelas*	+	-	+
Flathead chub	*Platygobio gracilis*	+	-	+
Longnose dace	*Rhinichthys cataractae cataractae*	+	-	+
Creek chub	*Semotilus atromaculatus*	+	-	+
Northern river carpsucker	*Carpiodes carpio carpio*	+	-	+
White sucker	*Catostomus commersonii*	+	-	+
Shorthead redhorse	*Moxostoma macrolepidotum*	+	-	+
Black bullhead	*Ameiurus melas*	+	-	+
Channel catfish	*Ictalurus punctatus*	+	-	+
Stonecat	*Noturus flavus*	+	-	+
Northern pike	*Esox lucius*	-	+	+
Rainbow trout	*Oncorhynchus mykiss*	-	+	+
Brown trout	*Salmo trutta*	-	+	+
White bass	*Morone chrysops*	-	+	+
Green sunfish	*Lepomis cyanellus*	-	+	+
Pumpkinseed	*Lepomis gibbosus*	-	+	-
Orangespotted sunfish	*Lepomis humilis*	+	-	+
Bluegill	*Lepomis macrochirus macrochirus*	-	+	+
Smallmouth bass	*Micropterus dolomieu*	-	+	+
Largemouth bass	*Micropterus salmoides salmoides*	-	+	+
Black crappie	*Pomoxis nigromaculatus*	-	+	+
Iowa darter	*Etheostoma exile*	+	-	+
Johnny darter	*Etheostoma nigrum*	-	+	+
Yellow perch	*Perca flavescens*	-	+	+
Sauger	*Sander canadensis*	+	-	+
Walleye	*Sander vitreus vitreus*	+	-	+

are important game fishes of impoundments within the Grand River drainage (SDGFP 2005d). Rainbow trout and brown trout are present in the Grand River below Shadehill Dam.

Only the eastern edge of the Grand River drainage was glaciated. The entire drainage, however, lies relatively close to the glacial front and presumably only cold-water fish could have survived within the region during glaciation. Probably few fishes survived post-glacial droughts within the Grand River drainage unless streams of the Cave Hills or Slim Buttes moderated drought severity. Most of the modern day fishes probably colonized the Grand River drainage within the last 4,000 years via the main stem Missouri River. The dominance of creek fishes within the Grand River drainage indicates that the main stem Missouri River was suitable for their dispersal at least some of the time.

Little Missouri River tributary drainage

The Little Missouri River begins at Flat Iron Butte in northeastern Wyoming. It flows north across the southeastern corner of Montana and then across the northwestern corner of South Dakota. From there it continues north to the Missouri River in North Dakota. Slick Creek and Boxelder Creek are the only major Little Missouri River tributaries in South Dakota. Camp Crook is the largest South Dakota city that lies within the Little Missouri River drainage.

A total of 24 fish species, of which 19 were native and five were introduced, was documented from the Little Missouri River drainage of South Dakota during the last 150 years (Table 3.16). Of these, at least 18 fish species (14 native, four introduced) are currently present. Five native species including lake chub, brassy minnow, sturgeon chub, black bullhead, and Iowa darter were absent from recent (post-1990) collections.

Table 3.16   A total of 24 fish species (19 native, 5 introduced) is known from the Little Missouri River drainage in South Dakota. Native species were present at the time of settlement by Anglo-Americans. Humans transported introduced species to the Little Missouri River drainage. Recent species (18 total of which 14 are native and 4 are introduced) were present in Little Missouri River drainage fish surveys since 1990. The symbol + indicates yes and the symbol – indicates no.

COMMON NAME	SCIENTIFIC NAME	NATIVE	INTRODUCED	RECENT
Goldeye	*Hiodon alosoides*	+	-	+
Lake chub	*Couesius plumbeus*	+	-	-
Common carp	*Cyprinus carpio*	-	+	+
Western silvery minnow	*Hybognathus argyritis*	+	-	+
Brassy minnow	*Hybognathus hankinsoni*	+	-	-
Plains minnow	*Hybognathus placitus*	+	-	+
Sturgeon chub	*Macrhybopsis gelida*	+	-	-
Golden shiner	*Notemigonus crysoleucas*	+	-	+
Plains sand shiner	*Notropis stramineus missuriensis*	+	-	+
Fathead minnow	*Pimephales promelas*	+	-	+
Flathead chub	*Platygobio gracilis*	+	-	+
Longnose dace	*Rhinichthys cataractae cataractae*	+	-	+
Northern river carpsucker	*Carpiodes carpio carpio*	+	-	+
White sucker	*Catostomus commersonii*	+	-	+
Shorthead redhorse	*Moxostoma macrolepidotum*	+	-	+
Black bullhead	*Ameiurus melas*	+	-	-
Channel catfish	*Ictalurus punctatus*	+	-	+
Stonecat	*Noturus flavus*	+	-	+
Northern pike	*Esox lucius*	-	+	+
Green sunfish	*Lepomis cyanellus*	-	+	+
Largemouth bass	*Micropterus salmoides salmoides*	-	+	-
Iowa darter	*Etheostoma exile*	+	-	-
Yellow perch	*Perca flavescens*	-	+	+
Sauger	*Sander canadensis*	+	-	+

The Little Missouri River drainage was not glaciated except in its lowest reaches in North Dakota. If streams of the Black Hills supported fishes during glacial periods and extreme post-glacial droughts, then it is possible headwater streams of the Little Missouri River could have as well. The Black Hills comprise the largest 'island oasis' of the northern Great Plains but smaller uplifts, such as the Pine Breaks that lie within the Little Missouri River drainage, represent smaller oases (Kornfeld 2003) and perhaps some of them contained suitable refugia for a few fish species during prehistoric droughts. Nevertheless, most modern day fishes probably reached the

Little Missouri River from the main stem Missouri River within the last 4,000 years. Stream capture by the Cheyenne River drainage might also have enabled some species to colonize the Little Missouri River drainage.

Missouri River Valley

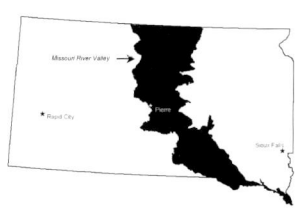

The Missouri River Valley includes central and extreme southeastern South Dakota. Waters of the Missouri River valley include four main stem Missouri River reservoirs (Lakes Oahe, Sharpe, Francis Case, and Lewis and Clark), two free-flowing stretches of the Missouri River (below Fort Randall Dam and below Gavins Point Dam), and small tributary drainages such as Spring Creek, Oak Creek, Swan Creek, Okobojo Creek, Medicine Creek, American Crow Creek, Bull Creek, Whetstone Creek, Choteau Creek, and Emanuel Creek. Mobridge, Gettysburg, Pierre, Chamberlain, and Yankton are the largest South Dakota cities that lie within the Missouri River Valley.

A total of 100 fish species, of which 68 were native and 32 were introduced, was documented from the Missouri River Valley of South Dakota during the last 150 years (Table 3.17). Of these, at least 85 fish species (59 native, 26 introduced) are currently present, including the only remaining South Dakota populations of the federally and state endangered pallid sturgeon, the state threatened sicklefin chub, the rare river shiner, the rare suckermouth minnow, the highfin carpsucker, the grass pickerel, and the burbot. The state threatened pearl dace and state threatened northern redbelly dace are present in some small tributaries. Other rare species present in the Missouri River Valley of South Dakota include American eel, silver chub, central quillback carpsucker, blue sucker, and golden redhorse. Eight native species including silver lamprey, sturgeon chub, blacknose shiner, silverband shiner, western blacknose dace, black buffalo, yellow bullhead, tadpole madtom, and plains topminnow were absent from recent collections. Channel catfish, northern pike, chinook

Table 3.17   A total of 100 fish species (67 native, 33 introduced) is known from the Missouri River Valley in South Dakota.   Native species were present at the time of settlement by Anglo-Americans.   Humans transported introduced species to the Missouri River Valley.   Recent species (86 total of which 60 are native and 26 are introduced) were present in Missouri River Valley fish surveys since 1990.   The symbol + indicates presence and the symbol – indicates absence.

COMMON NAME	SCIENTIFIC NAME	NATIVE	INTRODUCED	RECENT
Silver lamprey	*Ichthyomyzon unicuspis*	+	-	
Lake sturgeon	*Ascipenser fulvescens*	+	-	+
Pallid sturgeon	*Scaphirhynchus albus*	+	-	+
Shovelnose sturgeon	*Scaphirhynchus platorynchus*	+	-	+
Paddlefish	*Polyodon spathula*	+	-	+
Longnose gar	*Lepisosteus osseus*	+	-	+
Shortnose gar	*Lepisosteus platostomus*	+	-	+
Goldeye	*Hiodon alosoides*	+	-	+
American eel	*Anguilla rostrata*	+	-	+
Skipjack herring	*Alosa chrysochloris*	-	+	+
Alewife	*Alosa pseudoharengus*	-	+	-
Gizzard shad	*Dorosoma cepedianum*	+	-	+
Central stoneroller	*Campostoma anomalum anomalum*	+	-	+
Goldfish	*Carassius auratus*	-	+	+
Grass carp	*Ctenopharyngodon idella*	-	+	+
Red shiner	*Cyprinella lutrensis lutrensis*	+	-	+
Spotfin shiner	*Cyprinella spiloptera*	-	+	+
Common carp	*Cyprinus carpio*	-	+	+
Western silvery minnow	*Hybognathus argyritis*	+	-	+
Brassy minnow	*Hybognathus hankinsoni*	+	-	+
Plains minnow	*Hybognathus placitus*	+	-	+
Silver carp	*Hypophthalmichthys molitrix*	-	+	+
Bighead carp	*Hypophthalmichthys nobilis*	-	+	+
Common shiner	*Luxilus cornutus*	+	-	+
Sturgeon chub	*Macrhybopsis gelida*	+	-	-
Shoal chub	*Macrhybopsis hyostoma*	+	-	+
Sicklefin chub	*Macrhybopsis meeki*	+	-	+
Silver chub	*Macrhybopsis storeriana*	+	-	+
Pearl dace	*Margariscus margarita nachtriebi*	+	-	+
Golden shiner	*Notemigonus crysoleucas*	+	-	+
Emerald shiner	*Notropis atherinoides*	+	-	+
River shiner	*Notropis blennius*	+	-	+
Bigmouth shiner	*Notropis dorsalis*	+	-	+
Blacknose shiner	*Notropis heterolepis*	+	-	-
Spottail shiner	*Notropis hudsonius*	+	-	+
Silverband shiner	*Notropis shumardi*	+	-	-
Plains sand shiner	*Notropis stramineus missuriensis*	+	-	+
Topeka shiner	*Notropis topeka*	-	+	-
Mimic shiner	*Notropis volucellus*	-	+	+
Suckermouth minnow	*Phenacobius mirabilis*	+	-	+
Northern redbelly dace	*Phoxinus eos*	+	-	+
Bluntnose minnow	*Pimephales notatus*	+	-	+
Fathead minnow	*Pimephales promelas*	+	-	+
Flathead chub	*Platygobio gracilis*	+	-	+
Longnose dace	*Rhinichthys cataractae cataractae*	+	-	+
Western blacknose dace	*Rhinichthys obtusus*	+	-	-
Creek chub	*Semotilus atromaculatus*	+	-	+

Table 3.17 continued.

COMMON NAME	SCIENTIFIC NAME	NATIVE	INTRODUCED	RECENT
Northern river carpsucker	*Carpiodes carpio carpio*	+	-	+
Central quillback carpsucker	*Carpiodes cyprinus hinei*	+	-	+
Highfin carpsucker	*Carpiodes velifer*	+	-	+
White sucker	*Catostomus commersonii*	+	-	+
Blue sucker	*Cycleptus elongatus*	+	-	+
Smallmouth buffalo	*Ictiobus bubalus*	+	-	+
Bigmouth buffalo	*Ictiobus cyprinellus*	+	-	+
Black buffalo	*Ictiobus niger*	+	-	-
Golden redhorse	*Moxostoma erythrurum*	+	-	+
Shorthead redhorse	*Moxostoma macrolepidotum*	+	-	+
Black bullhead	*Ameiurus melas*	+	-	+
Yellow bullhead	*Ameiurus natalis*	+	-	-
Blue catfish	*Ictalurus furcatus*	+	-	+
Channel catfish	*Ictalurus punctatus*	+	-	+
Stonecat	*Noturus flavus*	+	-	+
Tadpole madtom	*Noturus gyrinus*	+	-	+
Flathead catfish	*Pylodictis olivaris*	+	-	+
Grass pickerel	*Esox americanus vermiculatus*	+	-	+
Northern pike	*Esox lucius*	-	+	+
Muskellunge	*Esox masquinongy*	-	+	+
Rainbow smelt	*Osmerus mordax*	-	+	+
Cisco	*Coregonus artedi*	-	+	+
Lake whitefish	*Coregonus clupeaformis*	-	+	+
Cutthroat trout	*Oncorhynchus clarkii*	-	+	+
Coho salmon	*Oncorhynchus kisutch*	-	+	-
Rainbow trout	*Oncorhynchus mykiss*	-	+	+
Kokanee salmon	*Oncorhynchus nerka*	-	+	-
Chinook salmon	*Oncorhynchus tschawytscha*	-	+	+
Bonneville cisco	*Prosopium gemmifer*	-	+	-
Brown trout	*Salmo trutta*	-	+	+
Lake trout	*Salvelinus namaycush*	-	+	+
Burbot	*Lota lota maculosa*	+	-	+
Western banded killifish	*Fundulus diaphanus menona*	+	-	+
Plains topminnow	*Fundulus sciadicus*	+	-	-
Brook stickleback	*Culaea inconstans*	+	-	+
Mottled sculpin	*Cottus bairdii*	-	+	+
White bass	*Morone chrysops*	-	+	+
Striped bass	*Morone saxatilis*	-	+	+
Rock bass	*Ambloplites rupestris*	-	+	+
Green sunfish	*Lepomis cyanellus*	+	-	+
Pumpkinseed	*Lepomis gibbosus*	-	+	-
Orangespotted sunfish	*Lepomis humilis*	+	-	+
Bluegill	*Lepomis macrochirus macrochirus*	+	-	+
Smallmouth bass	*Micropterus dolomieu*	-	+	+
Largemouth bass	*Micropterus salmoides salmoides*	-	+	+
White crappie	*Pomoxis annularis*	-	+	+
Black crappie	*Pomoxis nigromaculatus*	-	+	+
Iowa darter	*Etheostoma exile*	+	-	+
Johnny darter	*Etheostoma nigrum*	+	-	+
Yellow perch	*Perca flavescens*	+	-	+
Sauger	*Sander canadensis*	+	-	+
Walleye	*Sander vitreus vitreus*	+	-	+
Freshwater drum	*Aplodinotus grunniens*	+	-	+

salmon, white bass, smallmouth bass, and walleye are important game fishes of main stem Missouri River impoundments (SDGFP 2005a). Paddlefish, rainbow trout, and sauger are additional sport fishes that are more prevalent in free flowing river reaches of the main stem Missouri River.

The Missouri River Valley was mostly glaciated so if any fish were present during glaciation they must have been cold-water species. Impacts of post-glacial drought periods on the Missouri River and its fishes are unknown. Most fishes probably colonized the Missouri River Valley of South Dakota from the Missouri River to the southeast (downstream) because even if any fish survived in waters to the northwest (upstream) during the period of glaciation, they would have been cold-water fish that are not native to the Missouri River Valley in South Dakota. However, these cold-water species might have used the Missouri River during prehistoric cold periods. For example, the lake chub, longnose sucker and mountain sucker might have used the Missouri River during a cold period to reach the Cheyenne River drainage and colonize streams of the Black Hills. These species presumably colonized other river drainages at the same time. For example, lake chubs colonized the Crow Creek drainage at some point and mountain suckers reached the White River drainage of Nebraska. In any case, subsequent climatic warming restricted the range of cold-water fishes to only a few South Dakota streams.

The small tributary drainages of the Missouri River valley support typical creek fish faunas. This illustrates the fact that the Missouri River must have served as a dispersal corridor at some time in order for streams of the small tributary drainages to be populated by creek fishes. Fish biologists sometimes overlook this fact. One continuing debate is whether or not walleyes were native to the Missouri River Valley and western tributary drainages. It is impossible to know for sure without fossil evidence because walleyes were introduced to the region before fish surveys that documented the native fish faunas were conducted. The argument that walleyes are not native partly depends upon the viewpoint that the historical Missouri River was not suitable walleye habitat. Whether or not this is true, it is certain that many fishes used the main stem Missouri River for dispersal at one time or another even though they are not typical inhabitants of the Missouri River itself .

Fossil records documented the presence of unidentified gar, sucker, shiner, chub, catfish, madtom, sunfish, and darter species on the Missouri Coteau of north-central South Dakota during glaciation (Martin 1973). Post-glacial fish fossils from the Missouri Coteau include brassy minnow, western banded killifish, brook stickleback, and yellow perch from the Siebold Paleolake, North Dakota, (Cvancara et al. 1971, Newbrey and Ashworth 2004) and (tentatively) finescale dace, creek chub, and white sucker from the Prophet Mountain Site, North Dakota (Sherrod 1963). These species were probably present in all western South Dakota river drainages because they most likely reached the Missouri Coteau via the Missouri River (Clayton 1967) and presumably dispersed among all suitable waters. Post-glacial droughts apparently eliminated western banded killifish and yellow perch from western South Dakota. Brassy minnow, finescale dace, creek chub, white sucker, and brook stickleback possibly survived somewhere in western South Dakota throughout post-glacial time. At least creek chub and white sucker survived post-glacial drought periods in aquatic refugia of the Black Hills (Martin et al. 1993). Whatever species survived in refugia may have recolonized western South Dakota from them as droughts subsided. Otherwise they must have recolonized western South Dakota more recently via the Missouri River.

Discussion

The diversity of native fishes in South Dakota is partly due to the diverse geography of the state, which includes a variety of aquatic habitats that suit different fish species. This is shown by the fact that 71 typical creek fishes have inhabited the state. No individual creek ever maintained all 71 species but the combination of a wide variety of creek-like habitats throughout all regions of South Dakota provided suitable environments for them all. The importance of creek-like waters for maintaining fish biodiversity in the state is evident because creek fishes are the largest component of the present day fish fauna of all river drainages (Figure 3.5).

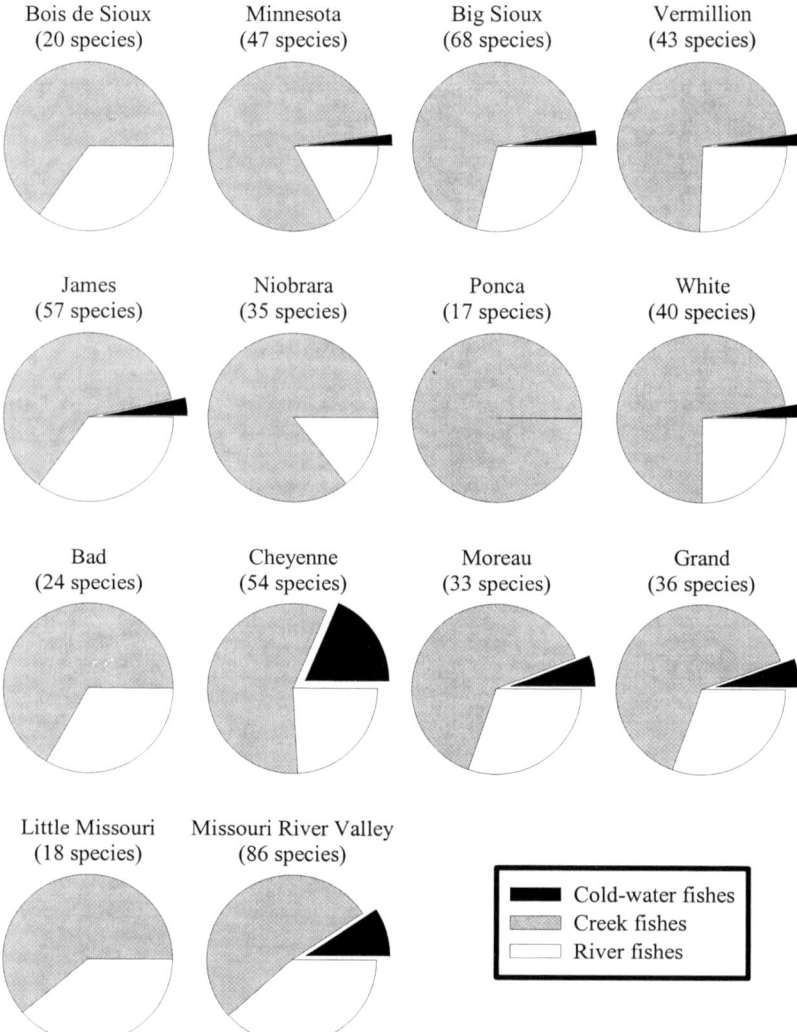

Figure 3.5   Pie chart showing the composition of creek fishes, river fishes and cold-water fishes (see text for definition of fish types) in 15 major South Dakota river drainages, based on collections made since 1990. The Crow Creek drainage was not shown because there have been no post-1990 collections.

The fishes of South Dakota used all four aquatic routes to colonize the waters of the state. Direct stream connections were presumably the dominant aquatic route because they are always present. Trans-divide connections between headwater streams were most important east of the Missouri River and in the Nebraska Sandhills because river drainage divides were low in both areas. These connections were only usable when conditions were wet enough to flood drainage divides. Stream captures were most important west of the Missouri River because of more active stream erosion within a high-relief, unglaciated landscape. Human made connections have become increasingly important due to intentional and accidental fish introductions, particularly within human made habitats, such as reservoirs.

The introduction of nonnative fish species has added to the fish fauna of South Dakota. All river drainages have introduced species and South Dakota has more fish species now than it did at the time of settlement. Some introduced species are widespread but others are restricted to man-made habitats such as reservoirs or reservoir tailwaters. More species have been introduced to river drainages in the Great Plains than to river drainages in the Central Lowlands. In fact, many of the species introduced to the Great Plains were already native to the Central Lowlands. Most of the introduced species that were not native to any portion of South Dakota are cold-water fishes. These species are restricted to Black Hills streams, cold-water impoundments of the Black Hills and Missouri River, and cold impoundment tailwaters. Under the current climatic conditions these species would have never reached cold-water habitats of the state without human aid. Impoundment and tailwater populations could not have become established unless humans had constructed large dams.

Nevertheless, the primary reason that so many fish species inhabit South Dakota is that southeastern South Dakota has a relatively mild and moist climate and a wide variety of aquatic habitats including the Missouri River Valley. The majority (95%) of fish species documented within the state are present either in the Missouri River Valley or in river drainages to the east. Because of the milder climate, eastern South Dakota river drainages have more species than similar sized western river drainages. For example, the Big Sioux River drainage includes 10,056 km^2 within southeastern South Dakota and currently supports 68 fish species while the somewhat larger White River drainage includes 13,274 km^2 within southwestern South Dakota but currently supports only 40 fish species.

Based on an analysis of historic versus recent collections, all river drainages except the Moreau River drainage have lost native species. Nine native species, silver lamprey, bowfin, mooneye, skipjack herring, blackchin shiner, silverband shiner, northern hog sucker, black buffalo, and slenderhead darter, have disappeared from South Dakota within their native range (skipjack herring are now present in the Missouri River where they are nonnative, Cross and Huggins 1975). Five of the missing species formerly occupied only one of the river drainages in the state, but four (silver lamprey, silverband shiner, northern hog sucker, and black buffalo) were known from two or more drainages. Presently, 15 fish species, pallid sturgeon, lake chub, sicklefin chub, hornyhead chub, river shiner, carmine shiner, southern redbelly dace, highfin carpsucker, longnose sucker, mountain sucker, grass pickerel, trout-perch, burbot, northern plains killifish, and logperch, are each restricted to one river drainage. Nine of these species formerly occupied more than one. Species limited to only one river drainage within South Dakota presumably have a relatively high risk of extinction from the state because they are more susceptible to localized impacts (i.e., all their eggs are in one river drainage).

Native fish declines were greatest in the Minnesota, Big Sioux, and Vermillion river drainages and in the Missouri River Valley. The nine missing fish species were mostly distributed among these river drainages. The majority of the 15 fish species that are presently restricted to only one of the river drainages within South Dakota are also found among these river drainages. Thus, the Minnesota River, Big Sioux River, and Vermillion River drainages along with the Missouri River Valley are critical for conserving fish biodiversity within the state. This makes sense because these waters are within regions that have relatively equitable climates that are capable of supporting a relatively high number of fish species.

The fish fauna of South Dakota is constantly changing. Some species found in the state during prehistoric time have never been collected in historic time. Some historical species have disappeared during the last 150 years. Many species have been introduced to new river drainages and several have been introduced to the state from elsewhere. Presently, invasive species such as grass carp, silver carp, and bighead carp are expanding their ranges in the state, while sensitive native species such as pallid sturgeon, lake chub, and blacknose shiner are declining. We can expect additional changes to the fish fauna of the state in the future as humans continue to alter aquatic habitats and introduce foreign species while the climate continues to change.

Acknowledgements

The authors thank the many students and river biologists whose studies (see references) were the basis for this chapter.   ❖

Chapter 4

Missouri River Fisheries—Past, Present, and Future

Kent Keenlyne

" Towns born on the banks of the river and nourished at the breast of the river were also ravaged by the wild, raging river during spring and early summer floods. Man and the river have had a love-hate relationship ever since. "

— Chuck Sowards (about the Missouri River)

The Missouri River represents the largest single body of water in South Dakota both in volume and surface area. From the headwaters in Montana, the Missouri River flows south easterly and bisects the state into a western Great Plains Region and an eastern Prairie Pothole Region. Its total length within South Dakota is about 484 miles; a distance longer than the state is wide.

Today, Lewis and Clark would barely recognize the Missouri River of their epic journey 200 years ago. The once winding river with myriads of shallow, braided channels, sandbars, and islands that these adventurers recorded in their journals, has been replaced in South Dakota with four reservoirs and highly altered river channels below the huge earthen dams that cross the valley floor. Fully, 372 miles (76.8%) of the 484 miles of Missouri River in South Dakota now lies under reservoirs resulting in a tremendous change in river dynamics, sediment movement, river flows, water chemistry, and the subsequent fishery. Most people are unaware, though, that physical and biological change is still occurring on the river as a result of the dams and that change will continue for many years to come as the river attempts to reestablish a new state of equilibrium from the alterations made by humans.

Missouri River Fishes

No thorough fish inventories existed for the Missouri River when the state became "South Dakota" some 100 years ago. The earliest surveys were quick assessments conducted in the late 1800s to see what kinds of fish occurred at specific locations. Often done at the mouths of tributaries or at railway crossings, researchers gathered information by pulling seines along shorelines, in backwater chutes, or in oxbows to see what species of fish could be found. These shallow-water surveys often failed to capture the larger fish that could escape the nets and those that lived in deeper and faster water, so species lists were occasionally supplemented by local fishermen's reports or through other observations. These early collections suggest a predictable large river ichthyofauna of catfishes, sturgeons, suckers, buffalo fishes, and an assortment of shiner and chub species (minnows).

The number of each kind of fish caught was seldom recorded in early studies so the relative abundance of fish during the early years is difficult to assess. It was not until 1928 that Professor E. P. Churchill, University of South Dakota, and W. H. Over, Curator at the University of South Dakota, along with some of their students, conducted surveys of the major rivers in the state including the Missouri River. Their statewide work added 32 more species to the existing 35 species of fish recorded in the state at the time. It was not until the passage of the cost-sharing Federal Aid in Sport Fish Restoration Act in 1950 (commonly known as the Dingell-Johnson Act or D-J Act) that the state was able to finally obtain sufficient funds to begin systematic statewide surveys on fishery resources. Reeve Bailey, University of Michigan, joined Martin Allum from South Dakota to establish a standardized statewide fishery sampling program in 1952 which included sampling stations on the Missouri River.

Dam Building Era

The intensive sampling program begun in 1952 was timely for establishing a better picture of the ichthyofauna in the river before it became altered by humans. In 1944, as World War II was coming to an end, federal legislation was passed to put returning soldiers to work while stimulating economies in stressed rural areas. The Flood Control Act of 1944 contained two such features, irrigation and dam development on the Great Plains, an area that still suffered economically from severe agricultural losses during the Depression of the 1920s and the Dust Bowl Era or Dirty Thirties. The legislation included a Missouri River Basin development plan commonly referred to as the "Pick-Sloan Plan," which combined a plan by the U.S. Army Corps of Engineers (Colonel Pick, Division Engineer for the Army Corps of Engineers, Omaha, Nebraska) for flood control on the river and another plan by the U.S. Bureau of Reclamation (Glen Sloan, Regional Director for the Bureau, Billings, Montana) for mass development of federally supported irrigation projects in the arid West. Both plans called for the construction of large dams on the upper Missouri River. While the two plans would seem to be compatible, it has become apparent that there is not now, nor ever has been, sufficient water in the river to support all the expectations of massive irrigation in the Plains along with commercial navigation flows in the lower basin for which the water is now used. Though the massive irrigation proposed

Figure 4.1 MISSOURI RIVER POLITICS—Cartoonist J. Darling depicted the national debate about public works projects that spent millions of dollars on dams in South Dakota instead of improving housing in cities. It seems that there has always been political controversy surrounding the waters of the Missouri River, from planning in the 1950s that pitted urban interests vs. rural interests to operating the system of dams today that pits upper basin states vs. lower basin states. (Courtesy of the "Ding" Darling Wildlife Society)

in the "Sloan Plan" has never materialized, construction of the "Pick Plan" dams began on the river in 1946 (Figure 4.1).

Construction of Fort Randall Dam, the first main stem dam in South Dakota, and second in progression upriver, began near Lake Andes, South Dakota, in 1946. The dam, however, was not closed until July 1952. The ensuing reservoir, Lake Francis Case, is approximately 107 miles long and ends at the base of the next upstream dam, Big Bend Dam. The lake has a maximum pool of 102,000 acres with 540 miles of shoreline and a maximum depth of 140 feet. Maximum storage is 5.6 million acre feet (MAF) of water. One acre foot of water covers an acre with one foot of water, so Lake Francis Case holds a water volume equal to the amount that would cover all of South Dakota with water about one-inch deep. The average lake depth is 50 feet and the lake serves as a storage reservoir for downstream navigation with some flood storage capacity. It is normally drawn down 18 to 20 feet in the fall as a cushion for flood control. The normal exchange rate (pool turnover rate) during the winter drawdown period is three months and nine months during normal summer pool operating levels. The major tributary to this pool is the White River and the average sediment deposition rate into the lake is 16.6 thousand acre feet, enough sediment to cover $25mi^2$ with sediment that is one-foot deep. This amount of sediment input easily explains the build up of sediment where the White River enters the Missouri River.

Construction of the lowermost dam, Gavins Point Dam, near Yankton, South Dakota began in 1952 (Figure 4.2). The dam was closed in July 1955, and formed Lewis and Clark Lake that extends 25 miles to Niobrara, Nebraska. It is the smallest of the main stem reservoirs and serves as the flow regulation structure for the navigation project that begins downriver at Sioux City, Iowa. At maximum pool, the lake is only 31,000

Figure 4.2 Fisheries scientists and students sampling fish with a beach seine on a sandbar downstream from Gavins Point Dam. (photo by C. Berry)

acres, with 90 miles of shoreline, and a maximum depth of 45 feet. With a storage capacity of less than half a million acre feet, turnover rates are high and vary from three to 10 days during the year depending on whether releases are being made in support of the nine-month navigation season downriver. The pool is relatively shallow with an average depth of 16 feet and receives considerable sediment from its primary tributary stream, the Niobrara River. The lake receives an average sediment load of 2.5 thousand acre feet per year.

The next main stem dam to close on the Missouri River in South Dakota was the Oahe Dam (Figure 4.3) located about six miles upriver of Pierre, South Dakota. Construction began in 1948 but the huge dam was not closed until August 1958. At maximum pool, the resulting reservoir, Lake Oahe, covers 373,000 acres with a shoreline of 2,250 miles. The dam is 200 feet deep and the lake extends north into North Dakota.

Figure 4.3   Fishing boats and water intake structures near Oahe Dam. Anglers use downriggers, trolling sinkers, snap weights, in-line sinkers, planer boards and lead-core line to present lures to salmon and other species in the deep, cold waters near the Dam. (photo by C. Berry)

Oahe is the lowermost of the three large storage dams on the river and holds 23.3 MAF of water. The lake has an average depth of 63 feet and thermally stratifies over the lower two-thirds of the lake usually by the end of June. The turnover rate is not a factor and the lake has developed into a clear, cold-water environment. Water elevations fluctuate about 20 feet annually but the lake is drawn down by 40 feet or more during drought periods, leaving largely barren shorelines for much of the year. On an annual basis, pool levels peak in December or January as upstream releases are made for power production at Oahe Dam. Levels then decrease for power peaking needs at Oahe with more rapid drawdown in March or April to support downstream navigation needs. Depending upon local winter snow conditions on the Plains, water levels may rise starting in April or May with rising conditions into early summer as mountain snowpack melts in the Missouri River headwaters in Montana and Wyoming. The major tributaries to Oahe arise from the Plains to the west and include the Cheyenne, Moreau, Grand, and Knife rivers. Approximately 34,000 acre feet of sediment are accumulated in the reservoir each year.

The last main stem dam constructed on the Missouri River in South Dakota under the Pick-Sloan authorization was begun in 1959 about 21 miles upriver of Chamberlain, South Dakota. The Big Bend Dam was closed in July 1963, and forms Lake Sharpe, which backs water 80 miles to the vicinity of Pierre. The dam holds 78 feet of water depth, with a maximum pool of 61,000 acres and a shoreline of 200 miles. This reservoir has an average depth of 34 feet and the pool holds about 1.9 MAF of water that has a 27-day turnover rate. It is the most stable pool in the system, normally fluctuating less than two feet throughout the year. The Bad River flows from the west into the extreme headwater of the lake and has caused problems at the upper end of the lake that contribute to ice jamming when releases are increased in winter for power production at Oahe Dam. Approximately 4,400 acre feet of silt accumulation occurs in Lake Sharpe annually.

Changes After Impoundment

The nature of the Missouri River fishery was already documented by state biologists before the main stem dam system was filled in 1968 and a water management operational plan was instituted by the Corps of Engineers, operators of the dams, and associated power plants. By the time the dams were built, the native river fishery had been fairly well characterized. Before development, the river contained about 45 fish species that were adapted to the Missouri River aquatic ecosystem. The predominant predators consisted of species well adapted to capturing prey under the turbid water conditions of the river. The top predators were the channel catfish and pallid sturgeon. Other common larger species were the shovelnose sturgeon, which fed primarily on bottom dwelling insect life; the sauger, which fed on a variety of minnows; and the paddlefish, which filtered plankton and zooplankton from the water column. A variety of chub and shiner species also were found occupying the continuum of niches along the river, including oxbow lakes, backwater marshes, sidechannel sloughs, and shallow bays. The most abundant species in the river was the flathead chub, which had a sensory system well adapted to turbid water conditions. Other abundant small species included the silvery minnow, plains minnow, and brassy minnow along with sicklefin and sturgeon chubs. A variety of vegetation and detritus feeding "rough fish" were also common in the river including the river carpsucker, bigmouth and smallmouth buffalo, freshwater drum, white sucker, and redhorse sucker. Backwater sloughs, bays, and oxbows would hold quillback suckers and goldeyes along with longnose and shortnose gar and other species better adapted to less turbid conditions.

Researchers were able to document the changes in the fish fauna as the reservoirs filled (Box 4.1). As in most new impoundments, the vegetation-spawning species thrived in the new environment as the rising waters flooded tributary embayments and the bordering prairie and riparian vegetation. The newly flooded vegetation and rich organic soils provided an abundance of nutrients previously unavailable to the river. Under natural conditions, rivers must obtain nutrients from bordering riparian zones during flooding. By its turbid and flowing nature, the river produces relatively little energy. The basis of the food chain (the carbon, nitrogen, and phosphorus elements) are obtained by the river from tributaries and floodplains during out-of-bank flows. Thus, the newly flooded lands provided excellent habitat for some of the species which used vegetation for spawning and fed close to the bottom of the food chain.

In the early years of reservoir filling, the big and smallmouth buffalo, carp (an adapted introduced species), and drum populations exploded leaving researchers with the impression that the new "Great Lakes" could become an impressive commercial fishery. Once filled, however, the reservoirs no longer provided the conditions for these species to successfully spawn and their populations gradually declined in the ensuing years. In the large Oahe Reservoir, where filling

4.1
Dr. James Schmulbach

Dr. James Schmulbach, with his recent Ph.D from Iowa State University, joined the Zoology Faculty at the University of South Dakota in 1959 to take over the fish and ecology courses from the well-known Professor Dr. Edward P. Churchill. Colleagues described Doc Schmulbach as "quite a go-getter" for his efforts to increase fisheries research at the University.

Doc Schmulbach was a charter member and President of the Upper Missouri River Chapter of the American Fisheries Society that was started to "take a look at some of the things that were happening to the Missouri River." He was the first to see changes in the river and its fishes after the dams were closed. Doc wrote "Continuing reduction in the size of the aquatic backwater habitats portends deterioration of the Missouri River fishery. The entire degradation-aggradation cycle of the sediments has been changed by the dams...the future harvest rates of this fishery will remain modest or probably decline."

Doc Schmulbach retired from USD in 1993 after a 32-year career. One student was Wayne Nelson-Stastney, who is now the Missouri River Coordinator for the U. S. Fish and Wildlife Service. Fifty years after his major professor wrote about the loss of backwater habitat, Wayne is overseeing projects to restore them.

continued for nearly 10 years, the "rough fish" populations continued to thrive and larger predator species, notably the northern pike, also thrived. Although more adapted to the less turbid backwater bays and oxbow lakes where it could more readily see prey, the northern pike thrived in the new environment and became a noted sport fishery in Lake Oahe (as well as in Lake Sakakawea in North Dakota), attracting anglers from all parts of the United States to catch this prized trophy fish. Once the lakes filled, however, and the early spring flooding of vegetation no longer occurred, the northern pike also declined in numbers and remains so today.

Several other species also responded to the changing environment of the new reservoirs. The white and black crappies, also relatively rare species in the natural river ecosystem, were able to use the flooded riparian areas and the newly inundated brushy draws as ideal spawning sites, thus providing a popular fishery in the upper reaches of the reservoirs for several years after reservoirs filled. Once the lakes began their annual fluctuations during normal operations, however, shorelines began to erode from bank wash, removing woody vegetation from shallow areas and embayments. With spawning areas reduced, crappie species also began to decline and their abundance remains low in the smaller reservoirs. The predatory niche of the crappie gradually filled with the better adapted broadcast spawning species such as the goldeye and white bass, which continued to thrive for another 20 years.

As the reservoirs began their annual fluctuations, and wave wash pulled soils away from shallow shorelines, areas of gravel and boulders were left along the fluctuating lake boundaries. This was especially prominent along the eastern shores of the reservoirs that coincided with the western and southern margins of past glaciation along the Missouri River trench. Several gravel and boulder spawning species found the conditions conducive for their needs, including the walleye. Although the sauger was more abundant than the walleye in the pre-dam Missouri River, the sauger used gravel spawning sites in flowing water conditions while the walleye could utilize gravel and rock areas with little or no current. The walleye soon became the predominant sight-feeding predator on the maturing lakes as northern pike populations declined and it remains the most dominant, native sport species in the river.

The areas downstream of the dams were also changing. With the increased releases from the dams, the river channel downstream eroded with a subsequent lowering of the channel bed. Downstream of Gavins Point Dam, for example, the channel has degraded about 18 feet. The loss of backwater channels along the river has reduced the water table in bordering floodplain lands, and adjacent oxbow lakes and marshes have disappeared from lack of subsurface water and periodic flooding that maintained them in the past. While the reservoirs had already eliminated nearly 77% of the floodplain, many of the remaining shallow water areas disappeared due to channel incision and the lack of flooding. Native chub and shiner species, including the pre-dam dominant flathead chub, have virtually been eliminated from the Missouri River due to destruction of their habitat. Species like the sicklefin and sturgeon chubs, once fairly common species, are now candidates for federal listing as threatened or endangered. The larger predatory species that once fed on these species have been impacted; the pallid sturgeon for example, is now listed as Endangered by the federal government.

River development also impacted other larger native river fishes. Impoundments significantly changed water chemistry in the river including water temperatures downstream of the dams. Flow patterns were severely altered and vary drastically from week to week, day to day, and even hourly. Those species adapted to natural river temperatures and flows for cues to spawning, including the shovelnose sturgeon, paddlefish, burbot, and blue sucker (besides the above mentioned pallid sturgeon), declined precipitously when they failed to reproduce or had limited success with only periodic reproduction. There is no evidence of shovelnose sturgeon, once a common Missouri River species in South Dakota, reproducing in the state. The only paddlefish remaining in the reservoir are those that were stocked or remain from when the dam was closed.

Stocking New Niches

As the reservoirs filled, fishery workers realized that many native species were disappearing and that the new lake system represented an entirely new set of environmental conditions that many of the native species could not adapt to and survive. They began to search for species that could thrive in reservoir environments and the cold-water habitats created in the larger and deeper reservoirs like Oahe. Some of the earliest stockings were rainbow and brown trout from existing trout hatcheries in the Black Hills. Although Oahe remains cold enough for trout survival, the necessary spawning areas do not exist to allow trout to maintain populations. Attempting to fill the new cold-water niches, several other cold-water species have also been stocked in Oahe, including lake herring, lake whitefish, coho salmon, lake trout, cutthroat trout, Kokanee salmon, steelhead trout, Bonneville cisco, and chinook salmon. Although the stocked fish may continue to survive into adulthood, none of the stockings have produced self-maintaining populations due to a lack of suitable natural spawning areas. The chinook salmon has been maintained in Oahe for the last 20 years through a put-and-take system of gathering eggs from adults, raising the fry in hatcheries, and restocking to maintain year classes of salmon.

Fishery workers also attempted to find new species that could fill prey species niches. Spottail shiners are the only species to have become marginally successful but seem to concentrate along shorelines and, thereby, live in an environment not conducive to providing a prey base for the large cold-water species. Rainbow smelt were stocked in the Lake Sakakawea in the early 1970s and have since become established in the Oahe Reservoir. In the winter months, when water temperatures are at their coldest in the river, smelt occasionally pass through the Oahe Dam power plant and become dispersed downstream. The lower lakes, however, become too warm in the summer for smelt to establish permanent populations downstream. Like many newly successful species, the smelt enjoyed several years of rapid expansion in Oahe before its population crashed. At present, the fish are much smaller and experience population ups and downs related to spawning success and reservoir conditions.

In the last 20 years, smallmouth bass have also been introduced in both the South Dakota reservoirs on the Missouri River and in the riverine reaches downstream of the dams. The smallmouth has been successful in finding areas to spawn and enjoys scattered populations throughout the Missouri River system in South Dakota, but they are most plentiful in the reaches downstream of the Gavins Point and Fort Randall Dams. There are currently no plans to stock additional species in the Missouri River in South Dakota as the stocking of fish has been considerably reduced in recent years with the establishment of self-maintaining populations of sport fish preferred by anglers.

Changing Riparian Vegetation

The reservoir and river environment continues to change today even though most people think of them as static environments. The remaining riverine sections have seen major changes due to channel incisement and the loss of backwater channels, marshes, and oxbows along with the elimination of most in-channel islands. The riparian areas, however, continue to change and these changes are reflected in the adjacent river environment. At one time, American elm and cottonwood were the dominant trees along the rivers in South Dakota. The elm has largely been extirpated due to the spread of Dutch elm disease in the last 20 years and the cottonwoods will also soon disappear from the river floodplain. Cottonwoods require spring overflows onto the floodplain to reestablish and, since the river no longer escapes its banks, cottonwoods, which have not replaced themselves since the dams were closed some 50 years ago, are gradually dying out due to age. The dominant tree species remaining are green ash, Russian olives, red cedar, and other much smaller trees. Where once the large elms and cottonwoods overhung eroding banks, the banks are now largely unshaded and do not contribute logs and large branches to the riverine food chain. The encroachment of homes along these presumably flood-free sections also poses a threat from additional clearing of land and pollution from septic systems built in the porous floodplain sands. These flat areas along the river are seen as prime agricultural areas and landowners pay taxes on the full potential of the land, which discourages the owners from allowing the land to remain wooded. The loss of woody material to these riverine areas, the loss of overbank flooding which provided nutrients into the river, continued channel incisement, and the loss of streamside vegetation will continue to reduce the energy base that feeds the river, thus further reducing river productivity.

When the reservoirs first filled, thousands of acres of trees were also flooded and their trunks and limbs provided a substrate for a wide variety of aquatic insect life. These important substrates remained viable for 30–40 years in some places and gradually rotted away or were demolished or uprooted by ice flows. These "stump fields" continue to deteriorate and will eventually be lost from the system along with the important contribution they provide as a substrate to some of the food chains.

Reservoir Sedimentation

Sedimentation is an ongoing process. By acting as sediment "traps", the reservoirs retain sediments where flows dissipate or are insufficient to retain the material in suspension. In the roughly 40 years since the reservoir system has been in operation, large deltas have begun to form in the headwater of Lewis and Clark Lake from sediment from the Niobrara River, in Lake Francis Case from the White River, and at the headwater of Lake Sharpe from the Bad River. The delta from the White River is currently large enough to affect reservoir elevations upstream when the Corps attempts to push large volumes of water through the lake. Since little change is observed from viewing the lake surface, few people grasp the amount of change that is going on under the surface by sediment deposition. Since deeper areas of the lake do not experience the beneficial act of currents sweeping silt away, bottom substrates are gradually being buried in fine sediments with resulting changes in benthic aquatic life and oxygen balance in the deeper waters.

The changes, however, are not necessarily uniform. In Lake Oahe, for example, the bulk of sediments entering the lake come from its western tributaries. Since Oahe is so deep, most of these river mouths were inundated and the resulting bays are now gradually being filled with sediments. Unfortunately, tributary streams often serve as primary spawning and nursery areas for many naturally reproducing fish species. As the bays continue to become filled with sediment and cover woody and gravelly points, catfish and gravel-spawning species will be adversely affected. The delta at the head of Oahe also continues to expand creating broad mud flats and a spreading river channel which can block fish passage into the upper riverine sections during low flow periods. This process will continue into the foreseeable future with increasing impacts to the local ecology.

It is projected that the reservoirs will be able to accommodate sediment deposition for many years into the future. While this may be true, the calculations are misleading in relation to the welfare of the fishery. Earlier, it was mentioned that Lewis and Clark Lake, the smallest, and lowermost reservoir on the system, has a turnover rate of only 3–10 days. This calculation was based on storage volumes available 40 years ago and the turnover rate could be much greater today. As sediments continue to accumulate in the reservoir, turnover rates increase, resulting in faster flushing of juvenile fish through the system and adding to quicker eutrophication of the lake. Francis Case had a turnover rate from 3–9 months (depending upon lake elevation at different times of the year) 40 years ago but has already formed a large delta off the mouth of the White River, thus reducing storage capacity of the lake and increasing turnover rates. Lake Sharpe had a turnover rate of 27 days but now has an extended delta forming in its headwaters from sediment from the Bad River. As turnover rates increase, lake productivity is affected and relative abundance of insect and zooplankton species change which, in turn, affect the growth and welfare of the fish community.

New Reservoir Syndrome

One of the more significant findings of the early reservoir researchers was a decline in the growth rate of many fish species after the lakes filled and became operational. In a natural large river like the Missouri, very little energy is actually produced in the river itself. Instead the river goes to the bordering floodplain to capture nutrients and energy during periods of high water or flooding. Marshlands, sloughs, and oxbows along the river contain aquatic vegetation and provide conditions which capture energy that the river can use when flooding. The preponderance of energy, however, comes from the woody vegetation that grows on the floodplain floor along with the large volumes of detritus in the form of leaves that are produced by the riparian forest canopy. Most riverine fauna time their reproductive cycles to take advantage of this natural river "flood pulse." Even the floodplain vegetation synchronizes with the normal flooding cycle. Floodplain succession is a result of the flooding pattern and readjustment of floodplain features from erosion and deposits associated with changing river flows. Once these processes are eliminated, the entire ecology of the riverine system is interrupted, including the energy transport system.

When first inundated, the adjacent lands released great amounts of energy and nutrients into the new aquatic system. As outlined earlier, fish species that use vegetation as a spawning substrate found the new conditions beneficial and their populations expanded. It was the accompanying increase in nutrient release and the resulting bloom in phytoplankton and zooplankton that allowed the populations to expand so rapidly. This phenomenon has been documented repeatedly in the succession of newly created reservoirs. It has also been repeatedly documented, however, that there are subsequent declines in growth of these early successional species once the large flush of energy and minerals has been used in the food chain, and new resources are no longer available to maintain an increasing biomass of fish. At that stage, the large populations developed under the near ideal spawning and growing conditions start to experience resource shortages and growth (and populations) begin to decline. Sometimes the decline is very rapid.

The same phenomena is common for species occupying the second trophic level in new reservoir environments, although the process is slightly delayed since they depend upon the first trophic level for their link in the food chain. These largely predatory species often show phenomenal growth rates and large populations until prey species begin to naturally decline due to decreased energy in the system. Populations of the second successional tier follow the rise and fall of the primary consumer population explosion. The once trophy northern pike fishery on Oahe and Garrison reservoirs is an excellent example of this process.

Fisheries Management Challenges

Reservoir productivity is a continuing problem for fishery managers on the Missouri River in South Dakota today. Because most of the reservoirs fluctuate considerably during the year, their shorelines experience loss of topsoil due to wave wash and are relatively sterile environments that support little, if any, plant life. In Oahe Reservoir and in Lake Francis Case, normal annual fluctuation can be 20 feet per year leaving much of their shorelines largely devoid of vegetation. This prevents the establishment of stands of shoreline aquatic vegetation even though water clarity would allow the growth of plant beds were water levels more stable. The consequence is that the normal growth of plant life that occurs along the margins of natural lakes does not develop in the Missouri River reservoir system. Consequently, a significant amount of energy available to the lake does not materialize, greatly reducing the productivity of the water body. Even though the reservoir may have hundreds of miles of shoreline, fishery carrying capacity is severely restricted.

The primary source of energy to the Missouri River and Missouri River reservoirs in South Dakota now comes from tributary streams. In a Plains state like South Dakota, this creates difficult long-term management problems for fishery managers because they cannot control either the inflow from tributary streams or the water flow pattern through the reservoir system. In large reservoirs like Oahe (or Garrison in North Dakota), most energy is quickly captured in the food chain near the source of entry and very little passes out of the system past the downstream dam. This explains what anglers on Oahe already have observed, that the most productive fisheries lie in the upper half of the reservoir in habitat areas close to major tributaries (the Cheyenne, Moreau, Grand, and Knife rivers). The proclivity of smelt and salmon to suspend near the thermocline in the heat of summer has helped extend fishing into the lower third of the reservoir and assists in dispersing some of the fishing pressure, but most fishery pressure follows the natural capture of energy in the vicinity of tributary streams or in the bays of intermittent streams.

It is difficult for Missouri River fishery managers to maintain good distributions of fish species year classes in reservoirs. Part of the problem is the timing of water level rises during the critical spawning periods of species that normally spawn in shallow water areas. The vagaries of energy inflows to the system, which feed both the primary and secondary trophic levels of the aquatic food chain, are also beyond the control of the managers. Particularly in drought years, the arid plains of South Dakota do not capture enough snowmelt in spring to bring the nutrients and energy to the river and reservoir environment or to the major tributaries that feed the system. Even if spawning conditions are good, a shortage of energy at the critical fry stage may cause a collapse in the food chain and a weak year class results. Reservoirs like Lake Sharpe are particularly vulnerable to such scenarios; the absence of permanently flowing tributary streams and the intermittent streams that enter into Lake Sharpe are particularly vulnerable to a shortage of winter snowpack on the plains. Consequently, few, if any, nutrients reach the lake in such years. In these cases, fish can suffer from a shortage of food which severely affects growth rates and body conditions resulting in extremely thin fish. Attempting to maintain a good distribution of year classes of sport fish under such conditions is difficult to say the least.

The role of energy as a limiting factor in the Missouri River will continue to challenge fishery managers in future years. Researchers now know that it is likely that carbon has become a limiting factor in both the reservoir and river environments. It is a fairly simple remedial measure to artificially augment failed year classes of preferred species to try to maintain a balanced fishery but, if food is already a problem, little can be done to correct an energy deficiency for the food chain.

Better water level management could greatly assist the Missouri River fishery managers in maintaining balanced fish populations, but fishery concerns are identified as subservient to all other uses in the present management criteria for the Missouri River as established by the federal agency that presently manages river flows. Recent recommendations to offset the balancing of the upper three large storage reservoirs in Montana, North Dakota, and South Dakota to assure alternate fishery production in at least one reservoir each year has been a step forward, but still would only occur provided the action did not interfere with flood control, navigation, power generation, irrigation, or water supply concerns that are listed as higher priorities for river management. The question of Missouri River management criteria has received considerable scrutiny over the last 15 years, including a recent report by the National Research Council (NRC 2002), and will likely continue to be an issue for years to come. The Missouri River is characterized as the common thread that holds the Region together but the seam that also tears the Region apart.

The future holds another aspect that will continue to complicate the lives of fishery managers. Rivers are known to do two things. One is to carry water and the other is to transport sediment to the ocean. During the "Big Dam Era" of the 1950s, water planners attempted to control the first of these processes by developing huge dams on most of the large rivers in the country followed by elaborate operational plans to use the stored water for benefits other than fish. The second aspect, sediment transport, has still to be addressed and the process is continuing to be ignored by managers oblivious to its existence. Sediment deposition is already recognized as a problem on the Missouri River on Lewis and Clark Lake, Lake Francis Case, and Lake Sharpe. This is after only 50 years of reservoir operation—and the problems will continue to grow at accelerated rates in the future as storage capacities are reduced. For the fishery manager, this means shortened turnover rates in the reservoirs and probably changes in the kinds and numbers of fish in the reservoir. Access problems will continue to become more acute as storage is reduced and storage lakes, of necessity, will exhibit more fluctuation to supply flows downriver. Under severe draw down conditions (as would occur in a prolonged drought similar to the 1930s), Oahe would not develop a summer thermocline and would cease being a cold-water fishery with a resulting crash in the present cold-water fishery.

Sediments will continue to settle out at the head of Lewis and Clark Lake, at the mouth of the White River in Lake Francis Case, the head of Lake Sharpe, and from the western tributaries and at the headwaters of Lake Oahe. In future years, these deltas will continue to spread creating marshy areas with braided channels. In Lake Francis Case and Oahe, where reservoir elevations fluctuate several feet annually, judicious placement of woody barriers on the deltas would enhance the formation of low islands that could succeed to woodlands and, once again, add nutrients to the aquatic environment. Structures could also be used to create greater channel depth variability in the delta rivulets and provide substrates for riverine benthos. It is unlikely, however, that these areas would revert to ideal spawning areas for many of the chub and minnow species that have been lost to the system because the sediments lack the gravels and course sands that would revert to gravel bars or riffle complexes important to sustaining many of these species.

The future of the Missouri River fisheries depends largely on the water management scheme implemented in the future and the eventual priority given to fisheries in the operational management of the system. Without a higher status in management decisions, the fishery will, by necessity, require substantial artificial support to maintain as a sport fishery.

The walleye will continue to be the predominant sport fish for the foreseeable future. The lower reservoirs will gradually convert to greater proportions of the so-called "rough fish" and bass species will gradually increase proportionately as "sport fish." If river management continues as currently established, pallid sturgeon, shovelnose sturgeon, and paddlefish will disappear as will many of the native riverine chub and minnow species that are already severely reduced in numbers. Fishery managers will have to continue to be innovative in setting harvest regulations and may eventually have to resort to restrictive quotas on various sizes of the more heavily fished sport species or begin to become more restrictive on season lengths or closures in order to maintain a quality sport fishery while retaining sufficient stocks to continue self-maintaining populations.

Acknowledgements

The editors and the author of this chapter would like to acknowledge the many biologists who have worked on the Missouri River through the years, particularly Larry Hesse (Hesse 1987, Hesse et al. 1989). Pete Carrels, a South Dakota writer, has helped the public understand our love-hate relationship with the river (Carrels 1999). The National Research Council has issued a good, readable overview of the history and prospects for recovery of the Missouri River (Berry 2003, NRC 2002). The Doc Schmulbach photo was from Cummins et al. 1982. ❖

Chapter 5

Small Impoundments

Steven R. Chipps, Stephen K. Wilson and David O. Lucchesi

66 Water is the pioneer which the settler follows, taking advantage of its improvements. 99

— Henry David Thoreau

Impoundments vs. Lakes

To the casual observer, small impoundments can appear very similar to natural lakes; both are "lake-like" environments that provide angling opportunities, waterfowl hunting, wildlife watching, and water-sport recreation. What then, is the difference between a small impoundment and a natural lake? The purpose of this chapter is to 1) compare and contrast the habitats and fisheries resources in small impoundments with those found in natural lakes and 2) discuss factors affecting habitat quality and fisheries production in South Dakota's small impoundments.

Unlike natural lakes, small impoundments ***are not*** natural features of the landscape. Most small impoundments in South Dakota were constructed during the 1930s as part of the Works Progress Administration (WPA) or the Civilian Conservation Corps authority (CCC). These programs, established by President Franklin D. Roosevelt shortly after taking office, provided employment to many young men (between the ages of 18-25) to work in the National forests, parks and rangelands. Many of the small impoundments constructed during this era were less than 150 acres in size, and are now currently owned or leased by the state.

Aside from being "artificial", at least two important features distinguish small impoundments from natural lakes: water residence time and the ratio of watershed area-to-surface area. Water residence time, usually measured in years, is the amount of time it takes to replace the entire volume of water of a lake or impoundment. In wet years, water entering impoundments comes mostly from the inlet and can be offset by water output through spillways or other overflow structures near the dam (Figure 5.1). Because excess water can flow through an impoundment, water residence time in the impoundment tends to be much shorter than in natural lakes. In contrast, many natural lakes have no outlets; as a result, water level continues to increase during periods of above average precipitation, as we've seen in the mid-1990s. During dry years, water is lost primarily through evaporation and/or ground water recharge and water levels in both natural lakes and impoundments can become low.

Outlet   Inlet
Dam

Figure 5.1 A typical South Dakota small impoundment showing the inlet, dam and outlet structures.

The ratio of watershed area-to-surface area (WS) reflects the amount of landscape drained per surface area of a lake or impoundment. For impoundments, the average WS ratio is much higher than that for natural lakes. This is related to two factors; first, because impoundments are designed so that excess water flows downstream, they can be much smaller in size than natural lakes (Table 5.1). For a given watershed area, a small surface area results in high WS values. Secondly, unlike many natural lakes that were formed from glacial depressions, impoundments are usually constructed in areas with more topographical relief (i.e., narrow valley or draw).

In many ways, small impoundments are quite similar to natural lakes. Habitat inventories show that maximum water depth and water clarity are similar for South Dakota's impoundments and natural lakes (Table 5.1). Whether a small impoundment or a natural lake, seasonal patterns in water temperature usually depend more on the size and depth of the water body than its classification as an impoundment or lake. Small shallow lakes and impoundments tend to be well-mixed in summer months, so that water temperature varies little with water depth.

Other habitat features, such as the composition and abundance of aquatic plant communities, differ between small impoundments and natural lakes. On average, impoundments tend to have more aquatic plant species than natural lakes

Table 5.1 Range of habitat characteristics for small impoundments and natural lakes. Data are from recent habitat and water quality inventories in 24 small impoundments and 15 natural lakes in South Dakota.

VARIABLE	SMALL IMPOUNDMENTS	NATURAL LAKES
Surface area (acres)	27–180	355–15,550
Ratio of watershed area-to-surface area	36–1401	2–67
Maximum depth (feet)	6–31	8–41
Fetch (miles)	0.3–1.3	1.4–9.0
Water clarity (feet)	1–7.5	1.5–8.0
% submerged vegetation	0–70	0–31
% sand	0–31	0.3–35
% silt	55–95	45–85
Algal biomass (ppm[1] chlorophyll a)	0.001–0.214	0.005–0.081
Nutrients (ppm phosphorus)	0.02–1.4	0.01–0.5

[1]ppm refers to parts per million.

(Figure 5.2). Similarly, the amount of submerged vegetation coverage is generally higher in small impoundments than in natural lakes (Figure 5.2). This may be due, in part, to differences in surface area and topographical relief between small impoundments and glacial lakes. Because glacial lakes are generally larger than small impoundments, they have a larger fetch (exposure to wind) and are more prone to wind and wave action. In near shore environments, wave action can reduce plant growth by uprooting young plants and/or resuspending sediments that reduces light availability. Small impoundments, on the other hand, generally have 1) less fetch, 2) more shoreline development (i.e., more bays and arms) and 3) occur in areas with more topographical relief, that tends to protect small impoundments from effects of wind and wave action. Some natural lakes, such as Enemy Swim and South Buffalo, are relatively protected from wind and wave action and as a result, have more submerged vegetation than typical, wind-swept lakes such as Poinsett and Thompson.

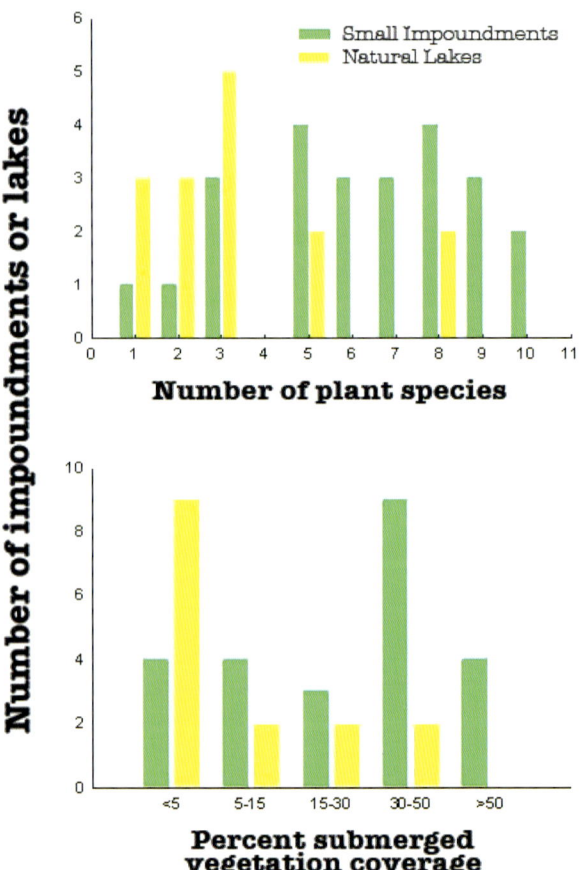

Figure 5.2 Number of aquatic plant species in South Dakota's small impoundments and natural lakes [top panel]. Bottom panel shows percent submerged plant coverage in South Dakota's small impoundments and natural lakes.

Habitat Diversity in Small Impoundments

South Dakota's small impoundments are generally distributed along a northwest-to-southeast cline and they occur both east and west of the Missouri River (Figure 5.3). Water clarity decreases in small impoundments from west to east in the state (Figure 5.4). Because row crop agriculture is more predominant east of the Missouri River, increased runoff (silt and nutrients) from the surrounding watershed likely contributes to increased turbidity. Water clarity also has an important influence on aquatic plants because turbid conditions can reduce light penetration and preclude aquatic plant

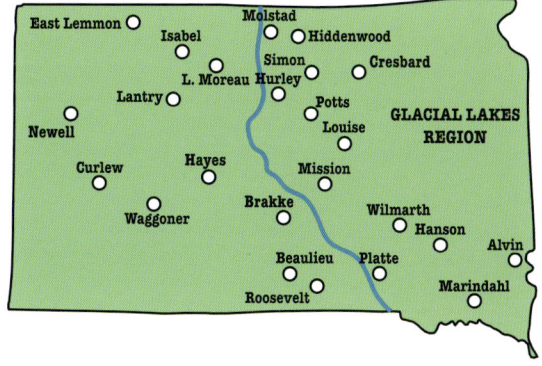

Figure 5.3 General location of some popular South Dakota impoundments. Habitat, water quality, and fisheries were inventoried in these 24 impoundments from 2000 to 2002.

growth. In general, small impoundments located in southeast South Dakota support fewer plant species (and lower plant coverage) than those in the northwest (Figure 5.4).

Like natural lakes, habitat features in small impoundments can be quite variable. In a 2000-2002 inventory of 24 small impoundments, maximum water depths ranged from 6 feet in Lake Platte to over 23 feet in Lake Alvin (Table 5.2). Surrounding land use also varies considerably from one impoundment to another and can have an important influence on water quality and fish production. In Mission Lake, agricultural land use constitutes only about 3% of the watershed, as compared to over 77% for Lake Alvin. Such large differences in land use patterns likely contribute to differences in water quality. Nutrient concentrations and algae biomass (measured as chlorophyll), for example, are substantially lower in Mission Lake (0.07 ppm phosphorus; 0.003 ppm chlorophyll a) than in Lake Alvin (0.15 ppm phosphorus; 0.04

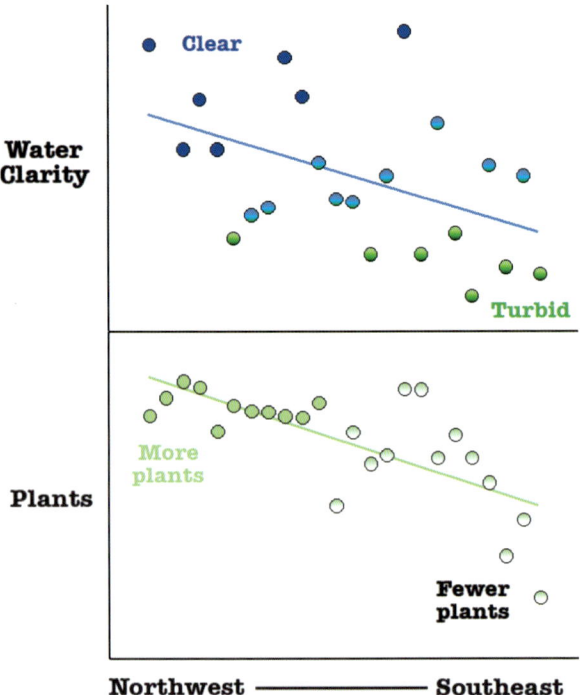

Figure 5.4 Geographical patterns in water clarity (top panel) and aquatic plant abundance (bottom panel) for small impoundments in South Dakota. Both water clarity and plant abundance decline from northwest to southeast.

ppm chlorophyll a). Hence, land use practices in watersheds surrounding small impoundments can have an important influence on water quality and fish production, particularly given the high ratio of watershed area-to-surface area that characterize many small impoundments.

Fisheries In Small Impoundments

Small impoundments provide important recreational opportunities in South Dakota. Although they comprise a small amount of the total surface waters in the state, small impoundments can receive relatively high angling pressure. Lake Alvin, located near Sioux Falls, SD, annually supports over 100 hours of fishing per acre of water. Compare that to large, natural lakes like Lake Madison or Lake Thompson with 10 to 30 hours of fishing per acre—and it's clear that small impoundments provide their share of angling opportunities.

The South Dakota Department of Game, Fish & Parks (SDGFP) classifies small impoundments as waters less than 150 acres. Water quality criteria are used to group impoundments into one of three categories; Class I, II or III waters. Most impoundments are classified as "warmwater permanent fish-propagation waters" (Class I), or "warmwater semi-per-

Table 5.2 Habitat and water quality characteristics in 24 South Dakota impoundments sampled from 2000 to 2002. See Figure 5.3 for general location of small impoundments.

IMPOUNDMENT	SURFACE AREA (ACRES)	MAXIMUM DEPTH (FEET)	WATER CLARITY (FEET)	AGRICULTURAL LAND (%)	NUTRIENTS (ppm[1] phosphorus)	ALGAE (ppm chlorophyll)
Alvin	104	23	1.5	77	0.15	0.042
Beaulieu	27	17	3.3	76	1.45	0.036
Brakke	146	17	4.3	71	0.06	0.029
Cresbard	69	16	2.5	26	0.66	0.026
Curlew	143	22	1.0	11	0.09	0.006
East Lemmon	180	16	4.6	66	0.11	0.014
Hanson	59	17	1.6	61	0.11	0.044
Hayes	74	16	3.1	46	0.14	0.009
Hiddenwood	27	18	3.3	11	0.2	0.012
Hurley	91	22	5.9	36	0.88	0.005
Isabel	143	23	2.4	39	0.19	0.214
Lantry	37	18	4.6	45	0.3	0.007
L. Moreau	37	20	3.0	37	0.07	0.035
Louise	156	21	5.2	21	0.45	0.004
Marindahl	151	31	3.9	63	0.04	0.007
Mission	59	15	2.0	3	0.07	0.003
Molstad	104	17	6.9	60	0.54	0.004
Newell	138	20	7.2	5	0.02	0.001
Platte	146	6	0.9	54	0.59	0.051
Potts	57	16	7.5	52	0.28	0.004
Roosevelt	91	18	2.0	56	0.26	0.104
Simon	77	14	3.9	47	0.78	0.007
Waggoner	99	16	5.8	7	0.09	0.021
Wilmarth	121	22	4.2	39	0.76	0.043

[1]ppm refers to parts per million

manent fish-propagation waters" (Class II). A few small impoundments are designated as "marginal fish-propagation water" (Class III) indicating that they experience frequent winter and/or summer kills.

Good shore fishing access, a lack of other nearby fishing opportunities and proximity to population centers make many small impoundments popular places to fish. Generally speaking, small impoundments are managed primarily for panfish species such as bluegills, sunfish, white & black crappies, and yellow perch (Figure 5.5). Important game fish species also include largemouth bass and northern pike. Although walleyes and other popular sportfish (e.g., channel catfish) are

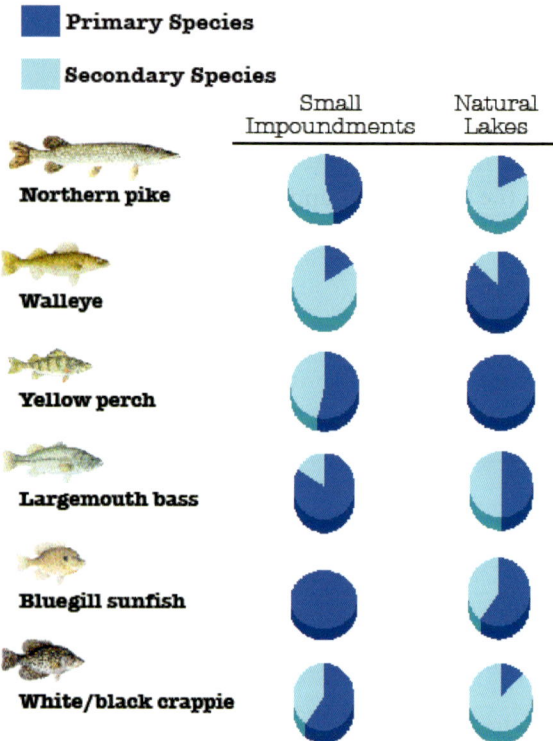

Figure 5.5   Common game fishes managed as primary or secondary species in South Dakota's lakes and small impoundments.

found in many small impoundments, they are generally managed as "secondary" species (Figure 5.5). Several state records have come from small impoundments, stock dams, farm ponds, or other artificial lakes in South Dakota; included are state records for bluegill, largemouth bass, green sunfish, white and black crappies, and muskellunge. State records for most other warm- and cool-water fishes, including yellow perch, walleye, and northern pike have been taken from natural lakes, rivers or the large Missouri River reservoirs.

Management Challenges

Providing quality fisheries in small impoundments can be a challenging task. In a recent survey by SDGFP biologists, watershed degradation was listed as one of the primary factors affecting fish production in small impoundments. Other issues contributing to low abundance and/or reduced growth rates of fish include poor habitat quality, low productivity, lack of vegetation, and an overabundance of rough fish. Fortunately, a variety of tools are available to fisheries managers to help increase game fish abundance, improve panfish size, and reduce rough fish abundance. Minimum size restrictions, creel limits, predator stockings (i.e., largemouth bass), and rough fish removal are just a few of the techniques used by SDGFP biologists to increase the quality of fisheries in South Dakota's small impoundments. In many small impoundments, however, maintaining quality fish populations over the long term will require large-scale efforts to improve conditions in the surrounding watershed (i.e., reduce silt and nutrient inputs).

Acknowledgements

We thank Sam Stukel (Stukel 2003) of South Dakota Game, Fish and Parks and Dr. Michael Brown, South Dakota State University, for kindly providing information on South Dakota's glacial lakes. ❖

Chapter 6

Eastern Natural Lakes

David W. Willis, Sam M. Stukel and Michael L. Brown

66 A lake is the landscape's most beautiful and expressive feature. It is earth's eye; looking into which the beholder measures the depth of his own nature. 99

— Henry David Thoreau

The state of South Dakota has been blessed by a tremendous resource of natural lakes in the eastern part of the state, with most concentrated in the northeast (Figure 6.1). These lakes, created by glacial activities, provide a resource quite distinct from the impoundments found in central and western South Dakota.

A wide variety of habitats occur within these natural lakes. Some lakes are little more than extremely productive duck sloughs, prone to winterkill; however, they can at times provide substantial sport fisheries. On the other end of the spectrum, some true jewels occur, such as Enemy Swim

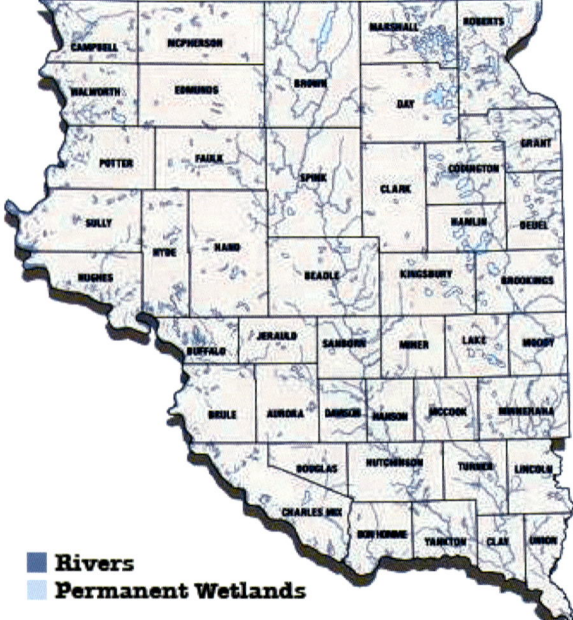

Rivers
Permanent Wetlands

Figure 6.1   Most of the South Dakota natural lakes of glacial origin are located in the northeastern part of the state.

Lake (Figure 6.2). This lake has rocky habitat, islands, aquatic plants, and one of the most diverse fish communities in any natural lake in the state. It also was one of the few lakes in the state to still contain water and retain a fish community during the Dust Bowl of the early 1930s.

The purposes of this chapter are to provide background information on the geologic origin of these natural lakes, review the variety of habitat characteristics for the lakes, summarize the current lake classification system used by South Dakota Department of Game, Fish and Parks (SDGFP) biologists when managing these lakes, and provide an overview of the water level dynamics that are characteristic of these lakes.

Geologic Origin of South Dakota Glacial Lakes

The landscape of eastern South Dakota was structured by glaciation during two ice ages: the Illinois Glaciation Period (Pleistocene Epoch) approximately 400,000 years ago, and the Wisconsin Glaciation Period (Holocene Epoch) about 10,000 years ago. During both periods, massive glaciers crept south, gouging the terrain and transporting enormous amounts of debris. These two glacial functions are responsible for the creation of eastern South Dakota's most distinguishing physical characteristics: the Prairie Coteau and the myriad of wetlands found there.

The Prairie Coteau is a 200-mile long plateau running northwest from northern Iowa to near the northern border of South Dakota. At its northern tip, the Prairie Coteau is geologically unique. Here it abruptly rises up several hundred feet above the bordering lowlands to the north and east. The westward slope to the James River basin is much more gradual—almost imperceptible in some places. The northern section of the Prairie Coteau is strewn with water-holding depressions ranging in size from less than 1 acre to over 15,000 acres.

The Prairie Coteau

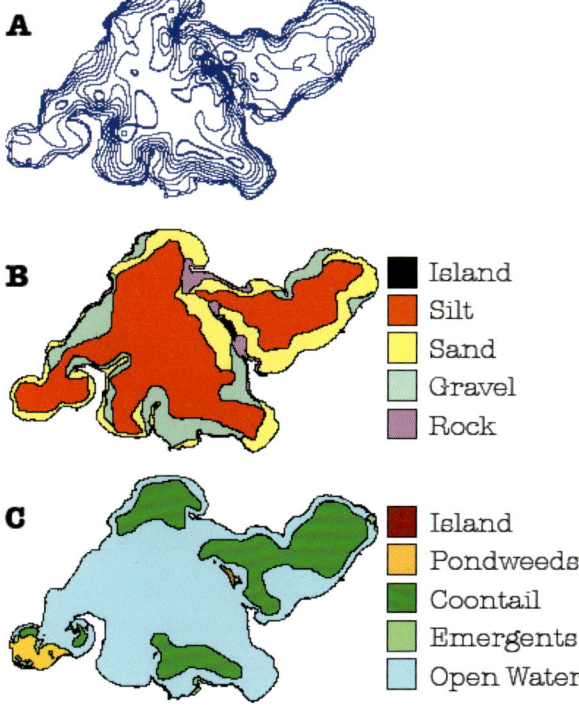

Figure 6.2   Contour (A), bottom substrate (B), and vegetation (C) maps for Enemy Swim Lake, South Dakota, August 1996. The depth contour map is in 1-meter (3.3 feet) intervals, which indicate the complexity of this particular lake basin.   Maps courtesy of Dr. Brian Blackwell.

began as an irregularity in the Cretaceous shale bedrock of the region. The shale under the modern Prairie Coteau has a relief of up to 650 feet above the surrounding bedrock. The Illinoian glaciers apparently completely covered this bedrock structure. Glacier advance and retreat likely increased the abruptness and height of the Prairie Coteau. Following a 200,000-year interglacial stage, the first of six advances of the Wisconsinan glaciers entered South Dakota. Ice sheet thickness declined with each advance, and only the first had sufficient mass to override the high ridge left from Illinoian glaciation. This initial Wisconsinan advance left the ridge covered with up to 80 feet of glacial till. Advances II and III were primarily responsible for shaping the Prairie Coteau into what it is today. The Prairie Coteau split these ice sheets like an enormous wedge as they passed southward. As the glacier advanced down both sides of the wedge, massive ice blocks and vast amounts of till were heaved onto the high ground. The second advance left an average of 100 feet of clay-rich till that was derived from shales in western Canada, along with granite particles of all sizes collected from the northern Pre-Cambrian Shield area. As each advance began to cease and recede, rivers of melt water carried even more glacial drift (termed outwash) onto the Prairie Coteau, burying the thousands of ice blocks left by the advancing glaciers. As they flowed, these melt-water rivers "sorted" their payloads based on particle size. This resulted in large sand and gravel deposits. Some outwash deposits in Day County, South Dakota, have been found to cover several square miles and to be up to 125 feet thick.

The drift and ice stranded on the Prairie Coteau at the end of the last glacial ages (10,000 years ago) formed the lakes, wetlands, and topography that characterize the area today. Wetlands on the Prairie Coteau were created one of two ways. Many of the smaller wetlands (and a few larger lakes) are located over the site of isolated ice blocks heaved onto the Prairie Coteau and buried under drift. Because they were buried, and therefore insulated, these blocks slowly melted over thousands of years. As they did, they left depressions in the landscape known as "kettles" or "potholes." Lakes Madison, Brant, and Herman (Lake County) apparently were formed in this way.

The majority of South Dakota's large lakes are termed "outwash" lakes. That is, they were formed as drift was swept forward off the glaciers in melt water. The outwash was sorted by size and deposited in a fan shape preceding the glacial margin. Natural depressions in this gravelly, porous deposit became lakes. Typically, these lakes are well drained and accumulate little dissolved solids, creating our best lakes in terms of water quality. Pickerel (Day County), Enemy Swim (Day County), and Poinsett (Hamlin County) are examples of such lakes. All of these lakes have extensive aquifer connections due to the porous nature of their substrate. As a result, their water levels are relatively stable over time.

Waubay and Bitter lakes (both in Day County) were also formed by outwash; however, they are located over impermeable till near the end of a deposit. This till essentially "seals" the lake bottom. That, plus the fact that they are closed basins, influences dissolved solid levels. Minerals flowing into these lakes are essentially trapped there. In other outwash lakes, such substances would slowly leach into the groundwater or flow out of the lake.

Diversity in Habitat Characteristics of the Eastern Glacial Lakes

Lakes can be small and shallow, such as Lake Goldsmith in Brookings County (280 acres, 8-foot maximum depth), or large and shallow, such as Lake Byron in Beadle County (1,749 acres, 10-foot maximum depth). Lakes can be very simple in basin shape, such as the nearly circular shape of Lake Brant in Lake County, or more complex and irregular such as Enemy Swim Lake in Day County (Figure 6.2). Lakes can have little submerged vegetation plant growth, such as Lake Poinsett (Hamlin County) or Lake Kampeska (Codington County), or contain extensive beds of rooted, underwater aquatic plants as in Lake Cochrane (Deuel County) and Enemy Swim Lake. Lakes can be overly productive (hypereutrophic; see Box 6.1) such as Lake Hendricks (Brookings), Lake Herman (Lake County), or Lake Kampeska, or only moderately productive (mesotrophic) such as Pickerel Lake (Day County) and Enemy Swim Lake.

A study completed by one of the authors (S. Stukel) quantified the available aquatic habitats in 15 eastern South Dakota glacial lakes using Geographic Information System technology to create aquatic habitat layers (Table 6.1). Aquatic habitat types were surveyed during the summer months of 2000, 2001, and 2002 and found to be highly variable among lakes. Some of the lakes were recently expanded due to above-normal precipitation, while others had been relatively stable. Hard substrates (e.g., sand, gravel, or rock) covered from 2.4

6.1
Lake Trophic Status

Ecologists use the term trophic status to classify the nutrient (primarily nitrogen and phosphorus) content of water bodies. In simplest form, there are three trophic states for lakes depending on lake production. The term production means the total biomass (living tissue) that accumulates over one growing season.

TROPHIC STATUS	LAKE CHARACTERISTICS
Oligotrophic	Lakes with low production have low nitrogen and phosphorus concentrations. Such lakes tend to be deep and clear (transparent) because algae (microscopic floating plants) are scarce.
Mesotrophic	Lakes with intermediate levels of nutrients.
Eutrophic	Lakes with high production have high nitrogen and phosphorus concentrations. Such lakes tend to be shallower and have low transparency because algae are abundant.

There is a *continuum* from oligotrophic to eutrophic as nutrient content increases. Lakes naturally change from less productive (mesotrophic) to more productive (eutrophic) over time as water depth decreases because of sedimentation and because nutrients accumulate. So, eutrophication is a natural process that occurs over long periods of time (e.g., centuries). When eutrophication is accelerated by man's activities, it is called "cultural eutrophication."

About 23 feet of sediment has accumulated in Lake Pelican (Codington County), and about 10 feet in Lake Kampeska. Enemy Swim Lake is the least productive lake in Eastern South Dakota and it is classified as *mesotrophic*. At the other end of the scale and perhaps *off the scale*, are many shallow water bodies with excessive nutrients that are classified as "hypereutrophic," which is another step beyond eutrophic. For example, Lakes Kampeska and Poinsett are hypereutrophic, and have noxious blue-green algae that lake residents call "pea soup" during late summer.

Table 6.1 Physical characteristics, habitat characteristics, and water quality of 15 eastern South Dakota glacial lakes in 2001 and 2003. The shoreline development index (SDI) indicates shoreline complexity; a perfect circle would have an SDI of 1.0, with increasing values for lakes with greater shoreline length in relation to surface area. The trophic state (total phosphorous) index reflects productivity, with higher values indicating higher productivity.

LAKE	WATER-SHED AREA (ACRES)	SURFACE AREA (ACRES)	MAXI-MUM DEPTH (FT)	MEAN DEPTH (FT)	SDI	HARD BOTTOM (%)	LAKE AREA <6.5 FT DEPTH	SUBMERGED VEGETATION COVERAGE (%)	TROPHIC STATE INDEX (PHOSPHORUS)
Albert	248,894	3,696	12.8	8.9	2.1	10.2	4.9	2	92.6
Alice	5,210	1,116	12.8	7.9	1.7	18.6	13.9	4.1	57.4
Blue Dog	7,652	1,528	8.9	5.9	1.5	7.2	35.1	0.1	47.4
Brant	73,753	1,037	14.1	9.5	1.5	8.6	5.3	1.7	37.4
Clear	21,822	1,170	22.0	12.5	1.5	34.2	5.6	18.8	57.4
Cochrane	832	356	23.9	13.1	1.4	11.1	6.8	31	53.2
Herman	36,420	1,286	12.8	4.6	1.8	8.8	65	0.1	79.1
Madison	29,168	2,640	16.1	7.9	2.2	11.33	11.4	0.1	84.7
Pickerel	23,820	980	41.0	15.7	2.2	42.1	3.5	7.2	60.6
Poinsett	291,970	7,896	23.0	16.4	1.3	31.5	0.9	0.6	92.6
Roy	9,607	2,052	21.0	9.8	3.7	24.4	9	23.2	65.4
Sinai	5,652	1,719	33.1	17.0	2.4	2.3	3.6	3.6	79.8
S. Buffalo	15,988	1,788	14.1	7.9	3.8	25.6	32.7	31.1	57.4
Thompson	262,847	12,447	25.9	14.4	2.6	21.4	4	5	90.2
Waubay	195,854	15,538	35.1	16.1	4.5	8.38	2.3	0.1	69.1

to 42.1% of lake bottoms. Submergent aquatic plant coverage ranged from 0.5 to 31.1%. Maximum depths ranged from 2.8 to 4.9 m, while the amount of cropland in the lake watersheds ranged from 15.9 to 66.7%. Total phosphorous levels in water samples from the lakes ranged from 0.01 to 0.46 ppm, and water transparency (measured with a Secchi disk) ranged from 0.5 to 2.5 m. These characteristics resulted largely from the geology and land-use practices found in this region.

SDGFP Lake Classification System

SDGFP biologists manage the eastern natural lakes as part of the "Large Lakes and Reservoirs" program. Such waters are defined as those that exceed 150 acres in surface area, but not including the Missouri River mainstem reservoirs. The large lake and reservoir resource includes 117 lakes and impoundments, of which 70% (82) of these are natural lakes, and all but two are located east of the Missouri River. Management responsibility for nearly all natural lakes in the state falls to SDGFP Region III (Sioux Falls) and Region IV (fisheries office in Webster). These large lakes and reservoirs are primarily managed for warm-water and cool-water fishes, except for three Black Hills reservoirs that are managed for cold-water trout fisheries.

The lakes are all categorized as Class I, II, or III, depending primarily on their importance to the state as fisheries. Class I lakes are primary permanent and semi-permanent waters that are highly important. They require regular data collection, at least every three years, to effectively manage. Many are primary walleye lakes, but important multi-species waters are included. Class II lakes are secondary permanent or semi-permanent waters that are somewhat important to the state's fisheries. They have relatively stable multi-species populations and/or lighter angling use requiring sampling at least every five years. Class III lakes are marginal fisheries and other waters that are occasionally important. This category includes winterkill lakes (see Box 2), beaver ponds, and other waters that are managed as the opportunity allows. Surveys on these waters are conducted as needed to evaluate population status and effects of restocking, winterkill, commercial fishing, etc. Approximately 95% of the Class I and II natural lakes have at least one boat ramp; few of the Class III water bodies have ramps.

Ever-Changing Nature of the Eastern Glacial Lakes in South Dakota

Any attempt to communicate the diversity exhibited by the glacial lakes of eastern South Dakota must include the effects of varying climatic patterns. What is a lake with a sport fishery one year can be dry a decade later. White-tailed deer used the cattails in the bottom of Lake Thompson during the early 1980s. By the late 1980s, 20 feet of water allowed the development of one of the best walleye and northern pike lakes in eastern South Dakota. Only six lakes retained fish communities during the drought years of the early 1930s.

6.2
Winterkill

Winterkill refers to fish deaths because of low dissolved oxygen in water under ice. Other forms of aquatic life, such as zooplankton (microscopic animals), are also susceptible to winterkill. The longer ice and snow cover a lake, the more likely it is to winterkill. Shallow, eutrophic lakes (meaning productive lakes) at northern latitudes are most susceptible to winterkill. A productive water body produces algae and other plants that decompose during the winter. Bacteria cause decomposition and lower the dissolved oxygen levels. Dissolved oxygen is replenished by photosynthesis by *living* algae and other aquatic plants. When the amount of oxygen used during decomposition is greater than the amount of oxygen replenished during photosynthesis, *and* when the lake is capped with ice, then dissolved oxygen levels decline and fish begin to die. Photosynthesis is reduced when snow prevents light penetration through ice. Also, gases such as hydrogen sulfide (rotten egg smell) and methane can accumulate during decomposition, and these gases are toxic to fish.

Fishes differ in susceptibility to low levels of dissolved oxygen. Some fishes, such as trout, tend to be sensitive and die when dissolved oxygen levels reach about 2 parts per million (ppm). Other fishes, such as the walleye, have an intermediate tolerance to low dissolved oxygen levels and die when dissolved oxygen levels are between 1 and 2 ppm. Tolerant fishes, such as northern pike, yellow perch, and fathead minnow, do not begin to die until dissolved oxygen is below 1 ppm.

One Brookings County landowner (Obie Stime) reported that the only crops his family raised in 1933 (Dust Bowl years) came from the bottom of what is now Lake Sinai in Brookings County. As of 2003, this particular portion of the lake contained nearly 40 feet of water.

One example of the ever-changing nature of these lakes is Waubay Lake, a meandered lake located in Day County, South Dakota. Waubay Lake presently encompasses four previously distinct water bodies, including North Waubay, South Waubay, Spring, and Hillebrands lakes. Water levels in these four lakes rose substantially through the 1990s resulting in a single water body. From 1990 through 1993, the water elevation was less than 1785 ft (544 m) above mean sea level, but increased to more than 1800 ft (548 m) from 1998 through 2002 (Figure 6.3). The maximum depth of Waubay Lake peaked at about 31 ft (9.5 m) and the surface area of Waubay Lake at 15,540 acres (6,294 ha) in 2002.

Eastern natural lakes provide many hours of year-round fishing (Figure 6.4) and other recreational activities.

State parks, campgrounds, and good boat ramps are located at most lakes. The variety of lakes and their changing

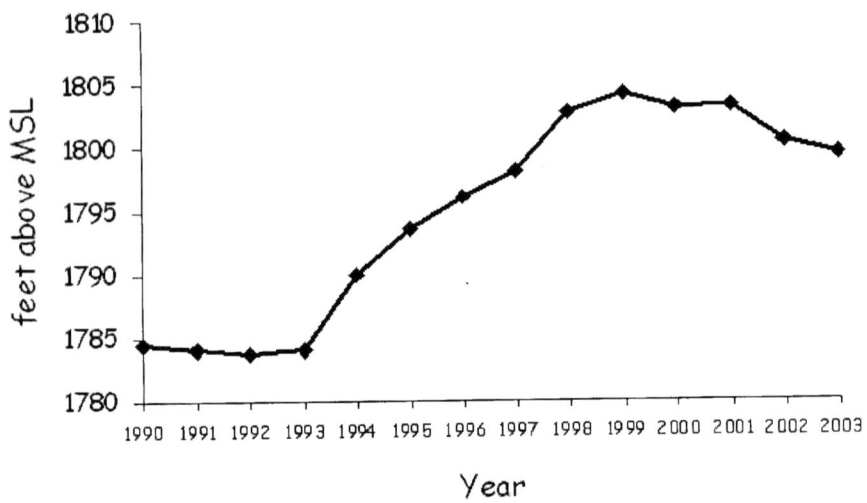

Figure 6.3 Approximate Waubay Lake water elevations from 1990 to 2003. MSL = mean sea level.

Figure 6.4   Eastern South Dakota lakes are fished year round from docks (a), boats (b) and ice (c). (Photos a and b by C. Berry, photo c from T. Holmlund)

nature require regular collection of data about both the fishes and anglers to insure good fishing in the future.

Acknowledgements

The South Dakota Department of Environment and Natural Resources (Gene Stueven and Bill Stewart 1996) completed a broad condition assessment of South Dakota lakes during 1995. References listed in the references section at the end of this book were important information sources for this chapter, particularly Blackwell 2001; Flint 1995; Leap 1996; Sando 1996; Schaap and Sando 2002; and Stukel 2003. ❖

Chapter 7

Past and Present Black Hills Water Resources

Ron Koth

"No fish except suckers & dace. Both take grasshoppers readily from the surface of the water. The suckers are better eating…"

— Colonel R.I. Dodge, 1875 expedition to the Black Hills, from Kime 1996

The Black Hills could be called an *"island on the plains"*. Sixty-two million years ago at a time when the Rocky Mountains were nearing the end of their formative process, the Black Hills uplift occurred. A dome-shaped region approximately 60 miles wide and 120 miles long was formed in western South Dakota and eastern Wyoming and for all practical purposes was a forested island in a sea of short and mid-grass prairie. From the surrounding grassland elevation of

3,000 feet above sea level, the Hills rise to a maximum elevation of 7,242 feet above sea level at the top of Harney Peak.

Documented history of the Black Hills was slight until 1874 when the Custer Expedition surveyed portions of the Hills on an expedition originating from Fort Abraham Lincoln across the Missouri River from present day Bismarck, North Dakota. The accounts of the Custer expedition led to the conclusion that the Hills were and are a unique feature in the prairie landscape. Further exploration and inventory of the Hills took place in the years following the 1874 expedition and led to not only the gold rush, but natural history observations of "clear, cool, and pure" streams as noted by Lt. Col. R. "Irving" Dodge in 1876 (Kime 1996). Today this forested island of hills is the primary component of the Black Hills National Forest.

Perennial cold-water streams and five natural lakes (Bear Butte, Cox, Mud, Mirror 1 and 2) found at the time of the Custer expedition were a marked difference from the seasonal drainages and warm-water rivers in the immediately surrounding plains. The combined effects of local geology and precipitation patterns allow the water resources of the Black Hills to be much different than those outside the "island." The geology of the Black Hills, though well documented, bears review here to understand how water resources begin, persist, and function, and how this relates to fisheries. Likewise, knowledge of precipitation is well known and documented for the Black Hills; however, a brief summary will help in understanding Black Hills water resources. The emphasis here will be on surface waters as they are linked to fisheries potential.

The Black Hills are erosion remnants of the dome-shaped granite uplift that pushed up the overlying sedimentary formations (Figure 7.1). From a fisheries view, this geologic development means that perennial stream flow is present in the regions of the Hills where the granite and/or metamorphic rocks are found. Here the underlying rock structure is relatively impervious, which allows for surface flow. In the high elevations on the western flank of the uplift, limestone is exposed. Precipitation recharge of ground water occurs through this porous limestone strata. Where the porous limestone intersects the impervious rock, springs flow to the surface and continue down slope.

Further down slope on the edge of the Hills, limestone is again exposed on the outer flank of the uplift and takes in surface flows and, under most conditions, eliminates surface flow of water and associated fisheries. The "loss zone" created by this limestone contact around the edge of the Hills means that except for streams such as Rapid Creek and Spearfish Creek with adequate base flow or flows controlled by impoundments (Rapid Creek downstream of Pactola Reservoir), fisheries downstream of the loss zone are

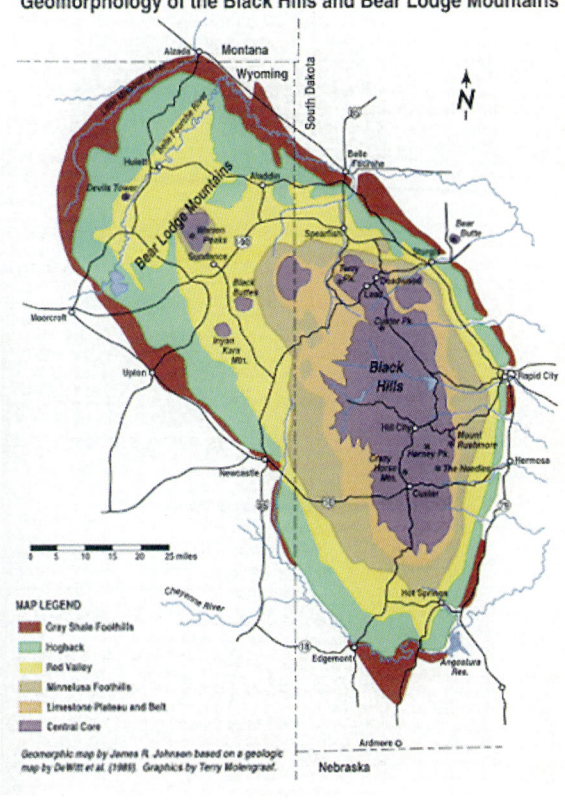

Figure 7.1 **Map showing geomorphology of Black Hills and Bear Lodge Mountains. Note the concentric rings of rock types around the central core of the Black Hills. Each plays a role in water conditions of streams; the high elevation limestone plateau is a recharge zone whereas the limestone around the margin of the Hills is a loss zone. (From DeWitt et al. 1989)**

temporary and are present only when adequate water is available to allow for surface flows once the loss zone of the limestone is "filled." To visualize this, imagine one-half of an egg-shaped dome with the long axis oriented north-south as the Black Hills. On the western one-half of the dome, limestone is exposed and collects water from precipitation. In the central portion of the dome the impervious granite and metamorphic rock allow for surface flow of water. Completely encircling the dome is a band of limestone that captures surface water and there is little stream flow except during wet periods. Specific situations in Rapid, Whitewood, and Spearfish creeks allow for surface flow beyond the loss zone and will be discussed later.

It's All About The Water

Water quality and stream habitat degradation was not mentioned as a problem in the 1950s (SDGFP 1959), prob-

7.1
Fire and water don't mix...or do they?

Fire plays an important role in the water budget of the Black Hills. How? First the facts about fire and water in the Hills.

Fire suppression has been successful in changing the fire return interval from 16 years to 42 years in the Hills. The end result is more trees.

Average precipitation in the Hills is 19 inches, but only 8% enters streams, wetlands and lakes. The remainder is cycled back into the atmosphere by evapotranspiration from vegetation. Ponderosa pine and other evergreens cover the majority of the Hills and photosynthesize all year long, thus transpiring huge volumes of water.

Now you have the facts to see why stream flow has been reduced during the years of fire suppression and increased forest cover, and why stream flow might increase after a forest fire. Fire, precipitation, vegetation, and geology are the primary driving forces for water availability for fisheries.

Fire and water really do mix when it comes to water for fish and fish habitat in the Black Hills of South Dakota (Anderson 1980, Fisher 1987).

ably because state agencies were preoccupied with the great growth and change in fish management after World War II. One of the changes was "the beginning of scientific game management and fisheries programs" (SDGFP 1959), and it didn't make sense to stock fish in polluted streams with poor fish habitat. Water quality degradation from mining, forest management, excessive grazing, and urban development was recognized in the early 1960s as threats to the success of trout fishing (Stewart and Thilenius 1964). By 1970, the Black Hills had lost 85% of its trout stream mileage (Glover 1975). Many of these "lost stream miles" have been reclaimed through modern stream habitat management, such as methods described by Hunter (1991).

Distribution and amount of precipitation in the Black Hills play a large role in availability of water for fisheries. In the northern portion of the Black Hills near Lead, average annual precipitation is about 29 inches, while in the southern Black Hills near Hot Springs, annual precipitation is only about 16 inches. A gradient from north to south generally applies with the most precipitation in the northern Black Hills in the high elevations. On the northern end of the Black Hills, the Butte County line is where average annual precipitation amounts drop to about 18 inches, which is the average annual precipitation for the entire Black Hills geographic region. The precipitation patterns coupled with the geology of the Black Hills as previously described, set the baseline conditions for water resources and fisheries (Box 7.1).

The geology and the high precipitation in the northern Black Hills watersheds combine to create the reliable, drought-resistant, perennial, cold-water streams that most people think of as Black Hills waters. In contrast southern Black Hills watersheds do not receive adequate precipitation to develop reliable perennial streams as easily as in the north. While other physical and climate differences between the northern and southern Black Hills play roles in water resources and fisheries potential, much is determined by geology and precipitation. Overall only about 8–10% of annual precipitation contributes to surface flow or groundwater recharge; the remainder goes back to the atmosphere via evapotranspiration.

Watershed Features

A north to south tour of the Black Hills will take you through 10 watersheds before reaching Custer State Park at the southern end of the tour. Primary water resource features in each watershed are discussed from Crow Creek in the north to Cascade Creek in the south.

The Crow Creek and Spearfish Creek watersheds are bounded by the Wyoming state line on the west, I-90 on the north, and generally State Highway 85 on the south and on the east by a line running north-south from Spearfish to a point just east of Cheyenne Crossing. Primary streams in the watershed are Crow Creek, Spearfish Creek, Little Spearfish Creek, and East Spearfish Creek. Ponds and impoundments include Mirror Lakes 1 and 2, Cox Lake, Iron Creek Lake, Yates Pond, Hanna Pond and Ward Draw Pond along with several small run-of-the-stream ponds on Spearfish Creek and Little Spearfish Creek. Water is relatively abundant in this watershed (Figure 7.2). Crow Creek is the largest spring-fed stream in South Dakota with a discharge rate of approximately 30–50 cfs.

The Spearfish Creek drainage produces an annual mean flow of approximately 54 cfs (cubic feet per second, USGS gauge #06431500). A particularly interesting feature of the Spearfish drainage is the "plumbing" developed by Homestake Mine, which has been in place since the late 1800s. The most visible feature of this plumbing system is the dry or nearly dry streambed from the south edge of Spearfish to the pond at Maurice approximately nine miles upstream. A large aqueduct draws all stream flow at Maurice and routes it to Hydro #1, a small hydroelectric plant in Spearfish. The aqueduct takes flow around the limestone loss zone and delivers nearly all of the stream flow to the hydro plant except for 2–5 cfs lost in transmission. This plumbing feature allows Spearfish Creek to flow from Hydro #1 undiminished by the loss zone, which would take approximately 20 cfs from the base flow.

Some professionals disagree regarding the potential for volume of loss to diminish over time or whether small springs down slope would reappear if a significant volume of water was allowed to flow in Spearfish Creek downstream of the Maurice intake.

In November 2003, Spearfish Falls on Little Spearfish Creek began to flow again on a regular basis for the first time since 1917. This event occurred as a result of the Homestake Mine closure, shutdown of Hydro #2 and a legal settlement of historic mining related pollution issues. The State of South Dakota now holds the water right for the entire flow of Little Spearfish Creek, which allows the falls to flow year-round. Spearfish Creek, from the confluence of Little Spearfish to the site of the old Hydro #2 plant, will have improved habitat conditions as a result of the changes in water management. Mirror 1 and 2, Cox, and Yates ponds are noteworthy as they all are primarily groundwater fed, meaning that little, if any, ice forms during winter and what does is not thick enough to allow safe access. Groundwater ponds generally have consistent temperature

Figure 7.2 **Examples of relatively high flows are found in Spearfish Creek (above) and Little Spearfish Creek (below) where a consistent stream originates from springs flowing out of the Madison limestone. (Photos by R. Koth)**

during the summer, making them desirable for cold-water fish, although high densities of aquatic vegetation sometimes makes fishing difficult. Fishing in this watershed, including the reach of Spearfish Creek flowing through the City of Spearfish, is excellent.

The Whitewood and Bear Butte watersheds are found in the historic and current day mining district of the Black Hills. The Whitewood watershed extends beyond the outer margin of the Black Hills onto the prairie towards the Belle Fourche River. Some mine drainage flows out of the watershed to the Spearfish watershed via the Annie Creek and Squaw Creek drainages, however most mining related drainage is within the Whitewood or Bear Butte watersheds. Mining-related pollution historically eliminated much of the cold-water fisheries in Whitewood Creek downstream of Lead and in Bear Butte Creek downstream of Galena, but today many of those adverse mining related impacts are gone or improved.

Soon after the start up of Homestake Mine in 1876, tailings from the gold process were released to Whitewood Creek. In addition to the heavy sediment load created by the volume of tailings, toxic substances including cyanide from the mine process also were a part of the tailings. This release of tailings to Whitewood Creek continued until 1977 when an impoundment dam was built to hold all tailings. Water was then recycled for use in the mine or treated prior to release to Whitewood Creek. Eliminating the discharge of tailings was followed closely by eliminating the discharge of untreated sewage with a modern wastewater treatment plant for Lead/Deadwood. These changes were significant improvements, but fishery recovery did not occur until after the discharge water was treated for removal of cyanide beginning in 1984. Today Whitewood Creek supports a naturally reproducing population of trout and other fishes, which is an impressive change from the days when the stream ran gray and lifeless. Whitewood Creek is also noteworthy as it and Rapid Creek are the only streams that regularly have flowing water all the way through the loss zone.

The Elk and Boxelder watersheds include large land areas but relatively small streams. Boxelder Creek and Elk Creek are the primary streams. Small ponds include Reausaw, Dalton, and Roubaix Lakes. The Elk Creek watershed is quite narrow in the Black Hills proper, encompassing only the immediate drainage along Elk Creek. This small area of the watershed has a large impact on water availability because it reaches into the higher elevations and higher rainfall area. Boxelder Creek above the small town of Nemo usually has adequate water and supports a good brook trout fishery. Just downstream of the Steamboat Rock picnic area along Nemo Road, Boxelder Creek contacts the loss zone and all of the flow usually goes underground. Only during periods of high rainfall or a prolonged wet cycle does Boxelder Creek flow far beyond the loss zone. This is due to the high loss threshold of approximately 50 cfs, the amount of water necessary in the stream before overland flows begin.

The upper Rapid Creek watershed includes the land area surrounding Deerfield Lake. Primary streams include North and South Forks of Castle Creek, Castle Creek, Ditch Creek, Slate Creek, North and South Forks of Rapid Creek, and Rapid Creek. In addition to Deerfield Lake, there is also Slate Creek Dam, a small impoundment on Slate Creek. Deerfield Lake is a 414-acre impoundment located at approximately 5,900 feet above sea level on the western edge of the central crystalline core of the Black Hills. Its location is very close to the high limestone headwater region. This location allows Deerfield Lake to experience very little fluctuation in water surface elevation because of the steady water supply from Castle Creek and the South Fork of Castle Creek that flows from the limestone region. Deerfield Lake has a maximum depth of 90 feet with an average depth of 30 feet. Improvements made to the outlet valve of the dam in 1995-96 allow for releases of 8–10 cfs (instead of only 2 cfs) during the winter period, thus improving conditions in Castle Creek downstream from the dam.

A much improved fishery has resulted after improvements in Castle Creek downstream of Deerfield Lake. In addition, three miles of fishery below Deerfield Lake are only accessible to walk-in anglers as a result of U.S. Forest Service rules. Deerfield Lake is the ice fishing destination of choice for many anglers. Deerfield Lake is managed as part of the water supply system for Rapid City and Rapid Valley irrigation interests. Fishing is generally excellent in the entire watershed, but somewhat diminished on Rapid Creek and Castle Creek below the confluences with the north fork of each stream.

Rapid Creek and Castle Creek flow through the naturally occurring "bog iron" region of the Black Hills, where surface exposure of iron deposits have either been disturbed by past mining or are naturally occurring, thus degrading water quality and overall productivity. Fisheries productivity in the affected stream reaches is reduced to below 20 lbs/surface acre compared to much more productive reaches in the Black Hills that can be over 200 lbs/surface acre.

The lower Rapid Creek watershed includes Rapid Creek just upstream of Silver City to the eastern edge of Rapid City in the Black Hills. Pactola Reservoir and Canyon Lake are the impounded waters within the watershed. Pactola Reservoir is the largest impoundment in the Black Hills at 785 surface acres with a maximum depth of 150 feet. Deerfield Lake and Pactola Reservoir form the storage component of the Rapid City municipal water supply and the Rapid Valley Irrigation District. Because of interest in maintaining as much water as possible in the reservoirs for potential use as municipal or irrigation water, flows in Castle Creek below Deerfield lake and Rapid Creek below Pactola Reservoir do not resemble the natural hydrograph.

The natural hydrograph has flushing flows in the spring that mobilize the bottom sediments and refresh the system, but flushing flows do not take place on a regular frequency that would normally occur with spring runoff. Similarly, flows released during the winter are much lower than what might normally be found below each reservoir. These impacts are more pronounced on Rapid Creek than Castle Creek. During times of low precipitation, however, both reservoirs are managed to keep water flowing in the stream reaches below. Rapid

Creek does not begin to lose water to the loss zone until midway through Dark Canyon, just upstream of the west edge of Rapid City, where it has a loss threshold of approximately 10 cfs (Figure 7.3). This loss only affects a short reach of Rapid Creek as the Jackson/Cleghorn springs complex supplies a net volume of 10–12 cfs of water year-round to the stream near the state fish hatchery on the west side of Rapid City.

A good trout fishery is present through Rapid City to about mid-town where stream gradient and temperature effects begin to diminish desirable habitat. The urban portion of the Rapid Creek fishery is accessible via the greenbelt running through the city. This was established following the major flood of 1972. Stream length through Rapid City has been significantly shortened from the original channel. Currently, approximately 10 miles of stream classified as cold-water fishery flows through the city, whereas the historical

Figure 7.3 An example of relatively low flow is found in Rapid Creek below the loss zone, where stream discharge drops from about 20 cfs (cubic feet per second) to about 12 cfs. The missing water (8 cfs) trickled into the porous limestone beneath the stream.

mileage may have been closer to 14–16 miles over a similar linear distance.

The Spring Creek watershed is located south of Rapid City and generally encompasses Spring Creek upstream of the Boy Scout Camp to the Stratosphere Bowl north of Rockerville. Within this watershed there are several impoundments including Sheridan Lake, Lake Alexander, Newton Fork Lake, Major Lake, Mitchell Lake, Sunday Gulch Pond, and Sylvan Lake. Primary streams include Tenderfoot Creek, Newton Fork Creek, and Horse Creek in addition to Spring Creek. This watershed often suffers from a scarcity of water so fish management options are limited.

Fisheries management includes a significant amount of put-and-take stockings based on water availability. Generally, once every 4–6 years a dry cycle impacts water availability to the degree that fisheries habitat shrinks along with existing fisheries. When water is available, productivity is high, leading to fast growth and rapid re-colonization of previously dry reaches of streams. Some headwater reaches of streams in locations such as Sunday Gulch or upper Spring Creek do not suffer from the same severity of flow fluctuations because of their linkage with the water-bearing limestone at higher elevations. Spring Creek below Sheridan Lake is one of the most visible reaches affected by water availability. During periods when water is available, it develops a naturally reproducing population of brown trout and is one of the most popular stocked rainbow trout fisheries in the Black Hills.

The Battle Creek watershed is located in the region surrounding Mt. Rushmore, south to Legion Lake in Custer State Park and east towards Hermosa. Small impoundments include Horsethief Lake, Lakota Lake, Legion Lake, and Center Lake. Primary streams in addition to Battle Creek include Iron Creek and Grace Coolidge Creek. This watershed is located further south in the Black Hills and does not have good contact with the high elevation water-bearing limestone. As such, water is a significant limiting factor and all streams are quite small and periodically lack flow. Fisheries potential therefore is somewhat depressed from more northerly watersheds, but does hold some bright spots where perennial water is found in the higher elevations of Iron Creek and Grace Coolidge Creek. Small impoundments are all managed as put-and-take fisheries, which is standard management for all impoundments in the Black Hills.

Custer State Park and the region to the south is not a true watershed. Primary streams included within this boundary are portions of Grace Coolidge Creek and French Creek. Stockade Lake is the largest of the impoundments within the boundary. All streams within the Custer State Park boundary are subject to periodic low-flow conditions. The upper reaches of Grace Coolidge Creek are least affected and support a good brook trout fishery. The remainder of streams in the park are managed as put-and-take fisheries due to the frequent recurrence of low flows and the high angling pressure found in the popular destination park. Stockade Lake, an impoundment on French Creek, supports a multiple-species fishery including trout, bass, crappie, perch, and northern pike. Numerous small run-of-the-stream impoundments and beaver ponds provide deeper water reservoirs when stream flow is diminished.

The Fall River and Cascade Creek watersheds are the southernmost watersheds in the Black Hills. Fall River is formed by the junction of Hot Brook and Cold Brook at the northern edge of the city of Hot Springs. Numerous hot springs enter Fall River, which is the shortest South Dakota river at only 7 miles long. Cascade Creek rises in Cascade Springs, the largest spring in the Black Hills. The Creek tumbles over 15-foot Cascade Falls on its way to its confluence with the Cheyenne River on the prairie below.

These streams are unique because of the supply of warm water from the numerous hot springs in their watersheds (Whalen 1994). The warm water and unique water chemistry fosters lush growths of aquatic plants, such as the white-flowered watercress (*Nasturtium officianile*), which is an exotic species that is harvested by locals for dinner salads (Larson 1993). Fall River and Cascade Creek are also habitat for several unique aquatic plants such as maidenhair fern (*Adiantum capillus-veneris*) and stream orchid (*Epipactis gigantia*). The fish assemblages in both streams are typical of warm water streams (e.g., creek chub, longnose dace) but there are some big goldfish (*Carassius auratus*) in Fall River where it courses through the Hot Springs. About 700 catachable rainbow trout are stocked in Cascade Creek to provide a fishery from January through March when water temperatures are cool enough to support trout.

A Look To The Future

Black Hills water resources are controlled by the overall geology and precipitation of the region and are much different than mountain streams found further west in the Rocky Mountains where snowmelt runoff is an important component of base stream flow. In the Black Hills there is a buffering effect of the groundwater-fed stream network as it takes time for the effects of a prolonged dry period to impact base flow as well as a time lag between a wet period and increased base flow volume. Depending on the beginning conditions and severity of a given dry period, 3–5 years can pass before significant changes to the base flow are evident.

Water resources in the Black Hills are resilient as they go through periodic drought or wet cycles; however, burgeoning development and cattle grazing present difficulties to all water resources today and may be even more important in the future. With the expansion of single-family homes into the Black Hills proper, a network of roads, wells, septic systems, and different land use patterns is changing the past history of the area. While cumulative effects of this development have yet to be seen throughout the region, groundwater in some areas has been adversely affected by high-density drain field development.

Free access by cattle to streams has impacted stream channel geometry and riparian zone health for many years. Even low intensity grazing may further degrade streams in the future unless grazing management programs are enacted by government and private landowners. With knowledge gained through the Black Hills Hydrology Study (Driscoll 1994), hydrologists learned that patterns of groundwater flow, surface water availability, and water quality are more interconnected than not. They also learned that conditions in geographically distant areas of the Black Hills may have considerable potential to affect water resources some distance away. These findings are important to the public for preserving, enhancing, and managing water resources to maintain the Black Hills fishery and the quality of life for Black Hills residents.

Acknowledgements

The editors and the author would like to thank the many fisheries biologists who have worked for the Department of Game, Fish and Parks on the streams of the Black Hills through the years (see Stewart and Thilenius 1964). Ron Glover did some of the first studies that focused on the physical and hydrological conditions in Black Hills streams (e.g., Glover 1975). Richard Ford continued the stream inventory (Ford 1988). Jack Erickson had much to do with the modern stream surveys (Erickson and Koth 2002), while the USGS conducted special hydrology studies (e.g., Carter 2002a, b). ❖

Chapter 8

Warm-Water Rivers and River Fishes in South Dakota

Charles R. Berry, Jr.

> "Oh winding Sioux!
> Oh winding Sioux!
> For many miles I've followed you
> Along your banks
> I long to stay
> Far from the dusty, traveled way."
>
> — Ole Rolvaag, *Giants of the Earth*

In Dakota Territory in the 1800s, the water and timbered banks of the Moreau, Cheyenne, and Grand rivers offered wintering places for cattle and horses and campsites for men of the range. Word of the abundance of grass and water brought wealthy financiers from Texas, Mexico, and Europe.

The City of Canton steamboat was launched June 25, 1892, and carried passengers on excursions up the Big Sioux

River nearly every night. The water was so clear that people could see the river bottom. They called it the Silvery Sioux.

In May 1895, the Aberdeen *Weekly News* touted the Rondell fishing grounds on the James River near Aberdeen as follows: "there is a fine stage of water at the present in the Jim and fish are reported quite plentiful…"

In the spring of 1905, there was trouble on the Julius Wickert homestead beside the Bad River. His big horse team, Buster and Babe, were behind a fence and the river was rising fast behind them. A neighbor cut the fence and saved their lives while Julius was in Ft. Pierre in a rowboat rescuing people caught in the flood.

South Dakota has 35 named rivers and hundreds of tributaries named creek or stream. Each river, stream, or creek has a special story that threads through the history of South Dakota like the rivers that thread through the landscape. The location of early settlements shows the importance of watercourses to South Dakota. Yankton, Vermillion, Rapid City, and Ft. Pierre were early river towns. Sioux Falls was founded because the falls promised power. At Vermillion, a steam-operated sawmill was set up in "heavy timber" near the river in 1859. The importance of rivers to stock growers is obvious. Indian camps, fossils, waterfalls, homesteader accounts, water wars, flood stories, and big fish catches are all part of the South Dakota river heritage.

Most of South Dakota's "rivers estate" is made up of warm-water rivers. The purpose of this chapter is to summarize information about warm-water rivers, particularly information about the four dimensions of a river, and river life with an emphasis on the fishes. Fishes are the icons of river wildlife, but rivers harbor many kinds of organisms. Humans depend on rivers for many services. Recent surveys of river habitat and fishes show that many warm-water rivers in the state are in good condition, but water quantity and quality are issues, as are best management and conservation practices. The Missouri River is a special warm-water river in South Dakota that is covered in an earlier chapter.

South Dakota's Warm-water Rivers Estate

When South Dakotans say they are "going to the river," they mean the Missouri River with its huge reservoirs that are national tourist destinations. South Dakota, however, has 14 other major river basins (Figure 8.1). We send water to other states from headwater streams of the Red, Minnesota, Niobrara, and Little Missouri rivers. Conversely, the James, Big Sioux, White, Cheyenne, and Grand rivers bring water from other states. The watershed map of South Dakota looks much different than the familiar state outline that is a political boundary. Three basins, the Bad, Moreau, and Vermillion, are all ours.

Rivers in western South Dakota flow west to east, whereas those in eastern South Dakota flow north to south. The last glacier left this pattern (Hogan 1995). Before glaciation, many South Dakota rivers, including the Missouri, flowed northward, as does the Red River today; however, the continental glacier blocked their flow. The rivers were forced to flow southward gouging the channel for the present day Missouri River.

South Dakota Watersheds

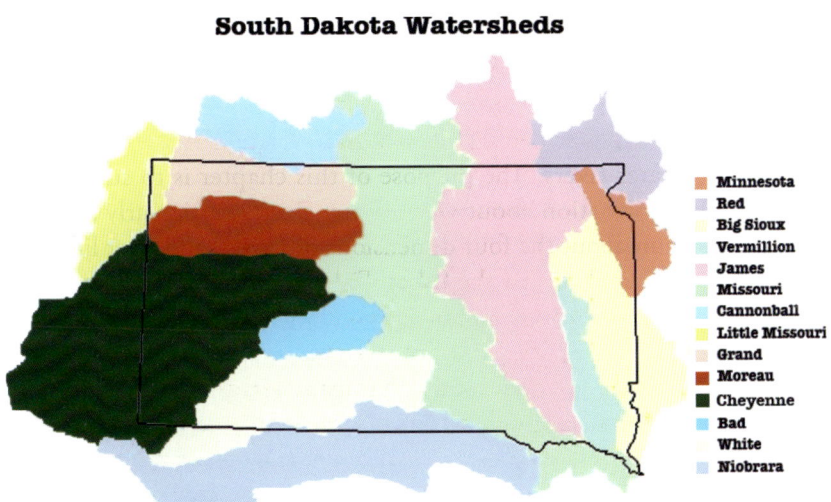

Figure 8.1 **Map showing the South Dakota borders and borders of major watersheds that are shared with neighboring states.**

Glacial melt waters gave birth to the James, Vermillion, and Big Sioux rivers with their north-to-south drainages.

Two other geological processes also help us understand the ecology of our eastern rivers. First, about one million ice blocks were left on the land that is now eastern South Dakota. These ice blocks melted and left depressions that we call wetlands (e.g., potholes, sloughs) and glacial lakes, which hold runoff and then meter water to the rivers as surface flow or underground flow. Wetland management is important to river management. An area adjacent to the basins of the Big Sioux, James, and Vermillion rivers (Figure 8.1) is called a "noncontributing area," meaning that runoff is trapped in lakes and does not enter the rivers.

Secondly, some of the glacial meltwater pooled in ancient lakes such as Lake Dakota in the upper James River Valley. This 100-mile-long lake had an outlet in Spink County that drained into a river that we now call the James. After Lake Dakota dried up, the James River was left winding across the flat lake floor. This geological history gives the James River two interesting features. The river is prone to flooding because the gradient is only a few inches per mile making it one of the lowest gradient streams in the United States, and the river can sometimes flow upstream when the wind blows from the south or when a storm increases the flow of a tributary. Understanding these facts is important to our attempts to manage James River floods.

What Is A River?

Philosophers say that we are drawn to rivers, not simply for the tangible resources they afford, but because they communicate to us "the power of eternal forces relentlessly at work." Brookings artist Virginia Coudron paints river scenes. Her depictions of the pastoral Big Sioux River near Brookings and the more rowdy Cheyenne River near Wall, makes a good contrast between our eastern and western rivers (Figure 8.2).

Scientists are less poetic, saying that a river is a "conveyor of water and sediment." Most people define the role of rivers as "moving water," but few add "and sediment." The deltas at the mouth of the Niobrara, White, or Bad rivers tell the sedi-

Figure 8.2 Brookings artist Virginia Coudron's paintings of the Cheyenne River (above) and the Big Sioux River (below) capture the contrast between rivers in the farmlands of eastern South Dakota with those in the rangelands of western South Dakota.

ment-carrying story. From the smallest headwater streams to the large Missouri River, rivers have chiseled South Dakota's landscape into the shape we see today.

A river basin (also catchment or watershed) is the land that collects precipitation for the river. Watersheds have always been recognized, but now the term watershed is heard more and more during water quality planning and conservation efforts. We hear about watershed management as a means of river or lake conservation, or even to stop siltation of marinas. It has been said, "land health and water health are not two issues, but one."

There are four river dimensions and all of them—longitudinal, lateral, vertical, and temporal—must be understood to use, live with, manage, and conserve rivers (Ward 1989). Water flows in the longitudinal dimension from the hill slopes and wetlands to the first river component that is called a rill, meaning small channel. Rills are formed when the water is powerful enough to begin eroding a channel. Rills become creeks that become streams that become rivers. Understanding the longitudinal dimension helps us understand the importance of watershed management to "neighbors" downstream, and helps explain patterns in river aquatic life. Fishes use the longitudinal dimension when they migrate into streams in the spring to spawn. Since fishes are quite vulnerable during spawning, South Dakota regulations usually include streams under the regulation called "Spring Stream Closures." South Dakota State University (SDSU) students found white suckers spawning in Six Mile Creek on April 14, 2004 (Figure 8.3). The suckers had presumably migrated upstream from the Big Sioux River.

The lateral dimension includes backwaters, side channels, and the flood plain that are wetted when the river rises. Scientists term this event the "flood pulse," saying that rising water levels inundate backwaters that become fish nurseries, gather organic material that will become the base of the river food webs, and have many other vital river functions such as shaping the channel. An important feature in the lateral

Figure 8.3 White sucker is a species that migrates upstream in the spring to spawn. This female was captured with a smaller male white sucker and many smaller fishes by seining in Six Mile Creek on the South Dakota State University campus. (Photo by C. Berry)

dimension is the riparian zone or strip of land adjacent to the river. The riparian zone is a transitional area between uplands and river. Riparian zones filter overland and subsurface flows, protect riverbanks, and provide habitat for wildlife. Like watershed management, we hear more and more about the importance of riparian zone management for river conservation.

The vertical dimension begins in the earth under the river and ends in the air above the river. In between are river bottom materials, instream fish habitat (e.g., logs, submerged vegetation), and water. The river bottom is inhabited by many kinds of organisms, such as insect larvae that emerge as winged adults and fly upstream to deposit eggs for the next life cycle. There are many connections in the vertical dimension. Much of the water that appears as river flow comes from groundwater. River currents suspend silt, sand, and bottom-dwelling insects in the vertical dimension. The drifting insects are fish food and the suspended sediment is redistributed downstream to reform the channel.

Finally, there is the temporal dimension that refers to time. Most rivers do not have stable flows. Indeed, South Dakota rivers have great fluctuations in seasonal and annual flows. Instability is natural and aquatic life has adapted to the fluctuating flows. A hydrograph is a figure depicting the flows. The annual hydrograph for South Dakota rivers generally has a peak in the spring and a low point in the fall and winter. Some streams have perennial flows, but many are intermittent at low flow, meaning that deeper pools are not connected by surface water, but are usually connected to groundwater. Western rivers, like the Bad and Moreau, are often intermittent. Even in the east, where rain is more plentiful than in the west, rivers can be intermittent during droughts. The lower Big Sioux River was intermittent three times in the 20th Century.

Early Rivers Accounts

The first descriptions of South Dakota warm-water rivers were by explorers traveling the Missouri River. Lewis and Clark get most of the publicity, but a Frenchman named

Trudeau was the first to describe our rivers (Abel 1921). The reader of the Trudeau journals that were written in 1794-95 (Abel 1921) does not need to know French to understand the phrase "campe a la grand riviere des scious a la droite du missourie." The phrase "a la droite" means "on the right side of."

Trudeau then writes in 1794 about the Missouri River: "Forty-five leagues from there, on the same side, goes out the river Jacques [James River], a beautiful river very abundant with beaver and other wild animals. It has, according to the savages, a course of more than a hundred leagues and comes also from the north." He also wrote about the Bad and Cheyenne rivers as he followed the Missouri River upstream.

There are other early descriptions of the landscape, but the writings of John Fremont and Joseph Nicollet in the 1830s seem special as they approached the James River Valley, Fremont from the west and Nicollet from the east (SDHS 1920). Fremont and a guide followed Medicine Creek. Fremont wrote: "…a rolling prairie, partly covered by short, sweet scented and grateful verdure…silicious particles of the soil are blackened by the smoke of the vernal and autumnal fires of the prairies, …there are no springs to quench the thirst…whence the most magnificent spectacle presents itself extending over the immense hydrological basin of the River Jacques." Fremont wrote that his guide said, "Well, come now, you want geography; look, there's geography for you."

Approaching the James River from the east, Nicollet wrote "…the grand simplicity of the prairie is its peculiar beauty… came in sight of the Riviere a Jacques, its scattered wooded line stretching as far as the eye could reach…Riviere a Jacques is deemed navigable for small hunting canoes for between 500 and 600 miles…but below Otuhu-oju (Oaks) it will float much larger boats…encamped near the mouth of the Elm River…this river and its forks are well timbered…magnificent multitudes of grand buffalo herds…for three days we were in their midst…"

These descriptions tell us that the James River had a main channel that could be navigated. Perhaps the tributaries had many beaver dams and were so braided that they were invisible as the prairie grasses blended into wetland grasses. If so, then the watershed was acting like a "sponge," with many springs and clear, cool water in streams. The riparian vegetation was mostly grass with scattered trees on bends or islands where fires could not reach.

We get a hint of how the "sponge effect" works from the notes of James Audubon (1897) who boated up the Missouri in 1843. His expedition moored their boat in the mouth of the Big Sioux River in early May. He described the Big Sioux as "a clear stream that abounds with fish." At the same time, he described the Missouri as "turbid and rapid." Why was the Big Sioux clear and the Missouri muddy? Perhaps the "sponge effects" and the beaver dams slowed runoff and held the soils of the Big Sioux River basin in place, while the powerful Missouri River carried sediment and logs that Audubon said "struck our weather side and kept me awake."

Lewis and Clark were not the first to describe South Dakota rivers, but they were the first to describe the fishes (DeVoto 1953). At camp near the mouth of the Vermillion River, August 25, 1804, they wrote, "two of our men last night caught nine cats that would together weigh three hundred pounds." Clark also wrote "Muskeetors verry troublesome (sic)."

First Scientific Fisheries Surveys

The first scientific surveys of South Dakota fishes were in conjunction with government surveys of railroad routes. A physician, Dr. John Evans, traveling with the railroad surveyors, made the earliest scientific records in 1853–1855. Evans listed the names of a few fishes he collected near Ft. Pierre, among which were paddlefish, creek chub, flathead chub, plains minnow, silvery minnow, river carpsucker, northern redhorse sucker, and channel catfish. All but the silvery minnow are common today. The decline of the silvery minnow has been linked to the increasing clarity of Missouri River waters after the major dams were built in the 1950s (Hesse et al. 1993).

In 1876, Edward Cope searched for fossils in the Moreau River basin, but also collected prairie animals and 10 fishes. He found 10 species in the "chain of pools" that made up Battle Creek in October. All of the species are present there today.

Fish surveys were done on the Big Sioux River at Sioux Falls in 1889 and on the upper James River in 1896, but the most important early surveys were made as the government searched for places to build fish hatcheries. The surveyors, Barton Evermann and Ulysses Cox, listed 69 species of fishes in South Dakota—46 species that they collected and 23 collected by others (see Bailey and Allum 1962).

Even then, fish transplantation had obscured the native range of many fishes. Evermann and Cox said that "both species of *Pomoxis* (crappies) were being extensively introduced into the waters of Kansas, Nebraska, and South Dakota, and it is not easy to determine definitely the natural western limit of either." The native range of a species is usually only an issue for ichthyologists to discuss, except for the fish they called the "wall-eyed pike."

The wall-eyed pike, or walleye, is of course the state fish of South Dakota. But is it a native species? Probably. Walleyes were found in 1892 in Crow Creek near Chamberlain, in Rock Creek near Mitchell, and in Choteau Creek near Springfield. At the turn of the century it inhabited waters of eastern South Dakota, probably in smaller numbers than its close relative, the sauger. Today, millions of walleye eggs are taken from lake and reservoir populations, reared at the Blue Dog State Fish Hatchery, and stocked. Natural reproduction supplies rivers with walleyes and very little stocking has taken place in the Missouri River reservoirs.

In the mid-1990s, an intensive study was done on walleyes of the Big Sioux River (Fisher 1996). Tagged walleyes stocked at Flandreau Park were caught from Sioux Falls to Watertown by anglers. Walleye growth rate in the Big Sioux River was better than that in other rivers of the region. The number of walleyes was higher after years of flooding than after low water years. Radio-tagged walleyes sought off-channel habitats (e.g., tributary mouths, backwaters) and in-stream structures (e.g., logs, bridge supports) when main channel velocity was high. Walleyes are broadcast spawners, meaning that eggs are shed in the water and settle downstream on gravel. About 40% of the eggs that incubated in clean gravel hatched compared to 1% of the eggs that settled on sand and silt. These data tell the story of three dimensions of the Big Sioux River—longitudinal and lateral fish movement, benefits of flooding to year-class strength, and egg drift to bottom substrates.

The next fish inventory of scientific significance was made in 1926, when the Department of Game and Fish initiated a statewide fish survey under the direction of Drs. Edward Churchill and William Over from the University of South Dakota. They wrote the first publication exclusively devoted to South Dakota fishes, which listed 81 species including five sucker species, two species of buffalo fishes, several dozen minnows, six catfishes, nine sunfishes, the northern pike, the "wall-eyed pike," and some miscellaneous species (Churchill and Over 1938).

Churchill and Over called the paddlefish the "spoonbill cat," and wrote that the spoonbill cat stirs up the mud on the bottom with the large paddle in order to obtain food. Today the exact function of the paddle is still being debated. The paddlefish strains plankton from the water and new information indicates that the paddlefish can sense electric fields. Churchill and Over also stated that little was known about its reproduction. Today we know that paddlefish gather in groups near the mouths of rivers to spawn. Fish culturists at the Yankton National Fish Hatchery net the adults, take eggs and sperm, and incubate the eggs. The young are reared in ponds until about 12 inches long, after which they are tagged and stocked in the Missouri River. South Dakota paddlefish have been caught throughout the Mississippi River basin and as far away as Tennessee.

In 1949, South Dakota Game, Fish and Parks (SDGFP) expanded its fishery staff to include a fishery biologist, R. C. Gibbs, who began sampling fishes of the state, particularly in lakes, but Gibbs was also directed to gather information on

Missouri River fishes before the new reservoirs were formed behind Gavins Point, Fort Randall, Big Bend, and Oahe dams. In 1950, Marvin Allum, from SDSU, joined Gibbs to assist with the surveys. In 1952, Dr. Reeve Bailey, from the Museum of Zoology at the University of Michigan, assisted the South Dakotans in the new statewide survey, which resulted in the next edition of *The Fishes of South Dakota*.

The surveyors reported 93 species found at 137 sites. The 54 sites sampled on warm-water rivers and streams were usually near bridges. Some major rivers were not sampled (e.g., Belle Fourche) and some were sampled at only one or two sites (e.g., Bad, Moreau, Grand, Vermillion). Fishes were collected with seines and species were reported only as present at the site. Brief observations on the habitat were also recorded. For example, at the site on the South Fork of the Grand River, about six miles north of Bison, they collected 11 species on August 23, 1952: five minnow species, three sucker species, black bullhead, and stonecat. The stream was about 50 ft wide and 2 ft deep with moderate flow. The water was 74° F and turbid. The river bottom was sand, gravel, rubble, and silt.

Modern River Fish Surveys

In the 1990s, SDGFP asked the staff and students of the South Dakota Cooperative Fish and Wildlife Research Unit (SDCFWRU) to update information on South Dakota stream fishes. The SDCFWRU has been stationed at SDSU for 40 years and has had a long association with SDGFP. Currently the SDCFWRU is part of the biological discipline in the U.S. Geological Survey. Just as in the 1950s, when the stream fish inventory began because of impending habitat change in the Missouri River, the 1990s' study also began because of an impending habitat change—a proposed project to channelize the James River. The James River fish and fish habitat study was expanded to include similar studies of the Vermillion, Big Sioux, and Minnesota rivers (Schmulbach and Braaten 1993, Dieterman and Berry 1994, Dieterman and Berry 1998).

River studies in the 1990s were different than those in the 1950s. Modern studies use routine procedures to measure fish habitat, which includes channel shape, type of bottom materials, and riparian zone plants. River ecologists have learned that these habitat measures record the history of the watershed, much as fish scales record their life history. The new information also included quantitative data on the abundance of each species, not just its presence. Modern fishery biologists use statistical analyses to compare the catch of a species between two sampling periods, which is more powerful information than simply recording that a species is present or absent. Finally, analyzing the fish community (all populations) is called biomonitoring, which is another method for determining river health. The number and kinds of species of fish can produce an "index of biotic integrity" for the river that augments chemical data on water quality, hydrological data on water quantity, and physical data on habitat (Milewski et al. 2001, Shearer and Berry 2002).

In the Big Sioux River, for example, the species-rich fish community and the good population size and growth rate of walleyes in 1992 is evidence of a river that was much healthier in the 1990s than in the 1960s when the Big Sioux was listed as one of the most polluted rivers in the country. Much of the improvement was due to water pollution laws and clean-up funding from the Clean Water Act. In the 1990s, more clean water species were found and more species were found at more locations than in the 1960s. Walleye growth was faster in the Big Sioux than in other rivers in the region. Channel catfish were abundant and the river supported a "trophy fishery" for some 30-lb channel and flathead catfish (Kirby 2001).

SDSU researchers were able to make some temporal comparisons between the new data and that collected in the 1960s for eastern rivers despite the scarcity of historical fish data. Game fish populations were healthy, and most native non-game species had persisted over this time (Shearer and Berry 2003). However, a different picture emerged for the non-game native species after more robust analysis of longer time scales and fish distribution (instead of just fish presence). Since the earliest surveys, 17 species are missing from eastern rivers, and eight species have had their ranges reduced to one river basin (Hoagstrom 2006). Many of the species probably disappeared long ago when land use changed dramatically or drought gripped the landscape in the 1930s (Bailey and Allum 1962, Cross and Moss 1987).

The warm water rivers of western South Dakota have not been ignored (Berry 2000). As the study of eastern rivers was winding down, SDGFP directed SDCFWRU scientists to "go statewide" with the river habitat and fish research. Soon there were new data for the Grand, Moreau, Cheyenne, White, Bad, and Keya Paha rivers (Loomis et al. 1999, Fryda 2001, Harland 2004, Harland and Berry 2004, Hoagstrom 2006). These studies were funded by the Federal Aid in Sport Fish Restoration Act that authorizes taxes on fishing equipment. The "user pays" money is then distributed to states for sport fishing restoration and research. The main sport fish in the western tributaries is the channel catfish.

Channel cats in the western streams have smaller populations and no trophy-sized individuals compared to populations in eastern streams. Channel catfish growth in western streams is normal for the region in all rivers except the White River where turbidity and stressful temperatures may limit their growth. Like the walleyes in the Big Sioux, the channel catfish data from the Belle Fourche River told of the benefits of an annual spring flood because their growth was better in flood years than in dry years (Doorenbos 1998).

The status of freshwater fishes in South Dakota is much better than it is in other states. Only eight native species are missing from South Dakota waters today compared to historic surveys (Hoagstrom 2006). The Topeka shiner is an example of a species that is rare elsewhere but relatively common in South Dakota. It was listed as endangered in 1999 because it has mostly disappeared from its range in five other states (Wall and Berry 2004). The Topeka shiner had been recorded in 24 streams in eastern South Dakota before 1999 but it has been found in about two dozen new streams since then. The Topeka shiner and other species are signaling that the fish habitat in eastern South Dakota streams is relatively intact compared to that in other Great Plains states.

Good stream habitat remains because only 3% of the stream miles have been modified in the eastern part of the state (Johnson et al. 1997). An early river channelization project left a straight scar on the Vermillion River. Between 1911 and 1913, 20 miles of the main channel were dewatered and flows were diverted to a ditch. The levees blew out during the great 1993 flood leaving surrounding fields littered with mollusk shells. The planned channelization of the James River was never really begun.

Good stream habitat remains because only about 33% of the wetlands in eastern South Dakota have been drained. This figure sounds high, but compare it to 90% in Iowa and southern Minnesota (Dahl 1990). To understand the significance of wetland drainage, one needs a watershed perspective of stream conservation. Wetlands filter and hold spring run off and storm water, which they discharge slowly to the stream in quantities that can be accommodated by the natural stream channel. Excessive wetland drainage has many effects on a stream; the first noticeable effects are more flooding, channel down-cutting, loss of the riparian zone vegetation, and bank erosion.

Another reason for the persistence of South Dakota's native fishes is the relatively low number of introduced and exotic fishes. Introduced species can cause a decline of native species by predation, competition for food and habitat, and by introducing diseases. Twenty-two non-natives have become established statewide; 12 non-native species in eastern rivers and 22 non-natives in western rivers. Eastern South Dakota has a good supply of natural ponds and lakes that are managed for native fish (e.g., northern pike, walleye, yellow perch), rather than for introduced species. The increase in non-natives in western South Dakota reflects stocking to fill coldwater habitats in the Missouri River (reservoirs and tail races) and Black Hills streams, and transportation of sunfishes, largemouth bass and yellow perch from the eastern part of the state to western stock ponds.

The first sample of the 15-year-long study of South Dakota's river fishes was taken on the mainstem of the James

River in 1989, the last from Bear Butte Creek in September 2004. Sixteen SDSU graduate students wrote theses as part of the statewide river study. Once there were data from only 54 sites on warm-water rivers, now there are new data from about 600 sites. These data have been useful in many ways. New distribution maps were made for the nine species of catfish (Doorenbos et al. 1999). The game species populations seem healthy, populations of some nongame species were more abundant than previously thought, and new ideas were developed about the ecology of South Dakota streams (e.g., Hoagstrom and Berry 2005). The declining ranges of some native species may be a warning that habitat is changing.

A huge database is being synthesized from the many habitat measurements made when the fishes were collected. One conclusion reveals the diversity of river habitats. The rivers of South Dakota are made up of 6,200 segments. Each segment has a measure of average temperature, stream size, flow, gradient, and groundwater potential. While many segments are similar to other segments, overall there are 159 unique types of river segments. When this information is coupled with the fish distribution data, "hotspots" of fish diversity and habitat diversity can be identified. Applying conservation measures *now* in the watersheds, riparian zones, and streams in the biodiversity hotspots may improve the economy and efficiency of conservation efforts and avoid endangering fish species. This conservation method is called "geographic approach to planning" (Smith et al. 2000, Wall et al. 2004).

Other River Life

This chapter is mostly about warm-water fishes, but there are many other species that rely on warm-water streams—from bacteria to bats. While bacteria have received some bad press because their populations are high enough to be called "pollution," bacteria are a naturally occurring component in the stream ecosystem. They are the organisms that break down the organic matter that washes or falls into the stream and they are the base of the food web for fish. Too many bacteria, however, can harm the use of rivers for drinking water, recreation, and other uses.

Benthic invertebrates are another major type of river life. They crawl over and cling to the stones and woody in-stream habitat, and bury in the stream bottom (Schumacher 1995). The term "benthic" means bottom and the term invertebrates refers mostly to "bugs." Most aquatic benthic invertebrates are the immature forms of insects seen flying above the river after they metamorphose into a winged adult. These critters are the next link in the river fish food web because they eat bacteria and gather algae, prey on each other, and, in turn, are preyed upon by fishes. One special component of the benthic invertebrate group is the freshwater mussels.

In 1915, surveyors found 14 species of mussels in the James and Vermillion rivers, which greatly exceeded the number of species found in the Bad and Cheyenne rivers (Coker and Southall 1915). The James River, between Riverside and Lesterville, supported a mussel shell fishery (tons of shells were harvested each year) for pearls and for shells used in the pearl button industry. The dominant shell was the three-ridge mussel. Mussels have interesting names that relate to their shape (e.g., floater, heel-splitter, pocketbook, pimple-back, paper-shell, pig toe). Since then, some 35 mussel shell species have been recorded in South Dakota, although populations have greatly declined, especially in the Big Sioux River. Nineteen species *may have disappeared* from the Big Sioux River. Mussels are important for biomonitoring because they are sometimes the first to signal problems with water quality (e.g., toxins) or habitat degradation (e.g., siltation).

A number of vertebrates other than fishes also contribute to the biodiversity of our warm-water rivers. New information on amphibians, reptiles, and turtles show that there are 11 species of frogs and toads and seven species of turtles (Fischer et al. 1999, Bandas and Higgins 2004). Five of South Dakota's seven species of turtles are more widely distributed than previously reported, which is a signal from these animals that the state's aquatic habitat is being conserved.

Another component of river biodiversity is furbearers. Beavers are well known for their dams that sometimes back up streams and cause flooding. Mammologists at SDSU reported one beaver lodge per mile in the middle Big Sioux River (Dieter and McCabe 1989). South Dakota rivers also have mink and muskrat and experiments are being done to reintroduce the river otter. An otter released in Nebraska was captured on Crow Creek in Buffalo County, South Dakota. River otter populations may be recovering from overharvest in the past. Nine of South Dakota's 10 species of bats favor rivers and other aquatic resources as favorite foraging areas. Most widespread is the little brown bat that forages for emerging aquatic insects over water—another example of the vertical dimension of rivers. The little brown bat can capture about 3,000 insects per night (Higgins et al. 2000).

South Dakota warm-water rivers also provide bird habitat. Wood ducks nest in all counties in the state, especially along main river channels where flows are more stable than in tributaries (Coughlin and Higgins 1995). Along the forested Big Sioux River, wood duck broods use the wetlands outside the floodplain during drought, but use forested wetlands in the floodplain during high-water years. Wood ducks are so tied to rivers that counting them as they fly past a set point (bridge) may be a way to estimate population size. Colonial nesters, such as the great blue heron, frequently use riverine riparian zones. A heron rookery on the James River was the focus of one of the few intensive studies of this migratory species that begins arriving in early May to make colonies of as many as 80 nests. With an average of three nestlings per nest, the adult herons are constantly busy catching fish, frogs, and other prey from the river floodplain (Dowd and Flake 1985).

River Values To Humans

We use rivers in many ways—as a source of water for irrigation, drinking, and industrial processes. We depend on rivers to receive treated waste from sewage plants and industry and untreated waste from city streets and rural farms. We also use rivers for recreation. Boating, fishing, hunting, camping, swimming, and sightseeing along the Missouri River are popular with South Dakotans. The Missouri River recreational fishing industry is worth about $50 million per year to South Dakota. Although much smaller than the Missouri River, many warm-water streams also provide South Dakotans with numerous recreational opportunities. A Brown County resident wrote about the James River, "Because for the last 50 years the Jim River has been a source of good fishing, food, and recreation for the family, and in our retirement years, good comradeship with fellow fishermen. While the Jim may be the lowest of rivers we hold it in the highest esteem."

An angler from the Rocky Mountains wrote, "My first view of the Big Sioux River was from an old iron bridge near Brookings. This bridge would become one of my favorite fishing spots. My first night on the river resulted in landing six channel catfish over 5 pounds, including one of almost 12 pounds, one sauger, one bullhead, and two large snapping turtles. Any trip that includes a fish over 10 pounds and has steady action for 3 hours of fishing constitutes a successful trip in my book."

These statements are sentiments from people who find great value in small, warm-water rivers and streams. Although there are few studies that go beyond the sentiments to learn just what South Dakotans are doing on our small streams, there is information on how people use the Big Sioux River. A standardized survey of anglers was conducted from March through October, 1995 at 50 sites from Watertown to Sioux City (Doorenbos et al. 1996). At the same time, the number of people using the river for recreation was recorded. About 13,900 people used the river for 25 kinds of river-related activities. An estimated 87,000 visits were made to the river, mostly for fishing, but also for picnicking, exercising, relaxing, camping, and sightseeing. An estimated 20,000 anglers caught 153,000 fish of 21 different species (Doorenbos et al. 1996).

What do these data mean? During that year, the total visits to 37 state-owned nature and lakeside use areas ranged from 2,600 to 93,000 visits. The estimated 87,000 visits to the Big Sioux River are at the upper end of this visitation range. Consequently, the Big Sioux might be thought of as a long, skinny state park, operating without most of the cost of park management. In fact, management and monitoring of South Dakota warm-water streams has always taken a back seat to the attention given to lakes for fishing and to prairie potholes for hunting.

Stream Fish Management

In the past, stream fish management in South Dakota usually meant "Black Hills streams" and river management meant "Missouri River management." After all, the Black Hills streams, Missouri River reservoirs, and glacial lakes were the fishing jewels of the state. There was no routine biomonitoring in streams as there was for lakes; no annual reports from creel censuses or test nets were produced annually from each regional office. Even today in the annual Fishing Handbook under the title "East River Outlook," there is only information for large lakes, small lakes, and impoundments, even though the Big Sioux and James rivers produce some nice catches of many species, especially catfish.

A great change occurred in 1994, however, when SDGFP undertook a new system of management that included a Stream Fisheries Program and the term "stream" did not mean just Black Hills streams. Overall objectives of the new plan were to upgrade the beneficial uses of streams, establish instream flow reservations on some stream reaches, develop watershed-based aquatic resource management policies, conduct stream preservation and restoration projects, develop a stream fish and habitat database, increase public knowledge and involvement with streams, respond to stream health issues, and maintain and enhance all populations of aquatic special status species. To date, the agency has made progress on most of these objectives. It's imperative that progress be made because of the importance of river habitats and river recreation, and because of the growing list of issues related to streams.

Warm-water Stream Issues

From 1990 to 2000, the Sioux Falls *Argus Leader* had published 121 major articles about the Big Sioux River, which tells much about issues in all South Dakota rivers. There were 38 articles on pollution, 37 on flooding, 30 on parks and fishing, 8 on bridges, and 8 on accidents. When rivers make news in South Dakota, the subject is usually flooding. Since 1870, 84 damaging floods have occurred on the Big Sioux River. During the flood years of 1992 and 1993, 1,600 residences in 37 counties were affected by flooding. Data show that flooding is getting worse, in part because of how watersheds are managed. People like to live and work close to rivers, so naturally homes and businesses in the floodplain are prone to flooding. Also, new water from drained wetlands and paved parking lots is increasing flood peaks. If 62% of the drained wetlands in the Vermillion River basin were restored, flooding would be decreased markedly, probably eliminating the need for dams, levees, and channelization (Johnson 1997).

The Federal Emergency Management Agency (FEMA) analyzed the 1990s floods in South Dakota and provided a 25-point hazard mitigation plan. The plan includes structural ways (e.g., channel clearance, dams, levees) and nonstructural ways (e.g., wetland restoration, floodplain managers, watershed education, regulations) to lessen damages from the next flood. The idea is to focus *less on controlling* floodwaters and *more on reducing* the negative impacts of flood damages. Fish and wildlife benefit from this approach as opposed to manipulating river habitat by engineering measures (FEMA 1993).

Low flows are also a river issue. In fact, both flooding and low flows indicate that rivers are highly dynamic. One longtime agency worker said, "I've been on a flood task force and a drought task force *in the same year.*" It is difficult to manage fish and wildlife when the environment fluctuates so much. River ecologists label the large-scale effects "system controls" on the river. "System controls" usually mold the habitat and fish assemblage at a specific river site (Poff and Ward 1989). Low flow is usually related to low precipitation, but irrigation withdrawals can also lower flows. For example, irrigation appropriations in Nebraska have decreased flows of the White River entering South Dakota (Sando 1991). The consequences of low flow usually appear under river ice in winter when fish kills occur. The James River was plagued with winterkills during the droughts of the late 1980s; the Big Sioux River had winterkills in 2003 and 2004 when an estimated 50,000 fish died upstream from the Flandreau Dam. While the cause is usually a lack of oxygen, water quality analyses also indicate that pollution sometimes plays a role.

Pollution is another river issue. South Dakota has a total of 10,298 miles of rivers and streams, about 7,360 miles have been assessed for water quality. About 56% of assessed streams are supporting all beneficial uses (DENR 2004), meaning that 44% have some type of pollution. The pollution from industrial and municipal pipes has been solved and fish communities have improved; however, bacteria, suspended solids, and nutrients (e.g., nitrogen, phosphorous) are being noticed. These pollutants do not come from pipes, but from the runoff from rural and urban land. This widespread source is called "non-point source" pollution. A TMDL (total maximum daily load) Program is underway statewide to assess river water quality for non-point source pollution (DENR 2004).

Watershed management is a means of addressing river conservation issues and has been the subject of workshops in South Dakota (e.g., Milewski et al. 1997). Watershed management is the process of coordinating land and other resource use in a watershed to provide the desired goods and services without harming soil and water resources. Watershed management borrows from the 3,600-year-old advice of a Chinese Emperor who said, "To protect your rivers, protect your mountains." The Bad River basin is an example of how the watershed and the river are linked.

Watershed managers were faced with a history of improper grazing and row cropping that exposed fragile soils in the Bad River basin (Chaney et al. 1990). Increased runoff eroded and cut gullies and stream channels and produced prodigious amounts of sediment, which was transported out of the Bad River basin and eventually into the Missouri River. During one extreme rain on May 14, 1998, the Bad River discharged 949,300 tons of sediment. The Bad River delta that formed in the Missouri River has caused flooding problems in Pierre and reduced recreational fishing. The Bad River basin project was started on Plum Creek. By applying known conservation practices (e.g., planned grazing, erosion control structures, off-stream livestock watering, conservation tillage), the amount of silt delivered from Plum Creek was lowered from 83 tons to 10 tons per year. When the program was implemented throughout the basin with 90% landowner participation, the sediment discharge to the Missouri River declined by 40%.

During the past 15 years, the information on South Dakota warm-water rivers and river fishes has increased greatly and the news is mostly good. The physical fish habitat is relatively unaltered compared to other states. Water quality is generally good and programs are in place to conserve and improve water quality. Water quantity varies seasonally and is mostly beyond human control. The game fish communities appear healthy, but some native fishes are declining, a possible warning that rivers have changed. From the smallest to the mightiest, South Dakota warm-water rivers and streams make up a wetlands vascular system critical to our lives – biologically, esthetically, socially, and economically.

Acknowledgements

The South Dakota Department of Game, Fish and Parks (SDGFP) funded many of the studies used in the writing of this chapter. The South Dakota Cooperative Fish and Wildlife Research Unit is jointly sponsored by SDGFP, the U.S. Geological Survey, The Wildlife Management Institute, the U.S. Fish and Wildlife Service, and South Dakota State University. ❖

Chapter 9

Water Quantity and Quality: Rich History and Current Issues

Charles R. Berry, Jr.

> " A pint a day
> Is all I say
> To keep my whistle wet.
> But so much more
> I have to pour,
> Before my table's set. "
>
> — Anonymous

In the 1880s, while homesteaders were staking out 160-acre claims east of the Missouri River, Federal government geologist John Wesley Powell was saying that west of the 100th meridian (near Blunt, South Dakota) larger claims of about 2,500 acres would be needed because of scarce water and fragile land. A settler's rhyme described the dry conditions another way:

Fifty miles from water,
A hundred miles from wood,
To hell with this dammed country,
I'm going home for good.

Settlers soon learned the foolishness of the adage "rain follows the plow." They tried to solve water shortages by "mining" groundwater from hand-dug wells, by erecting windmills, or by building earthen dams across draws. The water gathering process shocked many settlers, even the word "dam." "I considered it a swear word," said one woman from the East, "but every homesteader had to have a dam." Early settlers would be shocked to see the huge dams built 70 years later, but the modern dams with names like Big Bend, Angostura, and Fort Randall serve the same purpose as did their tiny ancestors, which is to hold water in an arid landscape. Enough water is now stored in Missouri River reservoirs to cover the state with one foot of water (Hogan 1995).

Securing large quantities of water for people, crops, and livestock was successful because of the civil works projects by the Bureau of Reclamation and Corps of Engineers and farm programs of the U.S. Department of Agriculture. As the populations of humans and livestock grew, another water problem arose. Water pollution began to be noticed in formerly clean streams, rivers, and lakes. In the countryside, plowing the prairies increased sedimentation of streams, and in the cities, primitive sewer systems removed sewage from neighborhoods, but dumped it untreated into nearby rivers. The history of water pollution control in South Dakota follows the history of federal efforts to lead, assist, and ultimately *require* states to improve waste treatment, pollution monitoring, and other complex aspects of pollution control. South Dakota has been a leader in some areas (e.g., lagoons for sewage treatment in small towns; meat packing waste treatment) and there have always been state programs and university research on water pollution problems.

Fishes and other aquatic life were affected by developments that changed water quantity and quality. Some changes were beneficial, some changes were harmful, some changes were unavoidable, and some changes were unexpected. The purpose of this chapter is to review basic science relating to water quality and quantity, recall water history, and define water issues of the present and future in South Dakota.

Water Basics

To most of us, water is what fills a pond or stream. To an angler, water is fish habitat. To a chemist, water is H_2O, the water molecule formula with two hydrogen atoms and one oxygen atom. Water exists in three states: liquid (flowing stream), solid (ice), and gas (water vapor). It is the only substance that can appear in all three states at the same time. The endless cycle of evaporation into atmospheric water vapor, condensation as clouds, precipitation as rain, and runoff to lakes, streams, and oceans is the *hydrologic cycle*.

Water has many unusual properties. Water adheres to substances and dissolves them, even changing rocks into soil. Water is called the universal solvent. Most liquids contract when they freeze, but water expands so ice is lighter than liquid water. This unusual property enables ice fishing; but more important, it insulates lakes enough to keep them from freezing solid. Water has a high surface tension, meaning that it sticks to itself. High surface tension enables water to support objects heavier than water itself—from fishing boats to the insects that skate across the surface of a pond.

Water contains dissolved oxygen that supports fish and other aquatic life. Most dissolved oxygen comes from aquatic plant photosynthesis; some comes from agitation (e.g., waves, waterfalls); and some comes from diffusion from the atmosphere. The normal amount of dissolved oxygen is about eight parts per million, which equates to about eight marbles in a dump truck load of marbles. Fishes gather oxygen using gills and respiratory pumps (i.e., a complex of bones, muscles, valves, and chambers) that are specialized for capturing the rare oxygen molecules. Low dissolved oxygen can cause fish kills under ice, during hot summer weather, or where water is polluted with organic material.

The human body is about 66% water. As common as it is, water is the most precious substance on Earth. Most South Dakotans (96%) feel that water quality should be protected for human health reasons regardless of cost, and support is still high (93%) for protecting water quality for important game fish and wildlife (Gigliotti 1999).

About half of the rivers and 70% of the lakes in the state have not fully met water quality standards over the past few decades (SDGFP 1994, DENR 2004). Siltation, water diversions, and pollution are known problems. Pollution can come from pipes (called "point-source" pollution) or from runoff (called "non-point source" pollution). In lakes, non-point sources of nutrients (i.e., nitrogen, phosphorus) are problems that cause algae blooms that give many lakes a "pea soup" appearance in the summer. The scientific term for a pea soup lake is a "eutrophic lake." Eutrophic lakes are prone to under-ice fish kills caused by low dissolved oxygen and the build up of toxic substances such as hydrogen sulfide, the chemical that has a rotten egg odor. A government study recently classified Lake Kampeska as "hypereutrophic," meaning it exists in a state of extreme biological degradation (Holien 2001).

In rivers, point-source pollution causes water quality problems near the pollution source. While dilution downstream usually lessens the problem, "dilution is not the solution to pollution." Toxic contaminants accumulate in water, in bottom mud, and in fishes. There are three main types of contaminants: naturally occurring elements (e.g., selenium), metals from industry (e.g., mercury, arsenic), and pesticides and herbicides from agriculture. In the 1970s, mercury from Black Hills mines threatened fishing in Whitewood Creek, the Belle Fourche River, the Cheyenne River, and then the Cheyenne River arm of Lake Oahe (Ruelle et al. 1993). However, toxins are not a significant problem in South Dakota.

History of Water Quality Issues

Today, three state agencies share responsibility for developing and protecting water resources. The South Dakota Department of Agriculture (Division of Resource

Conservation and Forestry) funds local conservation districts to educate both urban and rural landowners about best management practices to protect water. The Department of Game, Fish and Parks (Division of Wildlife) monitors fish and other aquatic life. The Department of Environment and Natural Resources (DENR), however, is the agency most directly responsible for water quality and quantity. The mission of the DENR is complex, promoting both development and conservation of water resources (Table 9.1). In addition, the history of water pollution control in South Dakota is inter-

Table 9.1 History and current responsibilities of the Department of Environment and Natural Resources.

TIME PERIOD	AGENCY NAME		COMMENT
Prior to 1973	Dept. of Health Division of Sanitary Engineering	Water Resources Commission, Geological Survey	Dept. of Health employees deal with pollution problems of rapidly expanding population, new wastes, and new Federal legislation.
1973–1978	Dept. of Environmental Protection	Dept. of Natural Resources	Minor reorganization and name change, new Federal Water Pollution Control Act has many requirements.
1979–1991	Dept. of Water and Natural Resources		Focus is on water and soil, while sharing air and solid waste work with Dept. of Health, and mineral and mining work with Dept. of Agriculture.
1991–present	Dept. of Environment and Natural Resources		Activities under one umbrella organization with two divisions are: regulation, geology, resources management, water rights, administration, and technical services; share stewardship with Game, Fish and Parks and the Dept. of Agriculture.
	Programs deal with education, dams, drinking water, feedlots, fish consumption advisories, hydrology, lakes, maps, permits, streams, swimming pools, water monitoring, water use and appropriation, sewage treatment, operator certification, and well water.		

twined with, and sometimes driven by, federal programs and information.

With statehood in 1889, water quality was the responsibility of the new Department of Public Health because water quality was related to human disease (e.g., typhoid fever) and because public waters were used for drinking and as depositories of waste. Cities like Sioux Falls, Huron, and Rapid City first built sewers to move human and industrial waste to rivers and later added sewage treatment plants to partially control the pollution entering rivers. For example, between 1910 and 1920, the population of Sioux Falls doubled from 14,000 to 28,000. The amount of wastewater more than doubled because industries, especially the meat packing industry, added to the waste. The volume of waste flowing from Sioux Falls to the Big Sioux River sometimes exceeded the flow of the river. The river looked and smelled offensive for miles below the city; fish life was absent; and these conditions were reported to have "caused much just complaint from the riparian owners downstream" (Bradstad 1930). In August 1927, the first units of a sewage treatment works in Sioux Falls were in operation to treat about four million gallons of domestic and industrial waste per day. Other towns also began treating sewage at various levels of treatment termed primary (35% purification), intermediate (50% purification), and complete (85% purification), but the treatment of sewage always lagged behind the amount of sewage needing treatment. For example, in 1936, about half of the population had sewers and about half of the sewer waste was treated.

During this era, water pollution biology emerged as a science and trained biologists began to study the responses of aquatic life to pollution. The first nationwide study of 982 sites included collections of plankton, bottom invertebrates, and fishes from four sites in South Dakota—Big Sioux, James, Missouri, and Cheyenne rivers (Ellis 1937). The report stated that "the unsightly and noisome conditions due to pollution are encountered so often that they have come to be accepted by many as the usual order of things." Dr. Edward Churchill, a biology professor at the University of South Dakota, did an early (September 11–21, 1940) biological assessment of water pollution in the Big Sioux River from Renner to Klondike (Churchill 1944). He collected plankton and bottom organisms (but not fishes) to determine the impacts of the Sioux Falls Sewage Treatment Plant. Nine months later, the first short course for sewage works operators was held at South Dakota State College in Brookings. Operators from 16 cities learned about the workings of a sewage treatment plant and visited the Brookings plant, but the tour and training did not include information about pollution effects on the Big Sioux River (Carl 1941). South Dakota first passed a comprehensive water pollution law in 1935 that was administered by the State Department of Public Health.

South Dakota Congressman Karl Mundt tried to help the national pollution problem in 1947 by introducing water pollution control legislation for interstate rivers, which were a federal responsibility. He cited the failure of voluntary controls and the inadequacy of state laws to protect interstate waters. The idea of pollution control through federal legislation instead of state legislation, however, was controversial and without industry endorsement, Mundt's bill and other federal pollution bills failed or were weak when finally passed.

While Mundt worked on water pollution at the national level, two groups were working on water pollution in South Dakota. One was the Izaak Walton League, which pushed for conservation and is credited with forcing improvements in Watertown's waste treatment in 1933. The other group was the dedicated professionals belonging to the South Dakota Water and Sewage Works Conference, which was formed in 1934. Among the early leaders were Charles Carl, an instructor in sanitary engineering at South Dakota State College, and Wallace W. Towne, director of the Division of Sanitary Engineering of the State Board of Health.

Towne was the secretary of the Water and Sewage Works Conference that met annually to share information about water pollution control. At the 1940 meeting in Mitchell, Towne reported that they had the most successful meeting to date with 105 people registered, of whom 59 were actively associated with the operation and maintenance of water and sewer systems. Towne probably did not know that this would be his last meeting for a while as the winds of war were blowing. After returning from military service, Towne reported on the South Dakota situation at a national meeting as follows, "Due to the fact that our engineering staff was almost completely depleted during the war years, and are just now returning, we have not been able to do much." He went on to state that adequate, competent personnel were lacking (Towne 1946).

Towne remained an important figure in pollution control in South Dakota until the early 1950s. He then was employed by the U.S. Public Health Service to conduct studies at the new Robert Taft Sanitary Engineering Center in Cincinnati, Ohio. He did research on a variety of industrial wastes and waste treatment processes, including one idea that he had formerly promoted in South Dakota. Small towns needed an inexpensive means of treating waste and Towne oversaw the development of 32 sewage stabilization ponds (Towne 1957). A stabilization pond is a lagoon that treats small quantities of waste through biological action of bacteria and aquatic biota in a pond. A discharge from the pond is allowed once every 180 days. Use of stabilization ponds spread to neighboring states and Towne encouraged the Public Health Service to study ponds to improve efficiency. One study involved spiking the pond at Philip, South Dakota, with radioactive tracers to monitor under-ice conditions.

The decades of the 1950s and 1960s saw progress made in cleaning up waste from domestic and industrial sources. South Dakota was somewhat ahead of other states in the Missouri River basin with 68% of the state's population having some partially treated sewage compared to 53% basin wide. Still, more and more studies indicated pollution in the country's streams, rivers, and lakes. South Dakota Game, Fish and Parks Secretary R. A. Hodgins wrote, "Pollution comes in many forms and South Dakota waters have seen and are continuing to see this dreaded killer take its toll. Every lake and stream in South Dakota, at least to some degree, is affected by industrial, agricultural, or municipal pollutants" (Hodgins 1971). Hodgins may have seen data that showed severe damage to fishing from industrial waste below the town of Belle Fourche and data showing that the Big Sioux River had the highest levels of phosphorus found in the nation. Hodgins' statement foreshadowed other bad news to come in 1977 when the Big Sioux River was listed as one of the most polluted rivers in the country. Fish surveyors concluded that "the quality of the game fish population in terms of desirable species is adversely affected by poor land use practices on the watershed and by various forms of pollution" (Nickum and Sinning 1971).

A turning point in national concern for water quality came in the 1960s. Pollution problems like those in the Big Sioux River, reports of rivers that caught on fire, and a national conference where South Dakota Senator Francis Case headed a round-table discussion on water pollution were important, but it took Rachel Carson's vision of a "silent spring" to galvanize public demand for meaningful political action (Carson 1962). In 1969, Congress declared a national environmental policy in Public Law 91-190. The National Environmental Policy Act stated:

> *The purposes of this Act are: To declare a national policy which will encourage productive and enjoyable harmony between man and his environment; to promote efforts that will prevent or eliminate damage to the environment and biosphere and stimulate the health and welfare of man; to enrich the understanding of the ecological systems and natural resources important to the Nation; and to establish a Council on Environmental Quality.*

Supporting this policy were strong amendments to the Federal Water Pollution Control Act (Public Law 92-500) that set out procedures that are still central to pollution control today. The old South Dakota Public Health Service would change in response to this federal legislation and has undergone several reorganizations to become the modern-day

DENR (Table 9.1). While some state leaders believed that the new laws were a great burden in many ways, others saw opportunity.

The new federal laws promised funding for training and research on water pollution control. Two engineering instructors at South Dakota State College, Jim Dornbush and John Andersen, saw the coming changes and decided to seek doctorate degrees in the emerging science of water pollution control—Dornbush at Washington University, Andersen at the University of Wisconsin. They later teamed up to apply for federal funding for annual short courses in water treatment and they began university courses in environmental engineering. Dornbush supervised the research of dozens of graduate students on the use of stabilization ponds for municipal and livestock waste treatment. He was involved with South Dakota water issues for 50 years. A former South Dakota State University graduate student, Dr. Donald Hammer, edited a book on the modern use of constructed wetlands for wastewater treatment (Hammer 1989).

A key element in the new federal water pollution law was the idea that each stream should have designated beneficial uses and water quality criteria to protect those uses. This approach had been studied and debated for several decades and South Dakota had proposed stream uses and water quality standards in 1966 (Dornbush 1971). One unusual designation for the use of Whitewood Creek, however, received federal attention and got South Dakota a slap on the wrist (Ford 1970). South Dakota had designated Whitewood Creek for pollution *removal*; in other words, it was a legally polluted stream. This use did not make common sense when cleaning up water pollution was a goal, so Congress inserted the term "*beneficial* use" into the federal law and South Dakota went about designating beneficial uses to streams according to their natural conditions, which ranged from clean cold-water trout streams to small, intermittent, turbid warm-water streams. Eleven classes of beneficial uses are codified in South Dakota law; five of the classes depend on use by fishes and all

Table 9.2 Beneficial uses of lakes and streams in South Dakota. Specific water quality conditions are required by law to support each use. Permanent waters have no history of fish kills due to naturally occurring problems. Semipermanent waters have fish kills about every 10 years, and marginal waters have fish kills every five years.

CLASS	BENEFICIAL USE
1	Domestic water supply waters
2	Coldwater permanent fish life propagation waters
3	Coldwater marginal fish life propagation waters
4	Warmwater permanent fish life waters
5	Warmwater semipermanent fish life waters
6	Warmwater marginal fish life waters
7	Immersion recreation waters
8	Limited contact recreation waters
9	Fish and wildlife propagation, recreation, and stock watering waters
10	Irrigation waters
11	Commerce and industry waters

uses reflect value to humans, livestock, industry, and fish and wildlife (Table 9.2).

Designated uses are protected by water quality standards that are also codified in law as legally established limits for pollution materials. For example, the lower main stem of the Big Sioux River is classified for uses 5, 7, and 8 (Table 9.2). Among the criteria to protect uses 5, 7, and 8 are the following standards: 1) suspended solids shall not exceed 90 mg/l; 2) bacteria must not exceed 200/100 ml in five samples in one month; 3) total chlorine shall be less than 0.02 mg/l, dissolved oxygen shall be greater than 5.0 mg/l, and so forth for several other measurements of water quality. The DENR does routine monitoring to determine if water quality supports the designated use.

Every two years, the DENR publishes an important document that summarizes the condition of about 10,300 miles of state rivers and streams and 573 lakes and reservoirs. In 2004, about 56% of the stream miles supported all designated uses and 34% of the lake acreage supported all designated uses. Fish kills are rare (about 12/yr) and most (93%) are caused by natural conditions related to hot summer temperatures and low dissolved oxygen (DENR 2004).

When viewed in the historical context, progress has been made in controlling water pollution. The pollution problems and fish assemblages in the Big Sioux River have improved after 25 years of stream improvements funded by the Clean Water Act (Dieterman and Berry 1998). Of course there is more to do in every stream and lake. The State Water Plan is designed to ensure the optimum overall benefits of water resources for the general health, welfare, safety, and economic well being of the people of South Dakota. Water *quantity*, on the other hand, is in the hands of higher powers.

Water Quantity

Hypothetically, if all of the water in the world was held in a 55-gallon barrel, 54.6 gallons would be unavailable for our use because it is salty ocean water or locked in ice. Only about 1.5 quarts would be available for our use, and most of the 1.5 quarts is underground or in the atmosphere. An incredibly small amount (about 0.5 ounce) is in freshwater lakes and rivers.

The large demand for a limited supply of water imposes a serious strain on water resources, especially in arid and semi-arid regions. South Dakota is classified as semi-arid in the west to sub-humid in the east (Hogan 1995).

Understanding our water supply begins with understanding precipitation and runoff patterns. There never seems to be a normal water year in South Dakota, but the seasonal cycle is somewhat predictable (Figure 9.1). Most precipitation occurs as rainfall in the spring. Total amounts increase from about 14 inches annually in the arid west to about 24 inches annually in the more humid east. The Black Hills are a wet "island" on the prairie, receiving up to 29 inches annually in the northern high elevations.

The area of land that catches precipitation and sheds runoff to a particular stream or lake is called a catchment, watershed, or runoff basin. In a watershed, some precipitation infiltrates to become groundwater and the excess runs over land into surface waters such as lakes and streams. Everyone has a watershed "address." The big river is like your nation, the lake is your city, the creek is your street, and water flow begins on your property. Understanding water quantity issues begins with understanding the longitudinal and vertical dimensions of the streams in a watershed (Black 1996).

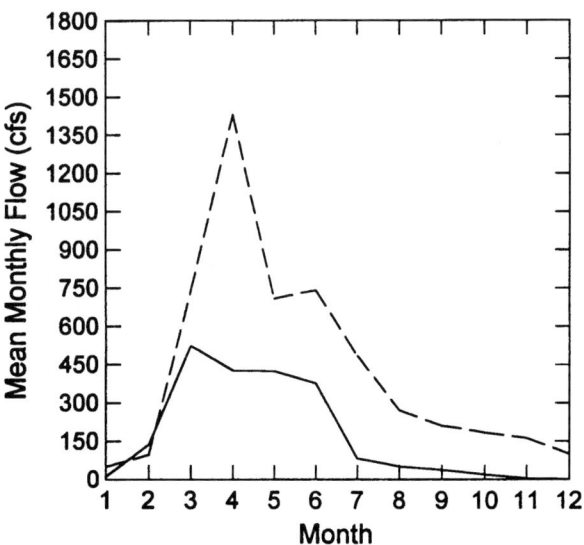

Figure 9.1   **Mean monthly discharge for the Bad River (solid line), years 1929-2000, and the Big Sioux River (dashed line), years 1949-2000.**

The longitudinal dimension means that rivers grow in size as they flow through the watershed. Wetland plants mark the saturated areas between hills where water flow begins to form the first stream, called a headwater stream. Headwater streams join to become larger streams, which join to become rivers. Headwater streams are the most abundant type of stream and are usually intermittent because precipitation is seasonal. Because headwater streams are numerous and are "the source" of the water supply, they are the focus of conservation and river restoration programs. For every mile of the Cheyenne River, there are 25 miles of headwater streams collecting water from the Black Hills, from the Badlands, or from the prairie.

In the vertical dimension, groundwater, surface water, and the atmosphere are connected. The connectedness on a grand scale is called the hydrologic cycle, and on a watershed scale is represented by groundwater recharge of springs and headwater streams. Wetlands and wet soils, called hydric soils, act like a sponge, soaking up precipitation and holding it as groundwater in aquifers to be metered out slowly to streams and lakes. Aquifers are water-holding beds of gravel, sand, and porous rock beneath the surface. Recoverable water in aquifers beneath South Dakota could flood the state with water to a depth of 1.5 feet. Irrigation uses about two-thirds of the water in South Dakota, whereas public water supplies, livestock, industry, and municipalities use the remainder (Amundson 2002). Every person uses about 160 gallons/day.

Flood control is the most important wetland service in the natural infrastructure of a watershed. Drainage has predictable consequences in the vertical and longitudinal dimension (Leopold 1994, Johnson 1997). Humans frequently try to hasten water drainage from a watershed by tiling and ditching wetlands, ditching headwater streams, and channelizing rivers. The changes are usually harmful to fishes by *reducing* water supply in small streams. (Less groundwater recharge leads to lower low flows or more periods of intermittency.) The changes are also harmful to downstream neighbors by *increasing* water supply in larger streams, thus causing higher high flows and more flooding.

History of Water Quantity Issues

"Whiskey is for drinking; water is for fighting over" so goes the quote attributed to Mark Twain, and there is a rich history of water conflict in the West. Two terms are central to any discussion of water quantity. First is the appropriation doctrine, a system of water rights that gives an individual the right to use a certain amount of water. This doctrine is different than the riparian doctrine that settlers from the East remembered. The riparian doctrine gave water to each riparian owner, which works well where water is plentiful and is used near the stream for livestock or households. In the arid west, however, water quantities were limited and developers wanted to move water long distances for irrigation. Court decisions in Dakota Territory began the shift from riparian water rights to appropriation rights (Wishart 2004). The appropriation doctrine gives water to a user on the date the use was established, which is often summed up in the phrase "first in time, first in right." Today, DENR's Water Management Board issues permits for water withdrawal, monitors stream flow, and issues "shut off orders" if stream flow becomes critically low.

Native Americans were "first in time" and their water rights are protected by the Winters Doctrine (Wishart 2004). Native Americans have "reserved" rights inherent when a reservation was created. Even if unused, an unquantified amount of Native American water is protected for economic goals on the reservation.

"Ownership" is central to a discussion of water quantity. Laws in 1877 gave water in navigable lakes and streams and the land under the water to the general public. When South Dakota became a state, it held in trust the water and beds of all navigable lakes and streams for the people. The trust lands were below the ordinary high water mark (OHWM), whereas land above the OHWM was private. The OHWM is a physical mark left by the action of water (DENR 2001). The term "ordinary" is not easy to define because of the variation inherent in periods of drought and floods and by human influence. The Water Rights Division of DENR finds the OHWM mark on public lakes. In navigable rivers, the practical mark is the "bank full mark" or the location on the bank where water spills onto the flood plain.

The U.S. Geological Survey (USGS) has monitored stream flow in South Dakota since 1903 (USGS 1987). Early gauging stations (1903-1906) were on streams in the Black Hills, on Indian reservations (1912-1920), and on the Missouri River (1928-1930). One Missouri River station has been in continuous operation since 1930. The USGS expanded stream gauging activities in the 1940s for the war effort and for Missouri River basin development. The state has cooperated with the USGS in monitoring water resources since 1944 and about 100 sites have records. Annual reports of river flows are available at libraries and on the internet (*www.usgs. gov*). A unique feature of the internet data is the real-time river discharges that are updated every 15 minutes at selected gauging stations. Besides compiling stream discharges, the USGS hydrologists have conducted many special studies on state water resources (e.g., Sando 1991, Williamson 2000, Heakin et al. 2006).

Water conflicts in the early West were less important because settlers used dry-land farming techniques. Dry-land farming meant conserving moisture where precipitation was scant. Farmers adjusted to the environment, in what today we would call an ecological sustainable approach. Dry-land farmers advised, "Don't try to change nature's laws to fit your notions and habits" (Carrels 1987). While their soil steward-ship retained fertility and moisture that produced crops, other settlers wanted bigger farms with irrigated cropland.

John Wesley Powell, director of the USGS, believed that large land grants and irrigation were best for the arid West and proposed regional water planning and forest management to produce water (Wishart 2004). Unfortunately, Congress did not understand these ideas that today we might call "water-shed management." Instead, Congress passed the Newlands Reclamation Act in 1902 to subsidize construction of huge irrigation projects. Like the debates on federal involvement in water pollution control, debates on government assistance for irrigation were bitter, but lobbyists promised a construction boom to be followed by drought-proof agriculture.

Early irrigation projects in South Dakota were small and confined to bottomlands of western rivers and creeks. The first big Reclamation Act project completed in 1917 was the Belle Fourche Project with 500 miles of canals to irrigate 90,000 acres. Engineers with little data or experience designed a reservoir that quickly went dry and made a mistake in soil classification that put the project on heavy soils that were unsuitable for irrigation (Reisner 1993). Most farmers went broke paying for the water and South Dakota Senator Chan Gurney criticized the project. The failure of other early irri-gation projects kept interest in irrigation low until the Dirty Thirties.

The Great Drought ushered in ambitious Bureau of Reclamation plans for irrigation. In South Dakota, the Bureau built dams in several canyons in the Black Hills (Table 9.3). An example is Pactola Dam on Rapid Creek, which is used for water supply and irrigation. Pactola Reservoir is cold and deep (maximum 158 feet) and has great recreational value.

Table 9.3   List of Bureau of Reclamation and Corps of Engineers reservoirs in South Dakota.

RESERVOIR	AREA (sq. mi.)	DATE DAM CLOSED	RIVER
Bureau of Reclamation			
Belle Fourche	12	1914	Belle Fourche River
Deerfield Lake	0.6	1947	Castle Creek
Angostura	6	1949	Cheyenne
Shadehill	10	1951	Grand River North Fork
Pactola Lake	1.2	1956	Rapid Creek
Corps of Engineers			
Lewis & Clark	39	1955	Missouri River
Francis Case	123	1952	Missouri River
Lake Sharpe	89	1963	Missouri River
Lake Oahe	487	1958	Missouri River

The Bureau of Reclamation reservoirs like Pactola created fishing and recreation opportunities. The U.S. Forest Service has jurisdiction over campgrounds and shoreline access areas. Game fish in Pactola are rainbow trout, lake trout, brown trout, and yellow perch. About 40,000 catchable-sized (9 inches long) rainbow trout are stocked annually because natural reproduction is insufficient to meet angler demands. Other species include white sucker, green sunfish, lake trout, and splake (purposeful brook trout x lake trout hybrid).

Keyhole (Wyoming), Angostura, and Orman dams in the Cheyenne River have changed river flows and created new fish habitat in reservoirs and tailwaters. In general, water quantity in tailwaters is stable, while water quality is cool and clear. Bottom materials are coarse rather than sandy because the erosive power of the river digs downward rather than creating lateral meanders across the flood plain. Some warm-water fish species declined, while others flourished. An example is the increase in smallmouth bass in a segment of the river downstream from Angostura Dam that now looks more like Fall River (clear, cool, stable flows) than the Cheyenne River (turbid, unstable flows, warm) (Hoagstrom 2006). Dams, which bring new waters and species for recreational anglers, also create conservation challenges for biologists who wish to conserve native fishes.

Many stock ponds that have been constructed for livestock water also provide good fishing. A west river stock pond is formed by building a dam across a ravine or draw that is subsequently filled by surface runoff or groundwater. Wetlands of South Dakota have been digitally mapped (Rieger et al. 2006, Johnson et al. 1997) allowing computers to count stock ponds and other wetlands. For example, about 1,129 stock ponds averaging about 5 acres, and 20 small reservoirs have been constructed on tributaries of the White River in South Dakota. The ponds may deplete river flow by impounding runoff. On the other hand, the stock dams may replace drained wetlands, keep cattle away from stream courses, and can be stocked with fishes for recreational purposes.

Stock ponds make up about 90% of the 100,000 constructed ponds in the state, of which 43,000 contain fishes. Under good conditions, constructed ponds provide high quality fishing. For example, ice fishing on one or more of the 400 small dams on the Fort Pierre National Grasslands can produce significant numbers of big bluegills, yellow perch, and largemouth bass. The ponds are usually very productive, which leads to fast fish growth. In the summer, casting for largemouth bass from a float tube is a good way to fish these ponds. South Dakota pond owners can follow some simple guidelines to properly manage their private ponds (Willis et al. 1990).

The typical farm and ranch pond east of the Missouri River is called a dugout because it is excavated into the shallow water table in flat terrain. Wetland inventory data shows that there are about 56,827 dugouts scattered across eastern South Dakota (Johnson et al. 1997). Dugouts have little impact on river discharge, but when located in the floodplain and flooded, fishes may migrate into them. As in other floodplain depressional wetlands (potholes), fishes may survive, spawn, and migrate back to the stream during the next flood, or poor water quality or predators may kill them. Such is the natural law of the flood pulse operating in the lateral dimension of a river.

No one federal water project transformed South Dakota as much as the 1944 Pick-Sloan Plan that authorized six dams on the Missouri River. A 1952 *Time* magazine applauded the Pick-Sloan Plan for changing "the most useless river there is" (Carrels 1987). By the 1970s, the Missouri River was transformed by the largest system of reservoirs in North America; four of the six reservoirs are in South Dakota (Table 9.3). The transformation has provided important benefits for Missouri River basin citizens, but it also significantly altered the river ecosystem. The decline of native species, combined with drought and flood events over the past decade, has led to a basin-wide debate on how to manage the waters of the Missouri River basin (CMRES 2002).

The consequences of Missouri River development to hunting and fishing were great. Drowned beneath the reservoirs were a million acres of fertile bottomland, many miles of prairie river, and great groves of cottonwoods. What had once been the most diverse ecological community on the northern Great Plains was swallowed up. On the other hand, towns like Chamberlain, Pollock, Mobridge, and Pierre have grown because of tourism on these reservoirs. The reservoirs and dam tailraces provide a great amount of recreation, hunting, and fishing that brings people from all over the world to South Dakota. For example, the fishery in the smallest reservoir, Lewis and Clark Lake, generates about 100,000 hours of fishing worth about $3.3 million annually (Wickstrom 2004). Upstream in Lake Sharpe, anglers fished about 400,000 hours in 2003, catching an estimated 550,000 walleyes and harvesting about 112,000, with others released. The hourly catch rate for walleye was an incredible 2.25 fish/hour. Other fishes caught by anglers were sauger, white bass, smallmouth bass, channel catfish, rainbow trout, and yellow perch (Lott et al. 2004).

Lake Oahe is the destination of most visitors lured by catches of big walleyes. Recreational fishing made a $20–24 million annual impact from 1995 to 1998 (Nelson-Stastny

2004). Lake Oahe fishing often dominates the news in South Dakota, but in the 1970s, a debate about water quantity and quality was in the headlines and some say the debate ended the career of a prominent politician. Aberdeen author Peter Carrels' book provides details (Carrels 1999).

The Oahe Diversion Unit was a large irrigation project that would move Lake Oahe water by canal 100 miles east to the James River Valley. A 1,200-mile network of smaller canals would then irrigate about 500,000 acres. Most politicians supported the irrigation project. Opposition came mainly from landowners in the path of the canals, from citizens wary that the irrigation return flows would degrade the James River, and from farmers who believed that irrigation would add salt to croplands. Overall, many South Dakotans had no desire to pay for a project they viewed as wasteful and destructive. Opponents formed the United Family Farmers and gained control of the local elected board overseeing development and promotion of the project.

The United Family Farmers were opposed by an organization called Friends of Oahe. The Friends had been organized by the South Dakota banking and construction industries and Senator George McGovern. After a series of heated (some say vicious) public hearings, the board asked Congress to deauthorize the Oahe Project, which was done in 1982. Instead of the massive Oahe Diversion Unit, the United Family Farmers promoted a smaller project known as the WEB (Walworth, Edmunds, Brown) water pipeline. The WEB eventually became the largest domestic water pipeline system in the United States, delivering treated Oahe Reservoir water to farms and communities in an area the size of Connecticut.

The threat of irrigation return flows and other developments required a number of fisheries surveys of the James River to assess impacts of the proposed developments. Consequently, there is more information on the James River fishery than on any other South Dakota river—11 fish surveys since 1975 (Shearer and Berry 2003). The most recent in 2000 showed that over the past 25 years, 50 species have been recorded, 9 more in 2000 than in 1975. One new species was the exotic grass carp; others were native species that had increased distribution and abundance (Shearer and Berry 2003). Recreational use of the river includes some 31 activities, with camping and fishing being the most popular (Hansen 1981).

Water quantity and quality have been at the center of an ongoing project termed the James River Restoration Project (Lange 1998). The history is almost as complex as that of the Pick-Sloan Plan on the Missouri River. Around 1900, springs and wetlands metered flows to the river and tributaries that were described as clear and vegetation choked. Today the same sites are described as turbid with a bottom of "clay, sand, and gravel." Water quantity seems to be too much or too little. Operation of the 90-foot high Jamestown (North Dakota) Dam has almost eliminated flooding to the state line. In 1975 and 1980, flow was nonexistent in parts of the upper river and fishes were concentrated behind hundreds of small dams (3–6 feet high) and rock crossings. These structures impede flow, impound water, and exacerbate flooding, as does drainage of about 40% of the wetlands in the watershed.

A 1970 plan suggested dredging the James River into a 400-mile-long barge canal for shipping agriculture products. Again in the 1980s, another James River canal dredging project was proposed (and briefly begun), this time to reduce flooding. Drought in the late 1980s fostered an "environmental initiative" that sought to provide minimum flows in winter to stop under-ice fish kills by releasing water from two proposed dams on tributaries. The Corps of Engineers and the James River Water Development District are still seeking ways to restore and improve the James River. Anglers hope the plans will maintain excellent fishing for trophy-sized catfish. The James River has been mentioned in national fishing magazines as a good catfishing river. Recent studies indicate that channel catfish and flathead catfish in the 30–40 pound range are present (Figure 9.2) in the James River, and the state record flathead catfish (63.5 pounds) was caught in the Jim River in 2006.

In the mid-1970s, the Environmental Protection Agency listed the James River as having intermediate to significant

Figure 9.2   Flathead catfish (42 lbs) caught in the James River during a research study of the channel and flathead catfish. In 2006, the state record 63-pound flathead catfish came from the James River. (Photo by J. Arterburn)

water quality problems. Phosphorus, nitrogen, and total dissolved solids exceeded the state's water quality standards. Water quality in both North Dakota and South Dakota portions of the river has improved in recent decades. Today, water quality on the main stem generally supports the designated uses of semipermanent fish propagation and limited contact recreation, but the degree of support depends greatly on the annual discharge. Water quality in the South Dakota portion of the James River has also improved, but the oxygen standard of 5 mg/l is sometimes violated when flows are reduced. Bacteria (fecal coliform) levels below cities and below Sand Lake National Wildlife Refuge can be two to six times higher than upstream of these sites. Although the watershed is dominated by agriculture, only minor amounts of pesticides (Aroclor, DDT, endrine) and metals (mercury) have been found.

From the Black Hills comes one recent development that is perhaps one of the most interesting stories of water quantity in South Dakota. Spearfish Falls is falling again after 85 years! In 1917, Homestake Mining Company diverted Little Spearfish Creek and stopped the flow of Spearfish Falls, the premier waterfall of the Black Hills. Evidence of its former glory is seen in many tourist pictures. Tourist trains often stopped so passengers could view the falls. When Homestake stopped mining operations, its water rights returned to the state, the falls returned to the scenic Spearfish Canyon, and the water added six miles of high quality trout stream to Spearfish Creek (Hunhoff 2004).

Water Conservation Future

The National Academy of Science has published advice (CMRES 2002) on how to manage the Missouri River. They proposed stakeholder involvement and agency programs that try "adaptive management." Adaptive management means trying a solution, monitoring the outcomes, and adapting the plan if the outcomes don't meet expectations. Central to their recommendations are restoring as much of the natural hydrograph and sediment transport (i.e., cut-and-fill alluviation) as possible. They concluded that Congressional action is the key to breaking a gridlock among basin states. A long-term assessment program was begun on the Missouri River in 2005, with sites in South Dakota included.

For other rivers, lakes, and streams, South Dakota has a State Water Plan (DENR 1993). The statewide goal is to achieve the optimum over-all benefits of the state water resources for the general health, welfare, safety, and economic well being of the people through conservation, management, development, and use of water resources. Among eight statewide objectives of the plan is an objective to enhance beautification and fish and wildlife benefits along with other beneficial uses. Among 16 general policies is the following: "Accomplish development of water resources in such a manner as to have minimal negative environmental impact."

Following well-crafted policies is one thing, but working out details is another as citizens discovered when factory farms and livestock confinement operations began to arrive. Livestock development is in our future. The cycle of "crops-to-cow-to-manure-to-cropland" within one watershed is appealing, practical, and possible; however, strict control of the manure in a holding lagoon and during land application is necessary so that groundwater and streams are not polluted. Initial proposals met with citizen reluctance partly because of the poor track record of this industry in other states (IWLA 1999). The industry will have to be a good neighbor and DENR will need programs and personnel to ensure that waters are not polluted. Besides dealing with large feedlots and dairy herds, the state will have to continue to help small operators upgrade waste management.

The DENR, the SDGFP, and SDDA should work together more closely on water conservation issues. It has been said, "land health and water health are not two issues, but one." While there have been few fish kills in South Dakota, investigation protocols and agency responsibilities are not clear when a kill does occur. Complaining about water pollution and reporting fish kills is a necessary public role. Each agency has an education program to improve public involvement and awareness of water issues. Anglers are one of the most important public groups (stakeholders) because anglers have insights into the conditions of rivers and lakes and the quality and quantity of fishes from these waters. Anglers need to be involved in preserving water quantity and quality. The Water Conservation Plan clearly informs local governments and citizens that they must be active and take leadership roles before seeking state assistance. This means participating in citizen reviews of water resource and solid waste management, being willing to participate in watershed management plans, and making water pollution complaints. Notifying DENR of a possible water pollution problem must be written and notarized, which requires citizens to be proactive.

Most important for the future of water conservation in South Dakota is an informed and active citizenry. The cross currents of politics, economics, and science require "advanced citizenship." The layperson can sort out the best course of action by asking the simple question, "How will this development influence the neighbors downwind, downhill, and downstream?"

Some streams and lakes do not fully support their beneficial uses (DENR 2004), so work needs to be done. Water conservation in the future will depend on thinking at the watershed scale, but also acting locally. Recent watershed management workshops have pointed out the abundance of funding, technical assistance, data, and expertise available to the public, if the public asks and participates.

Doing watershed management, however, is very difficult and requires skilled scientists and skilled public relations experts to help committed citizens, often in diverse groups, to work together. Human dimensions experts will be needed to resolve conflicts. Federal agencies offer short courses in watershed management, but there is little attention in universities or state agencies to produce or hire graduates with such training.

Foster Creek, a tributary to Lake Oahe in Stanley County, is a great example of a successful watershed management project (Boettcher et al. 1998). The project, funded by an EPA grant with assistance from DENR non-point pollution experts, shows how riparian vegetation can be restored, native range and wildlife habitat improved, and the economic well being of ranchers improved with proper grazing.

One way water information might be improved is to increase monitoring to find problems before a crisis occurs and costly clean up is necessary. Also, water pollution biologists should be asked for more biomonitoring as a means of determining the health of a water body and to assess the benefits of restoration programs (Larson 2001). Biota, such as insects and fishes, are there to tell the story of water pollution. Recall the fish story of water improvement in the Big Sioux after 25 years of Clean Water Act activities. Using a complex bioassessment approach, called an "index of biotic integrity," recent research has determined that the fish assemblage downstream from a poorly managed floodplain segment of the James River was different than the fish assemblage downstream from a healthy floodplain segment (Shearer 2001). The fish will tell the story of land management. Also, the science of TMDLs (Total Maximum Daily Loads) will be in our future. The TMDL process is a way to identify water quality problems, their sources, and solutions in each watershed.

South Dakota has few fish consumption advisories, but contaminants in water and fish may be a future topic. The mercury that has been appearing in some lakes probably comes from a natural phenomenon that occurs in new lakes, such as those lakes that expanded in size in the Northeast. Naturally occurring mercury becomes methylmercury when land is flooded and vegetation decays. South Dakota also receives mercury and other contaminants from air pollution, particularly from coal-fired power plants with inadequate emission controls. While this toxin accumulates in fish flesh, recent surveys by DENR indicated that fishes from most waters have only small amounts of toxic contaminants.

In conclusion, between the lines of each story of water history in South Dakota is the message that water is essential to agriculture, industry, recreation, and life. Water resources have been developed as much as possible. The task now is to use water wisely and to ensure that it remains unpolluted so that residents and visitors will have an ample supply of clean, fresh water for future use. It has been said, "quality of the environment, like freedom, must be protected and achieved anew by each generation." Many people hope that this is not true and hope that the public interest in water issues leads to actions by government agencies that protect water quality and quantity for the future generations of South Dakotans and all those who visit the state.

Acknowledgements

The South Dakota Department of Game, Fish and Parks (SDGFP) funded many of the studies used in the writing of this chapter. The South Dakota Cooperative Fish and Wildlife Research Unit is jointly sponsored by SDGFP, the U.S. Geological Survey, The Wildlife Management Institute, the U.S. Fish and Wildlife Service, and South Dakota State University. The State Library in Pierre was very helpful in finding information on Wallace Towne and on annual reports. ❖

Chapter 10

Cold-Water Fish Species

Rick Cordes

> " Alive without breath,
> As cold as death.
> Never thirsty, ever drinking,
> all in mail, never clinking. "
>
> — Tolkein

The cold-water fish species in South Dakota are limited in number and distribution because they must have temperatures of 65°F or colder. Most of the cold-water habitat in South Dakota is predominantly associated with the Black Hills, but it can also be found in a few Prairie Coteau streams flowing easterly to the Minnesota River drainage and cold-water springs of the Nebraska Sand Hills of south central South Dakota. Construction of the four reservoirs in the late 1950s and early 1960s on the Missouri River in South Dakota created additional cold-water habitat.

Before 1886, the cold-water fish species of South Dakota were limited to six native species that were present in the upper Missouri River basin, including the Black Hills, and isolated springs in western South Dakota following the last glaciation. Trout were first introduced to the Black Hills in 1886 by two local citizens of the Black Hills, Richard Hughes and Samuel Scott. Trout were stocked in many streams of eastern South Dakota from the early 1900s to the 1960s, but very few stocks survived. Trout, salmon, and cold-water forage fish species were introduced to the Missouri River reservoirs and tailwaters in South Dakota in 1956. Color pictures and a general description of the cold-water species can be found in Neumann and Willis (1994).

Black Hills

The Black Hills are a unique geological feature in the northern Great Plains region of North America and are often described as an island in the plains. Uplifting 60 million years ago created this northeastern isolated extension of the Rocky Mountains. The ancient mountain range consists of a central core of granite and metamorphic rock outcrops surrounded by bands of sedimentary rock, including limestone that has eroded over time. Today the Black Hills rise to over 4,000 feet above the surrounding prairies and the elevation contributes to higher annual precipitation than falls on the prairies surrounding the Hills. The porous sedimentary limestone continuously recharges the aquifers that in turn create the many springs and streams that flow with cold, clean water from the higher elevations outward to the prairies. Many reservoirs that have cold-water fish habitat were constructed during the last century to provide irrigation water, urban water supplies, and recreation.

The native cold-water fish species of the Black Hills are represented by three cyprinids (longnose dace, lake chub, and creek chub, Figure 10.1) and three catostomids (white sucker, mountain sucker, and longnose sucker). These minnow and sucker species were found in abundance in the streams of the Black Hills in 1892 and 1883 by Evermann and Cox (1896) who were investigating the feasibility of locating a fish hatching station. Mountain suckers are restricted to the cool mountain streams of the Black Hills. Lake chubs, historically believed to be abundant in many streams of the Black Hills, have recently been found only in Deerfield Reservoir on the headwaters of Castle Creek. The longnose sucker was not identified until collected in 1952 in Redwater Creek, a tributary to the Belle Fourche River (Bailey and Allum 1963). The

Figure 10.1 The creek chub (above) and five other non-game species were the native coldwater fishes of the Black Hills. Creek chub is a native stream fish throughout South Dakota; most are about 6 inches long, but some reach 12 inches. Brook trout (below) were stocked in 1886 and since then brook, rainbow, and brown trout have provided cold-water fishing opportunities in many Black Hills streams. Figures by Duane Raver Art, Freshwater Fish Collection, U.S. Fish and Wildlife Service.

other three species can be found in many waters of South Dakota.

Longnose dace are the most abundant cold-water minnow species in the Black Hills and are found in streams with moderate to swift currents. They prefer gravel substrates and can be found schooling in sheltered stream areas. Longnose dace feed on invertebrates dislodged from the substrate. Adults typically are three to four inches long. Sexually mature males and females appear in late May and spawn on sand and gravel riffles through early July.

Creek chubs are large minnows with adults reaching 8 to 10 inches long. Creek chubs in the Black Hills can be found in the lower reaches of many streams with moderate currents and gravel substrates. Creek chubs are opportunistic feeders, feeding on aquatic insect larvae, terrestrial insect drift, and small fish. Creek chubs spawn in late spring when water temperatures reach 55–65° F in gravel substrate. Males excavate a spawning pit by moving pebbles upstream with their mouths to form a ridge. Females then enter the spawning pit and deposit eggs, moving on to other nests or returning to the same nest to deposit more eggs. The male fertilizes the eggs and moves the gravel from the ridge to cover the eggs. After this, no further parental care is given to the eggs or hatched fry.

The lake chub population in Deerfield Reservoir is believed to be a relic of the glacial periods. Additional collections of lake chubs have been made in the upper Belle Fourche River outside the Black Hills. Evermann and Cox (1896) commented on the presence of numerous chubs in the tributaries of the Cheyenne River in the Black Hills. Lake chubs, which prefer cool streams and lakes, were once abundant in many streams in the Black Hills. Adults reach a length of six inches and form schools when in abundance. Lake chubs are opportunistic feeders and similar to other chubs, they search for food primarily by sight. Lake chubs spawn in late spring when water temperatures reach 55° F. They spawn near lake shores in the shallows on gravel substrate or they move into tributaries. Males chase females and stimulate them to spawn, causing the female to release small amounts of eggs that are immediately fertilized. The fertilized eggs settle into the substrate after which there is no further parental care.

Mountain suckers, the most abundant sucker (family Catostomidae) in the Black Hills, can be found in numerous streams wherever there is cool (55–65° F), clear, swift waters. Mountain suckers prefer bottom substrates with a cover of undercut banks, log jams, and boulders, and they usually occur in small schools. Mountain suckers are benthic feeders, using their cartilaginous mouth edges to scrape organic material from rocks. Mountain suckers spawn in the spring, but their preferred spawning habitat and behavior has not been documented.

White suckers inhabit the lower reaches of many streams in the Black Hills and are numerous in the many small reservoirs where they are often the most common species. It is speculated that the city of Spearfish derived its name from Native Americans who speared for white suckers as fish congregated in large schools in Spearfish Creek during the spring spawning migration. White suckers can reach 20 inches in length and weigh up to 3 pounds. They are benthic dwellers feeding on whatever they can remove from the substrate including aquatic insects, plant material, and detritus. Spawning occurs in early May with adults moving upstream or out of reservoirs into streams to congregate on gravel substrates. Males fertilize the eggs immediately after females deposit them over a gravel substrate; after that no parental care is provided.

The longnose sucker is the most widely distributed sucker species in North America. The small population in the Black Hills, found only in Crow Creek and Redwater Creek on the northern edge of the Hills, most likely represents another relic population from past glacial periods. They prefer streams with pockets of deep, clear pools with slow currents intermixed with riffle areas composed primarily of small gravel and sand. They often share habitats with white suckers. Longnose suckers are benthic feeders with diets consisting of insect larvae, plant material, snails, and crustaceans. Their spawning habitat and behavior is similar to that of white suckers.

Despite the introduction of trout into Black Hills streams and the degradation of habitat caused by mining, logging, grazing, and urban development, the populations of native minnows and suckers continue to reside in the waters of the Black Hills today. The most imperiled species is the lake chub, which has been extirpated from four of its historic drainages. There may be a decline in the population in Deerfield Reservoir that may be linked to soil erosion in the watershed or introduced piscivorous (fish eating) trout and bass.

The enterprising endeavors of Samuel Scott and Richard Hughes in 1886 resulted in the first introduction of brook trout (Figure 10.1) to the Black Hills. Both men recognized that the cold, clear water of the Black Hills streams should support trout. Fingerling brook trout were transported by buggy from Leadville, Colorado, across the plains of eastern Colorado and Wyoming to the Cleghorn Springs area west of Rapid City. In following years, the U.S. Fish and Fisheries Commission stocked brook, brown, and cutthroat trout in many of the streams in the Black Hills from 1890 to 1892. Evermann and Cox (1896) found a thriving population of brook trout in Spearfish Creek and reported the presence of brook and rainbow trout in Beaver Creek on a private ranch near Newcastle, Wyoming. The first rainbow trout were stocked in the Black Hills by the U.S. Fish and Fisheries Commission in 1896. Evermann and Cox's survey for a suitable site for a fish hatching station in 1892 and 1893 resulted in the selection of a site on Spearfish Creek for a hatchery. In 1898, the U.S. Fish and Fisheries Commission stocked its first brook, brown, and yellowstone cutthroat trout that were raised in this hatchery.

Churchill and Over (1938) reported populations of brook, brown, and rainbow trout in the Black Hills during their survey in 1927–28. They did not report yellowstone cutthroat trout, which would indicate that the introductions prior to 1928 were not successful. Historical stocking records indicate that the last yellowstone cutthroat trout were stocked in 1920. The stocking of brook, brown, and rainbow trout in the Black Hills waters continued throughout the last century, including numerous introductions into small reservoirs constructed by the Works Progress Administration (WPA) and the Civilian Conservation Corp (CCC) during the 1930s.

Since the initial stockings of brook, rainbow, cutthroat, and brown trout from 1886 to 1901, no new trout species were introduced in the Black Hills until 1977 when lake trout were introduced to Pactola Reservoir. The 1977 stocking produced numerous trophy lake trout, including several 20-pound plus lake trout in 2003. A Snake River strain of cutthroat trout was introduced to Pactola Reservoir in 1985, but stockings were discontinued in 1992 when the species failed to provide a recreational fisheries. Splake, a hybrid cross between lake and brook trout, were introduced in Pactola and Deerfield Reservoirs in 1988 and 1990, respectively.

The most abundant wild trout species in the Black Hills is the brown trout. It has populations occurring from the lower reaches to the headwaters of many of the spring-fed streams. They are most abundant in Spearfish Creek and Rapid Creek below Pactola Reservoir. Brown trout are more tolerant of water temperature changes and flow changes and will seek cover in deeper pools using undercut banks and tree snags. They are opportunistic feeders with a diet of aquatic and terrestrial insect drift and will prey on minnows and small trout. Brown trout spawn in the fall. Females can be seen digging beds in gravel substrates in late November where they deposit eggs. Attending males fertilize the eggs, which hatch the following spring and then receive no further parental care.

Brook trout have established wild populations in the small headwater streams of the Black Hills where water temperatures rarely exceed 65° F. The colorful brook trout thrive in these small streams, but it is unusual to see adult brook trout larger than seven inches. Brook trout feed on aquatic and terrestrial insect drift. They are fall spawners; females begin depositing eggs in shallow gravel nests in October, which are immediately fertilized by one or two males that accompany each female. Brook trout eggs hatch in early spring and receive no further parental care.

The only sustainable population of wild rainbow trout in the Black Hills occurs in a short reach of Spearfish Creek where a swift current is created by the higher gradient in Spearfish Canyon. In addition to the ideal habitat in Spearfish Creek, a nursery stream provides protection to young rainbow trout from piscivorus brown trout that do not inhabit the nursery stream but reside in Spearfish Creek. Rainbow trout prefer a diet of aquatic and terrestrial aquatic insect drift. Females spawn in the spring as the water temperature reaches 50° F. Females select a location with gravel to prepare a nest. Male rainbow trout become aggressive defending individual females and the nest, but when the female begins to deposit eggs in the nest, one or more males fertilize the eggs after which there is no further parental care. Eggs hatch 4-6 weeks later, depending on water temperature.

Management of trout populations in the Black Hills has evolved from a put-and-take stocking program from the early 1900s through the 1960s to a more diversified program of stocking and management of wild populations of brook, brown, and rainbow trout. Water quality and stream habitat degradation was not mentioned as a problem in the 1950s (SDGFP 1959), probably because state agencies were preoccupied with the great growth and change in fish management after World War II. One of the changes was "the beginning of scientific game management and fisheries programs" (SDGFP 1959), and it didn't make sense to stock fish in polluted streams with poor fish habitat. Water quality degradation from mining, forest management, excessive grazing, and urban development was recognized in the early 1960s as threats to the success of trout fishing (Stewart and Thilenius 1964). By 1970, the Black Hills had lost 85% of its trout stream mileage (Glover 1975). Many of these "lost stream miles" have been reclaimed through modern stream habitat management, such as methods described by Hunter (1991).

Cooperation among staff of the Black Hills National Forest, the South Dakota Department of Game, Fish and Parks, and sportsmen's groups has resulted in developing successful stream and small reservoir rehabilitation programs. One example is Whitewood Creek, which was a dead stream for over a century from mining activities. Evermann and Cox (1896) reported about Whitewood Creek as follows: "ruined by tailings from the numerous stamp mills. No fish were found in it…" Whitewood Creek was successfully rehabilitated when the Homestake Mine discontinued discharging untreated ore processing water into the creek in the 1970s. Today it supports populations of longnose dace, mountain suckers, and reproducing populations of brook and brown trout.

Missouri River

The Flood Control Act of 1944 authorized the development of the Pick-Sloan Plan for management of the water resources of the Missouri River Basin. The Pick-Sloan Plan included the construction of four reservoirs on the Missouri River in South Dakota. The Fort Randall Dam, completed in 1956, was the first of the four reservoirs with the potential of having cold-water habitats suitable for trout stocking. A total of 600 fingerling rainbow trout were experimentally stocked in 1951. The first tailwater stocking of catchable rainbow trout occurred at the Fort Randall Dam in 1956. The Fort Randall Dam tailwater is still stocked annually with brown trout. Lake Francis Case has received stockings of chinook salmon, rainbow trout, and cutthroat trout, but they have failed to provide recreational fisheries.

Oahe Dam was completed in 1963 and by 1964 an esti-mated 300,000 brown trout fingerlings and 600,000 rainbow trout fingerlings were stocked in the tailwaters north of Pierre. Fisheries biologists soon recognized that Lake Oahe had the potential to provide excellent cold-water habitat that could support a combination of cold-water predator and prey spe-cies. Kokanee salmon and lake trout were first stocked in the early 1970s, but failed to produce a sustainable fisheries. Rainbow smelt, a prey species, were observed in Lake Oahe shortly after their introduction in Lake Sakakawea in 1971. By 1980, they were well established as the most abundant prey species in Lake Oahe. Stocked rainbow trout fingerlings, however, failed to take advantage of the abundant rainbow smelt.

It was not until the introduction of chinook salmon in Lake Sakakawea that the right predator fish was found for the abundant rainbow smelt in Lake Oahe. The first reported catches of a chinook salmon by anglers occurred in Lake Oahe tailwaters in 1980 and Lake Oahe in 1981 and a highly pop-ular salmon fisheries was available in Lake Oahe as early as 1988. Today adult chinook salmon return in the fall to the Whitlocks Bay area on Lake Oahe to an artificial ladder and spawning facility operated by the South Dakota Department of Game, Fish and Parks. Collected eggs are reared to fin-gerling smolts and yearling salmon at Cleghorn Springs and McNenny state fish hatcheries for stocking in Lake Oahe.

Two steelhead varieties of rainbow trout from Lake Michigan and one wild strain of rainbow trout have had limited success in Lake Oahe, but with recent declines in both the reservoir water level and rainbow smelt abundance, the stocking programs have been discontinued. The fate of brown trout and cutthroat trout were similar. Rainbow trout still provide a popular recreational tailwater fisheries below Oahe Dam, as anglers of all types and ages try their luck at hooking a rainbow trout and the occasional trophy when the first signs of spring arrive in South Dakota.

Other Waters

Historical stocking records indicate that every county in South Dakota with some type of deep water, be it a natural lake, a man-made reservoir, or a stream, was stocked with at least one trout species during the last 100 years. The U.S. Fish and Fisheries Commission reported stocking lake trout in Lake Kampeska in 1891 and Lake Hendricks in 1898. Steelhead trout were stocked in Big Stone Lake in 1916. There is no evidence that these introductions were successful. Early fishery managers did not recognize that these prairie lakes did not provide suitable cold-water habitat. By the 1940s, it was known that fingerling and catchable trout could provide a temporary fisheries in winterkilled lakes or new impoundments in South Dakota until permanent populations of warm-water fishes were established. This practice continues today with about 15,000 trout stocked annually in prairie ponds.

Numerous streams in eastern South Dakota, including the Prairie Coteau streams that flow easterly to the Minnesota River drainage, have been stocked since 1929. Often these stockings only provided a temporary fisheries and required annual stockings of brook, brown, and rainbow trout. One exception, Gary Creek, which flows from the easterly hills of the Prairie Coteau in Deuel County, has maintained a wild brown trout population because of water quality improvements in the drainage during the last 20 years. Today, only occasional stockings are required to maintain the brown trout fisheries in Gary Creek.

The cold-water fisheries resources in South Dakota are limited because the amount of cold-water habitat is limited. At best, any new cold-water habitats will likely result from stream restoration projects in the Black Hills region and any future changes in the abundance or distribution of cold-water fish species will largely be dependent on new introductions. For the future, sustaining a cold-water fishery in South Dakota will require specific and specialized management efforts—and quite a bit of luck.   ❖

Chapter 11

Warm-Water Fish Species

Brian G. Blackwell

" Nothing makes a fish bigger
than almost being caught. "

— Author Unknown

Warm-water fish species currently abound across South Dakota, but this was not always the case. It is believed that only six warm-water fish species persisted in South Dakota during the glacial age. Following that period, however, many warm-water fishes were able to find pathways into South Dakota waters. The most common pathway was surely the Missouri River, since about 97% of the State's waters drained into the Missouri River after the glaciers melted (Cross et al. 1986).

Two minor pathways for post-glacial fish migration into South Dakota are in northeastern South Dakota, where the Minnesota River drains 2% of South Dakota and the Red River drains about 1%. One possible pathway was the hydrographic interchange at Lake Hendricks, a border lake in Brookings County, South Dakota and Lincoln County, Minnesota. Lake Hendricks flows east into the Lac Qui Parle River and ultimately into the Minnesota River; however, during high water periods the lake may also flow west into Deer Creek that is part of the Big Sioux watershed.

Another possible pathway for fish immigration was via ancient Lake Agassiz and River Warren. Melting glaciers formed Lake Agassiz, a 700-mile by 200-mile lake that covered much of Manitoba and parts of Ontario, Saskatchewan, Minnesota, and North Dakota. The ancient River Warren flowed out of Lake Agassiz through what is now Lake Traverse and Big Stone Lake, and at that time fishes in the Minnesota and Mississippi rivers probably moved upstream through the River Warren into Lake Agassiz. Lake Agassiz disappeared leaving the Red River and its tributaries. The 18 native species found in the Red River basin today probably used the River Warren as a pathway to colonize the Red River and its tributaries in the northeastern corner of South Dakota (see Hoagstrom and Berry 2006 for details about the origin of native South Dakota fishes).

An early account of South Dakota warm-water fishes can be found in the Lewis and Clark journals. During their expedition, fishes were a source of food along with wild game; unfortunately, most of the fish accounts described in the journals occurred outside of South Dakota. Captain Lewis did make an entry in his journal on September 21, 1804, while camped in Hughes County, indicating that catfish were numerous, but smaller than those they found further south.

The first scientific records of South Dakota fishes were made as railroad routes were surveyed in the mid-1800s. Fish collected during these surveys were shipped to the Smithsonian for cataloging and storage. Information concerning South Dakota fishes was limited until the late 1800s when Evermann and Cox (1896) listed 69 fish species from South Dakota. In 1933, Churchill and Over (1933) published

Fishes of South Dakota and reported that 81 fish species had been identified in South Dakota (Box 11.1). About 25 years later, a more thorough survey by Bailey and Allum (1962) found 93 fish species present in South Dakota. The number

11.1
Dr. Edward Churchill

Edward Churchill taught for 40 years at USD, leaving his stamp on the institution and on fisheries science in South Dakota. He was born in Iowa in 1882, received his B. A. Degree at the State University of Iowa, and his Ph.D. Degree from John Hopkins University in Maryland. He spent 4 years with the U.S. Bureau of Fisheries before becoming an Assistant Professor of Biology at the University of South Dakota (USD), where he became the first Head of the Department of Zoology in 1926. He remained in that position until 1948, and then returned to classroom teaching until 1961. He wrote an interesting little book titled *Three Thousand Coyotes and I* in which he recounted his life as a USD Professor.

He supervised 31 Master's theses in zoology but was especially interested in fish. Students said, "If you want to work with Dr. Churchill, select your problem before you go in or you'll come out with a fish." Miss Louella E. Cable was his first graduate student. Her study of the food of bullheads was the first Master's degree granted in zoology at the University. His first ever survey of fishes produced data on 81 species of fish and a book titled *Fishes of South Dakota*, which he coauthored with Dr. William Over. About the survey he said, "fishermen wanted to catch more fish and hoped for some magic from science, " and he also wrote "we were glad to have had a part in helping to make the people of the state more conscious of their resources in wild game and parks."

The black and white fish figures used in this chapter were drawn by Dr. Churchill's students and taken from *Fishes of South Dakota* (Churchhill and Over 1933).

of species now in South Dakota (see Chapter 3) is higher than in past years primarily because of fish introductions.

Defining the native ranges of fish species within South Dakota has proven difficult. South Dakota is prone to natural wet and dry periods. During the wet periods, fish are able to prosper and expand their ranges; however, South Dakota waters can be harsh environments for fish during drought years. During the drought of the 1930s, only six South Dakota lakes supported warm-water sport fish. All of those lakes occurred in the northeast—Big Stone, Cochrane, Enemy Swim, Kampeska, Pickerel, and Punished Woman.

Humans have been responsible for many of the changes observed in the distribution of fish species in South Dakota. Both native and exotic species have been stocked throughout the state. Fish stocking intensified following the appointment of a fish commissioner by territorial legislation in 1883. One of the duties of the fish commissioner was to stock fish fry (including common carp) received from the U.S. Commissioner of Fisheries. The amount of stocking made it difficult to define native ranges of many species, and as early as 1892, Evermann and Cox (1896) recognized the difficulty in defining the original ranges. For example, they found it difficult to define the western limit of black and white crappies because of stocking.

Besides fish stocking, other human activities have influenced fish populations and the assemblage of fishes in many waters. Agricultural practices have had a dramatic effect on fish and their habitats. Stream sediment loads increased with the onset of tillage agriculture and sediment deposition in lakes has reduced the water volume and clarity of many lakes. In the early 1900s, diversions from rivers into lakes (e.g., Big Sioux to Lake Kampeska) were created to maintain water levels during dry periods. The biggest problem caused by these diversions was the amount of silt that was deposited from rivers into lakes, thus decreasing lake water volume. For example, Punished Woman Lake, which did not experience winterkills during the 1930s, now frequently has winterkills.

Nutrient levels have also increased in South Dakota waters because of non-point source pollution. Fertilizers and other chemicals from home lawns and farmland enter streams and lakes. Also, cabins and houses adjacent to lakes often have unnatural shorelines, which can reduce the shallow littoral zone habitat that fish use.

Construction of stock dams to water cattle created new habitat for fish in many areas previously devoid of fish fauna. Much of central and western South Dakota is now dotted with stock dams, which provide warm-water fish habitat and recreational fishing opportunities.

Construction of the mainstem dams on the Missouri River dramatically changed the Missouri River from a turbid free-flowing river to several large impoundments. Fishes that evolved in the Missouri River have declined following construction of the dams while species more adept for living in a lake environment have prospered.

Today, many warm-water fish species occur in South Dakota. These species can be grouped into families based on various common characteristics. Several of the common families and associated species are described in this chapter. Color pictures of most South Dakota warm-water fishes can be found at *http://www.sdgfp.info/Wildlife/Education/CommonFishes.htm* (accessed on 11/22/06) or in Neumann and Willis (1994).

Catfish Family (Ictaluridae)

Catfish are represented by eight species in South Dakota (Doorenbos et al. 1999). Black bullheads and channel catfish (figure) are the two most famil-iar species. Catfish and bullheads 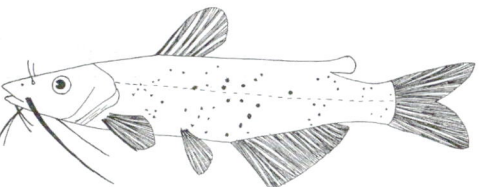 are easily recognized by their smooth, scaleless skin and bar-bels near their mouth that resemble cat whiskers. They have highly developed senses of taste and smell. Both the barbels and skin are capable of tasting, a feature that makes these spe-cies well adapted for nocturnal feeding. During reproduction catfish and bullheads build a nest and one parent remains with the eggs and young. One attribute many anglers have learned firsthand is the pain of being stuck by a catfish spine.

Bullheads

The black bullhead is the most common bullhead in South Dakota. Other bullheads include the yellow and brown bullhead. Black bullheads can be found in virtually all waters of the state and brown and yellow bullheads are found in limited numbers in eastern South Dakota. Many anglers today cringe at the thought of catching black bullheads, but that has not always been the case. Churchill and Over (1933) stated, "pound for pound for meat and food value, black bullheads are the most valuable fish of South Dakota due to the excellence of their flesh as food, their abundance, and ease of capture." Black bullheads continue to be commercially fished in many eastern South Dakota waters, and bullheads are exported or sold locally for community fish fries.

Bullheads are tolerant of poor water quality, thus enabling them to be successful in most habitats. They are opportunistic feeders eating insects, fish, and plant materials. Unfortunately, black bullheads often overpopulate and become stunted. Small fish are of little interest to anglers.

Catfish Species

Channel catfish and flathead catfish are the most common catfish in South Dakota. Blue catfish can be found in limited numbers, but they are a big river fish and their numbers declined with construction of the mainstem Missouri River dams. Channel catfish have a sharply forked tail, and their light-colored body is often marked with dark spots. Channel catfish are opportunistic feeders, so they eat a variety of organisms including insects, fish, and plant materials. Flathead catfish have a flat head with a protruding lower jaw and a tail that lacks the deep fork present in channel catfish and blue catfish. Flathead catfish are dark and often have a mottled color pattern. The adults differ from South Dakota's other catfish species because they are mostly piscivorous (fish eaters). Adult fish are often solitary and will return to a favorite resting spot during daytime hours and feed at night.

Channel catfish are found in all major river drainages in South Dakota. Flathead catfish are present in the Missouri River and its larger tributaries in southeast South Dakota. Overall, few South Dakota anglers fish for catfish, but may catch them when fishing for walleyes. Some of the angling public, however, are die-hard catfish anglers and pursue channel and flathead catfish during the summer months, probably because the flesh of catfish is considered excellent table fare.

Drum Family (Sciaenidae)

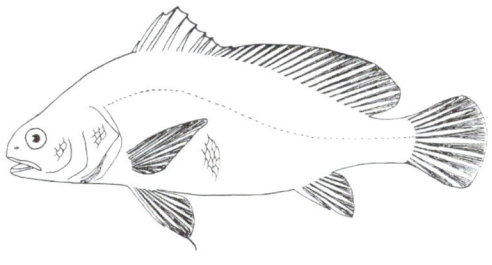

The freshwater drum (figure), also known as sheepshead, is the only member of this large family of fishes to occur in South Dakota and is the only member of the drum family that spends its entire life in fresh water. Freshwater drum are found throughout the Missouri River reservoirs and tributaries and in some lakes. Many anglers consider drum a rough fish; however, when hooked, freshwater drum are strong fighters, often battling much harder than walleyes twice their size.

The drum gets its name because of the drumming sound that males make when two special muscles move a tendon over the swim bladder. Females do not drum. Many anglers recognize drum for their "lucky stones" or ear bones (otoliths) that are unusually large, and sometimes used in jewelry making. The eggs of freshwater drum are different than those of most other fish because they float, which gives them the advantage of dispersing extensively in currents and wave action.

Gar Family (Lepisosteidae)

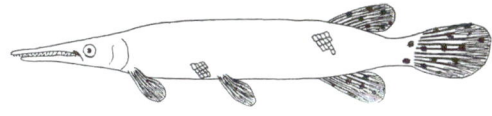

Shortnose and longnose gars (figure) are the only two members of the gar family that occur in South Dakota. Gars are primitive species with large diamond-shaped scales that form a protective armor. Gars are adapted for living in warm, slow-moving rivers and lakes and likely were more abundant in the Missouri River before the mainstem reservoirs were constructed. Their swim bladder is connected directly to the throat, allowing them to gulp air at the water surface.

The body of a gar is cylindrical and the dorsal fin is near the tail and the tail fin is rounded. The elongated jaws have many sharp teeth, but the bony jaws make them difficult to catch by hook and line. Gars are often found near the surface, making them favorite targets of archery anglers. It should be noted that gar eggs are poisonous to vertebrates and eating even a small amount will make a person sick.

Herring Family (Clupeidae)

Two members of the herring family occur in South Dakota—gizzard shad (figure) and skipjack herring. Skipjack herring were once found in Big Stone Lake and are occasionally found in the Missouri River below Lewis and Clark Dam. Gizzard shad are common in the Missouri River and its tributaries. South Dakota is on the northern range for gizzard shad. Until recently Oahe Dam acted as the northern limit on the Missouri River, but with the introduction of gizzard shad into Angostura and Shadehill reservoirs, they have moved down the Cheyenne and Grand rivers and into Lake Oahe.

In South Dakota, gizzard shad are prey for predatory fish. Walleye growth in Lake Francis Case and Lake Sharpe is influenced by the abundance of young-of-the-year gizzard shad. During winter, many gizzard shad perish because of cold-water temperatures. The survivors likely find warm-water refuge near artesian flows and survive to spawn during the following spring. Gizzard shad feed on microscopic plants and animals, and they may also eat decaying organic material.

Minnow Family (Cyprinidae)

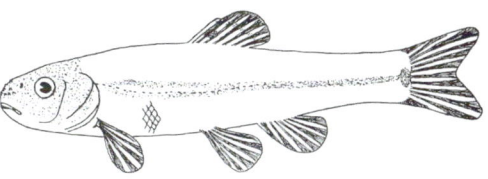

The minnow family has more fish species than any other family of fishes in South Dakota. We think of "minnows" as small fish, like the fathead minnow (figure), but members of the minnow family are some of the largest fishes in South Dakota. To describe all the minnow species would be a daunting task; thus, only a few key species will be discussed.

Common Carp

In the 1870s, R. Hessel of the U.S. Bureau of Fisheries brought common carp, an exotic species, to the United States, and following that, they were reared and distributed throughout the country. They likely arrived in South Dakota about 1885. Many anglers consider carp as rough fish and do not fish for them, but in many European countries carp are considered a game fish and are highly sought. Commercial fishers remove thousands of pounds of carp from South Dakota lakes each winter and most are shipped to markets in larger cities. Carp are opportunistic bottom feeders, but at times they will feed at the surface. They primarily eat insects and plant material.

Goldfish

Goldfish are native to Asia and Japan, but have been introduced throughout the world. In South Dakota, goldfish occur in Capitol Lake in Pierre and the Fall River at Hot Springs. Goldfish in the Fall River are thought to have originated from the Soldier's Home in Hot Springs. In South Dakota, cold-water temperatures likely limit goldfish since they need warm flows to sustain populations. It is illegal to use goldfish for bait in South Dakota waters.

Fathead Minnow

Fathead minnows, the most common minnow species in South Dakota, inhabit wetlands, lakes, and rivers. During spawning season, the male becomes dark around the head, hence fathead minnows were called blackhead minnows in the early 1900s. Anglers commonly use fathead minnows for bait. Bait dealers harvested an estimated 150,210 gallons of fathead minnows (about 2000 fish/gallon, Carlson and Berry 1990) from South Dakota waters during 2002.

Harvested minnows are sold in South Dakota and exported to other states with Minnesota and Wisconsin accounting for two-thirds of all exported minnows. This has changed from the early 1900s when fathead minnows were thought to have no economic importance because they were not good for bait since they were too dark in color and too sluggish to attract predators. In 2002, the sale of fathead minnows in South Dakota had an estimated economic value in excess of $2.2 million!

Golden Shiner

Golden shiners, native to eastern South Dakota, are now common across the state in lakes and ponds that are clear and vegetated. They feed on algae, vascular plants, and aquatic invertebrates. Golden shiners are a popular bait fish and commercial propagation of golden shiners is common in South Dakota. In 2002, bait dealers harvested 589 gallons of golden shiners from South Dakota waters. When using golden shiners for bait, anglers need to be sure that they have golden shiners and not European rudd (*Scardinius erythrophthalmus*) for the two species are similar in appearance.

The rudd is an exotic species and their introduction into South Dakota waters has the potential to disrupt native ecosystems. Rudd tend to have red median fins (dorsal, tail, anal fins) whereas median fins of golden shiners are silvery. Both species have a unique ventral keel behind the pelvic fins. The keel of the golden shiner does not have scales whereas the keel of the rudd is crossed by scales. Rudd have been found in 11 South Dakota counties, likely the result of anglers using them for bait when they thought they had purchased golden shiners.

Asian Carps

Asian carps are the newest fish species to call South Dakota home. The new Asian carp species in South Dakota include grass carp, bighead carp, silver carp, and black carp. All four of these species were originally introduced in the United States for use in aquaculture facilities in southern states. All four species have been found in the Missouri River below Gavins Point Dam at Yankton. The Dam has blocked the upstream movement of these carp, but anglers may move them upstream if Asian carp are accidentally mixed with other fish in bait buckets. There also is potential for these species to move up the Big Sioux and James rivers, ultimately entering lakes that occur within these drainages. Grass carp have been purposely stocked in a few small waters in South Dakota for control of aquatic vegetation. The bighead carp sometimes get attention because of their habit of jumping out of the water when a motor boat passes.

All of these carp species are capable of reaching sizes in excess of 30 pounds. Because they are not native to South Dakota, they have the potential to severely disrupt freshwater communities and habitats. Grass carp eat vegetation that is used by many native fishes as cover. Eliminating aquatic vegetation can disrupt and dramatically change a fish community. Both silver and bighead carp feed on plankton making them direct competitors with paddlefish, buffalo fish, gizzard shad, and all larval and juvenile fishes. Black carp feed on snails and clams.

Mooneye Family (Hiodontidae)

The mooneye family has only two members, the goldeye (figure) and mooneye, and both occur in North America and

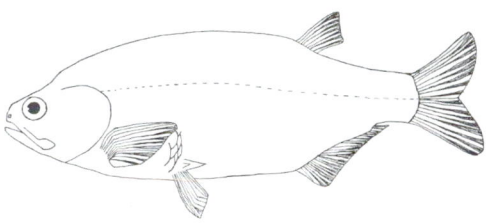

both occur in South Dakota. Records indicate that while mooneyes have been collected in South Dakota, goldeyes are more common. Goldeyes, commonly called skipjacks, are frequently found in Missouri River reservoirs and larger tributaries. Goldeyes are silver colored with a compressed body and a large gold-colored eye. Teeth are present on the jaws, roof of the mouth, and on the tongue. Young goldeyes are an important food for predatory fishes. They are an aggressive fish, often caught by anglers fishing for walleyes. Hooked goldeyes will often jump from the water, but tire soon after their initial fight. Because their flesh has numerous bones, they generally are not kept for consumption. In other areas of North America, goldeyes are an important commercial fish, especially when sold as a smoked fish product. The Winnipeg Goldeyes may be the only professional sports team (baseball) named after a fish.

Paddlefish Family (Polyodontidae)

Paddlefish, or spoonbill catfish, are a primitive river species. The paddlefish (figure) gets its name because of the long, flat bill or rostrum protruding from the snout. The mouth is large and the gills are covered with long filaments used to strain zooplankton from the water. Although they eat small zooplankton, they are a large fish with some weighing more than 100 pounds. Another unique feature of paddlefish is that their skeleton is composed of cartilage and not bone.

Paddlefish were once abundant in the Missouri River, but overfishing and the effects of river alteration following construction of the mainstem dams have led to a decrease in numbers. South Dakota currently has a limited season for paddlefish below Gavins Point Dam. In spring the South Dakota Department of Game, Fish and Parks and the U.S. Fish and Wildlife Service work together to collect and spawn paddlefish. Juvenile paddlefish are raised at Gavins Point National Fish Hatchery for stocking into the Missouri River reservoirs and other sites across North America. Tagged South Dakota paddlefish have been caught throughout the Mississippi River system (e.g., Tennessee River).

Perch Family (Percidae)

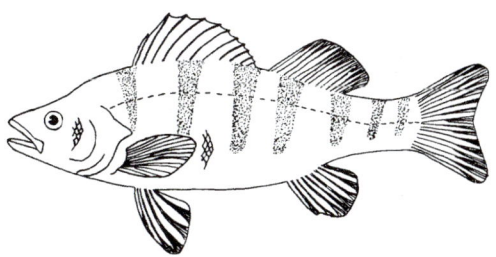

Fishes in the perch family, from the large walleye to the tiny darter, tend to have slender, cylinder-shaped bodies. The dorsal fin is separated into two parts—the front part having stiff spines and the latter part having flexible rays. All members of the family are carnivorous. Walleyes, saugers, and saugeyes (hybrid between walleye and sauger) all have large, canine-like teeth and eat fish, whereas darters eat invertebrates.

Members of the perch family are the most sought after species by South Dakota anglers because of their sporting qualities and their flesh, which is considered excellent table fare. In South Dakota, the walleye is king among most anglers, but yellow perch (figure) are a favorite of winter anglers. Fishing for members of the perch family annually brings in millions of dollars to the South Dakota economy.

Most people are aware of the larger members of the perch family, including the walleye, sauger, saugeye, and yellow perch, but few are aware of the small darters. Darters (e.g., Iowa darter, Johnny darter) lack or have a reduced swim bladder and thus sink to the bottom when they stop swimming. Darters usually live among bottom rocks of streams and rivers, and move along the bottom in quick dashes—hence the name "darter." Males build a nest beneath large rocks by cleaning away silt, and the female attaches adhesive eggs to the "ceiling" of the nest. The male guards and cleans the nest until the eggs hatch.

Yellow Perch

Yellow perch are probably native to eastern South Dakota from the James River eastward. Today they can be found across South Dakota, having been stocked in most waters.

Yellow perch are popular with ice anglers on many eastern South Dakota lakes. Anglers congregate on lakes where the "hot perch bite" is occurring, often establishing what appears to be small cities on the ice.

Yellow perch eat invertebrates and small fish, and generally as they increase in size, they eat more fish. They spawn during the early spring. Eggs are released in large skeins, not individually as with most other fish. Yellow perch spawning success tends to be greatest when spring runoff is high and the spring is warm and not windy. Young yellow perch are an important walleye food source in many lakes.

Walleye

Walleyes are thought to be native to South Dakota with early reports indicating that they were plentiful in many eastern South Dakota lakes and rivers. They were present in the Missouri River prior to the creation of the reservoirs, but were not nearly as abundant as they are now. Walleyes have been extensively stocked in waters across South Dakota.

Walleyes are aptly named because of their large opaque eyes. Each eye contains a structure (tapetum lucidum) that enables them to see during low light periods and also causes them to seek cover during high light intensities. Because of this, walleye fishing is generally better during dawn and dusk on clear water lakes. They have a large mouth containing many teeth for capturing and consuming smaller fish and sometimes invertebrates.

Because of its popularity with anglers, the walleye was named the state fish of South Dakota. Each spring the Department of Game, Fish and Parks collects and hatches in excess of 50 million walleye eggs to provide fry and fingerlings for stocking South Dakota waters. In the early 1900s walleyes already were valued by sportsmen, and tourists were attracted to waters where they were present. Anglers annually spend millions of dollars pursuing walleyes in South Dakota waters.

Sauger

Saugers typically are only found in the Missouri River reservoirs and larger tributaries. Saugers are more slender than walleyes and lack the white coloration on their lower tail fin. They are generally considered a river fish and likely were more abundant prior to construction of the mainstem dams on the Missouri River. Saugers tend to be more tolerant of turbid waters than walleyes.

Saugeye

Saugeyes occur naturally as hybrids in the Missouri River reservoirs and hatchery-produced hybrids have been stocked in various waters. Saugeyes do well in reservoirs that are turbid and have high water flow. Even though they are hybrids, all saugeyes are fertile. Distinguishing saugeyes from walleyes or saugers can be difficult and often cannot be done visually.

Johnny darter

Johnny darters occur throughout eastern South Dakota in lakes, streams, and rivers. Johnny darters are a small fish averaging less than 2.5 inches in length. Their bodies are long and cylindrically shaped with w- and x-shaped marks on the sides. Because they lack a developed swim bladder, they are almost always found on the bottom. Johnny darters can tolerate higher turbidity levels than other darters and can also tolerate some types of water pollution.

Logperch

Logperch are found along the eastern edge of South Dakota within the watersheds of the Big Sioux, Red, and Minnesota rivers. Clear water over sand and gravel is preferred by this species; areas of heavy silt are avoided. They are the largest darter in South Dakota, averaging 4-6 inches in length. Their body is similar to other darters, being long and cylindrical. They have a conical snout at the front of their heads above their mouths and dark vertical stripes on their sides.

Pike Family (Esocidae)

Within the pike family, the northern pike (figure) is the most common species found in South Dakota. Muskellunge and tiger muskellunge (a hybrid cross between female muskellunge and male northern pike) have been introduced into the state. The grass pickerel, a small member of the pike family, was found in Lewis and Clark reservoir in 1995. Grass pickerel are native to the Niobrara River, which enters the Missouri River from Nebraska at the upper end of Lewis and Clark reservoir.

Members of the pike family are easily recognized by their elongated body, a duck-bill snout, and a large mouth containing many sharp teeth on the jaws and on the tongue and roof of the mouth. Wise anglers use pliers to remove hooks from "the water wolf" or northern pike.

Northern pike

Northern pike are originally native to the Red and Minnesota river drainages in northeastern South Dakota and the Big Sioux and James river drainages. Northern pike have been stocked across South Dakota. Within the Black Hills, illegal stockings of northern pike by anglers have occurred in lakes managed for trout. These illegal introductions have caused changes in how some lakes are managed.

Northern pike spawn early in spring, often as the ice is going off the lakes. Their young grow fast in South Dakota waters, often exceeding 16 inches in their first year. Because northern pike can tolerate poor oxygen levels and have fast growth, they are often stocked in waters of marginal quality in the state.

Northern pike primarily feed on other fishes, but will also consume invertebrates and may eat small mammals and birds. Many anglers, becoming frustrated when northern pike refuse to bite in mid-summer, have suggested that pike lose their teeth. Research has shown that northern pike teeth are lost and replaced throughout the year; therefore, there is no seasonal tooth loss that keeps them from biting. Northern pike flesh is excellent table fare, but the many bones cause some anglers to shy away from eating them. When pickled, the flesh often tastes similar to pickled herring.

Muskellunge

The muskellunge is the largest member of the pike family. Food habits and spawning of muskellunge are similar to northern pike except that muskellunge spawn slightly later. Muskellunge are present in Amsden Dam and Lynn Lake in Day County, where populations are maintained through stocking. Amsden Dam has had a muskellunge population since 1975 and at Lynn Lake muskellunge were introduced in 2001. The goal of the South Dakota muskellunge program is to maintain a low-density muskellunge population (one 30-inch fish per five acres) that will allow an angler to catch a 36-inch plus muskellunge within about 20-40 hours of fishing.

Tiger muskellunge

Tiger muskellunge can be found in limited numbers in the Missouri River reservoirs and Orman and Angostura reservoirs in western South Dakota. They may occur naturally where there are both northern pike and muskellunge. Tiger muskellunge have characteristics of both parents and generally grow faster than northern pike or muskellunge. Because they are hybrids, tiger muskellunge are thought to be sterile; however, some females may be fertile.

Sturgeon Family (Acipenseridae)

Sturgeons are primitive fishes with armor-like scales that cover their body. Their head is pointed  and their mouth is on the underside. In South Dakota, the pallid sturgeon (figure) and shovelnose sturgeon are found in the Missouri River and lower reaches of its tributaries. Historically, sturgeons were an important commercial fish species in the Missouri River; however, because of their slow growth and late maturity, they are vulnerable to over-harvest. Construction of the Missouri River mainstem reservoirs altered riverine habitats and contributed to the decline of these species. Illegal harvest of sturgeons to make caviar from their eggs also has been a concern. Males reach sexual maturity at ages 5–7 whereas females began egg development at ages 9–12 (Keenlyne and Jenkins 1993).

The pallid sturgeon was listed as endangered on September 6, 1990 because catch records had declined greatly. For example, in the upper Missouri River, pallid sturgeon catches declined from about 50 fish/year in the 1960s to 6 fish/year in the 1980s. The Endangered Species Act requires the U. S. Fish and Wildlife Service to write a recovery plan for each endangered species and make other efforts to recover species (NAS 1995, Norris 2004). There are 15 "actions needed" listed in the pallid sturgeon recovery plan (Dryer and Sandovol 1993). Chief among the actions needed is the need to "restore habitats and functions of the Missouri and Mississippi river ecosystems while minimizing impacts on other uses of the rivers."

Other listed actions are: establish broodstock in a hatchery, increase public awareness, obtain more information about the fish, and stock young pallid sturgeon to augment existing populations. There are considerable efforts to increase their numbers, including research on propagation at Gavins Point National Fish Hatchery near Yankton. After much difficulty in learning how to culture pallid sturgeons, the first young fish were stocked near the confluence of the Yellowstone and Missouri rivers on August 11, 1998. Since then, more pallid sturgeons are raised than can be stocked so fish are available for genetics studies, laboratory tests, and other research, much of which has gone on at South Dakota State University (e.g., studies on diet, bioenergetics, and vulnerability to predation).

Because pallid sturgeons are endangered, anglers are not allowed to harvest them, nor can they harvest shovelnose sturgeon because they are similar in appearance to pallid sturgeon when small (less than 2 feet long). The pallid sturgeon is a good example of why we should save endangered species. If this native large sturgeon species were lost, the future possibility of a trophy fishery for wild fish or a commercial sturgeon farming industry would be lost also.

Sucker Family (Catostomidae)

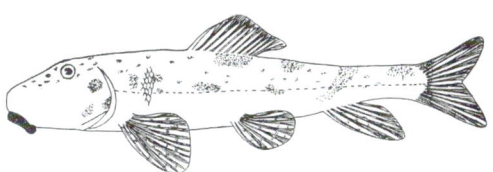

The sucker family has many fish species that are primarily restricted to North America. Members of this family, such as the northern hog sucker (figure), are characterized by fleshy sucker lips around a sub-terminal toothless mouth (opens downward), soft fin rays, a single continuous dorsal fin, and forked tails. Many of the species closely resemble members of the minnow family; some are deep bodied, similar to the common carp, while others are more cylindrically shaped, similar to minnows. As the family name implies, suckers feed primarily by sucking materials from bottom substrates or straining organisms from the water column. Because of their feeding style, few suckers are caught by hook and line anglers. Some are taken by spearing or with archery equipment. Juveniles of many of the suckers serve as food for predatory fish.

Blue sucker

The blue sucker is a rare fish in South Dakota occurring only in the Missouri River in limited numbers. The blue sucker is a fish of large rivers, doing best in areas having fast deep channels. Construction of the mainstem dams on the Missouri River likely reduced blue suckers. Their body is streamlined with the head tapering to a fleshy snout, an adaptation for living in areas of fast currents. Colors range from slate-blue to olive. Commercial fishermen once prized blue sucker flesh for its palatability; however, because its numbers have declined over most of its range, harvest by commercial fishermen has greatly declined.

White sucker

White suckers are common across South Dakota. They have a streamlined body with silvery to brassy coloration; males often become darker during the breeding season. Scales of white suckers become larger from the head to the tail. White suckers are commonly sold as "chubs" by bait shops in South Dakota. Bait dealers sold approximately 2,700 gallons of white suckers in 2002. In most waters the young are prey for predatory fishes such as walleyes and northern pike.

Bigmouth buffalo

Bigmouth buffalo fish are found in the Missouri River reservoirs and eastern South Dakota, where the species does well in many of the natural lakes. Large numbers of bigmouth buffalo fish, harvested each winter by commercial fishermen, are exported to markets outside of South Dakota.

The bigmouth buffalo is the largest member of the sucker family in South Dakota. Some 50+ pounders swim in South Dakota waters. The bigmouth buffalo is similar in appearance to the common carp, but it has a terminal mouth that lacks barbels and it does not have heavy dorsal spines like those on carp. The species often is found in large schools suspended at midwater or near the lake bottom. Predatory fishes consume young-of-the-year bigmouth buffalo, but because of their rapid growth, the young quickly become too large for most fish to eat.

Smallmouth buffalo

Smallmouth buffalo fish are found in the Missouri, Big Sioux, and James rivers. Like bigmouth buffalo fish, the smallmouths are similar in appearance to the common carp but without barbels or hard spines. Smallmouth buffalo fish differ from bigmouths in that their mouths are sub-terminal and they appear to prefer clearer water.

River carpsucker

River carpsuckers can be found all across South Dakota, except in the far northeastern corner. The river carpsucker is similar in appearance to the smallmouth buffalo except that their upper jaw extends beyond the front of the eye. When their mouth is closed the lower lip has a pronounced nipple-like projection. This species tends to congregate in large schools and prefers turbid waters. Their flesh is soft and rather tasteless.

Sunfish Family (Centrarchidae)

The sunfish family has over 30 species and all are native to North America. A laterally compressed deep-body form characterizes members of the sunfish family. The dorsal fin has hard spines in the front and soft rays in the back. The pelvic fins are located on the front portion of the fish almost directly below the pectoral fins. They find their food by sight and ambush it. Sunfish spawn during late spring to early summer when a nest is built and guarded by the male.

In South Dakota, anglers commonly fish for largemouth bass, smallmouth bass, bluegills, and crappies. Although sunfish do not attract as many anglers as the perch family, they are an important component of South Dakota fisheries. In western South Dakota, where small impoundments dominate the landscape, members of the sunfish family are the primary sport fish. The orangespotted sunfish is interesting because of its association with the endangered Topeka shiner (*Notropis topeka*). Like many sunfishes, the orangespotted sunfish sweeps gravel clean to make a circular nest, which the Topeka shiner also uses for spawning (Shearer 2003).

Orangespotted sunfish

Orangespotted sunfish were originally an eastern South Dakota fish, but likely were inadvertently stocked with other fish and now occur across the state. They are a small colorful sunfish that rarely exceeds three inches in length. Orangespotted sunfish occur in some lakes, but most often are found in slow-moving portions of rivers and streams. They feed on insects and other invertebrates.

Largemouth bass

Largemouth bass are native to the Red and Minnesota river drainages, but now can be found throughout South Dakota. They are aptly named because of their large mouth. To avoid confusion with smallmouth bass, anglers should observe the jaw. If the jaw extends past the eye it is a largemouth bass; if it does not extend beyond the eye, it is probably a smallmouth bass. Largemouth bass prefer clear standing water (lakes and ponds) and generally do best when there is submerged vegetation. They eat fish, invertebrates, and small land animals that may find themselves in the water.

Smallmouth bass

Smallmouth bass are native to the Minnesota River drainage in northeastern South Dakota, and have been extensively stocked throughout the state. In general, smallmouth bass prefer cooler water than largemouth and most often are associated with rocky habitat in clear water. They eat insects, invertebrates, small fish, and crayfish when available.

Bluegill

Bluegills originally were only present in the Minnesota River drainage in northeastern South Dakota, but have subsequently been introduced across South Dakota. When people talk about catching sunfish, they are generally talking about bluegills. Bluegills are a popular panfish species with many anglers because they are easily caught. They often occur in schools, increasing the chance of catching several at a time. Bluegills do best in lakes and ponds. In eastern South Dakota glacial lakes, they thrive in lakes that have areas protected from the wind with submerged vegetation. They are an important member of fish community in small lakes and ponds and most often co-occur with largemouth bass. When there aren't enough predators, bluegills may overpopulate and stunt at small sizes.

Green sunfish

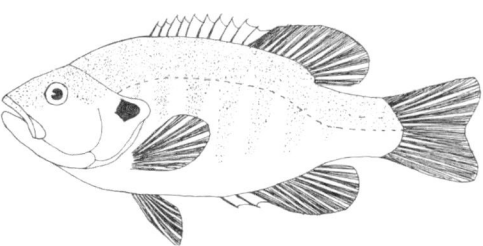

Green sunfish (figure) are found across South Dakota, except the northwest and northeast corners. They may have been inadvertently stocked with bluegill into some waters. Green sunfish can be distinguished from bluegills by their larger mouth size. Green sunfish are tolerant of poor water quality, including high turbidity, low oxygen, and high silt. Since green sunfish and bluegill spawning may occur at the same time and in a similar way, hybrids may be produced where the two both occur.

Pumpkinseed

Pumpkinseeds can be found in eastern and south central South Dakota. They were likely native to the Minnesota River drainage, but have been stocked in other locations. Where they are present, their numbers are not high and they often occur with bluegills. Pumpkinseeds are similar in appearance to bluegill and in waters where both occur they are often mistaken as bluegill. The easiest way to tell the two apart is that the gill cover lobe on a pumpkinseed is orange or reddish at the tip, while the bluegill gill lobe is entirely black. Pumpkinseeds feed on insects, but also eat snails and other mollusks.

Rock bass

Rock bass, found in northeastern South Dakota, were introduced into some Black Hills waters. As their name implies, they are often associated with rocky habitat. Rock bass have a large mouth allowing them to eat insects and small fish. Rock bass readily bite and often are a favorite of dock anglers because they are easily caught.

Black and White Crappie

Both black and white crappies are found in South Dakota. It is difficult to know if black crappie are native to South Dakota because as early as the late 1800s both species had been extensively stocked. An early record exists of white crappies being collected in Big Stone Lake. Crappies are popular sport fish with South Dakota anglers. Black crappies are often found in lakes having clear water, whereas white crappies are more likely to be found in more turbid waters. In South Dakota, black crappies are more abundant than white crappies. Black crappies have a speckled color pattern on their sides, whereas white crappies have dark vertical bars.

Both species feed on zooplankton, insects, and small fish. Crappies can be used as a panfish species with largemouth bass when managing small lakes and ponds. In some waters where predators are not sufficient, however, crappie populations may become overabundant and few fish will reach a large enough size to interest anglers.

Temperate Bass Family (Moronidae)

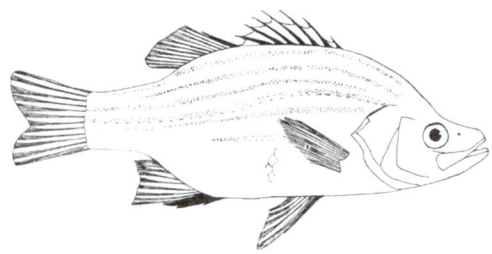

White bass (figure), commonly called silver bass, are the only member of the temperate bass family to occur in South Dakota. White bass, which originally occurred in the Minnesota and Big Sioux river drainages, have been introduced into many lakes across South Dakota. White bass are a schooling fish and generally occupy open water. Because of this, they tend to do best in large lakes and reservoirs.

In the spring white bass schools congregate in flowing water, often moving upstream before spawning. During years of high reproductive success, juvenile white bass become important food for walleyes where both species occur. White bass are primarily fish eaters. In the Missouri River reservoirs and southern North America, gizzard shad are important in their diet; however, in eastern South Dakota, where gizzard shad do not often occur, crayfish are a common source of food. Because white bass are carnivorous, they are readily caught by anglers, and they are good fighters when hooked.

Summary

Warm-water fishes currently abound in all types of wetlands throughout South Dakota; however, our waters can be a difficult place for fish to live because of drought periods. Many of our native fish species gained access to South Dakota following the retreat of the glaciers through one of three pathways—the Missouri River, Minnesota River, or Red River. Fish stocking and the invasion of exotics has increased the number of warm-water fish species present in the state today. Even though some cold-water fish species are highly sought after by anglers, the abundance and variety of warm-water fish species is, and will continue to be the primary fishery resource in the state of South Dakota for both anglers and commercial enterprises. ❖

Chapter 12

Threatened and Endangered Species

Douglas C. Backlund and Jeffrey S. Shearer

" Endangered and threatened species are of aesthetic, ecological, historical, recreational and scientific value to the nation and its people. **"**

— Endangered Species Act

Since the settlement of South Dakota, many changes have occurred to aquatic habitats statewide that have had profound effects on several fish species. Some changes have been beneficial to people, such as the construction of small dams that provide fishing opportunities and water for livestock, without being detrimental to native fish. Other changes, such as damming the Missouri River or polluting of streams and lakes with sediment and toxins, have been detrimental to native fish. The merits of such changes can be debated many ways, but the effects on native fish are documented and provide the best information on which to base current and future management decisions.

Biologically, the distribution of rare native fish species within a state like South Dakota can be defined in three ways: endemic species, disjunct species, and peripheral species. Endemic means the native fish species are confined to a specific region, range, or habitat, such as a single river system or a single mountain range. Endemics may or may not be rare. Endemics with a small range or occurring in rare or threatened habitats should be considered rare or worthy of special concern. Some endemic species may be widespread, but declining in overall distribution due to habitat loss. Some indigenous fish species of South Dakota that were formerly common and widespread, but are now rare, may now be categorized as endemics.

Disjunct species are populations of a species that may be common elsewhere, but in their disjunct habitat may be relics of a former larger distribution and therefore isolated from the main population. Such populations may be genetically unique if they have been isolated for a sufficient period of time or have been subjected to different selection processes. Over time if new species occur, disjunct populations can become endemic species. The Black Hills, due to a cooler climate and higher precipitation, harbors many disjunct species of plants and animals, including several fish species.

Peripheral species are those that are rare within the state due to their distribution pattern. South Dakota may be on the edge of their natural range. A peripheral species may be listed as a species of concern in South Dakota because it is scarce, even though it is common in the core area of its range. Some peripheral species, however, are rare or threatened throughout their range.

Rarity of species on a global scale is based on the status of the entire global population. If the entire global population of a fish species occurs in South Dakota and the species is common here, it could still be considered rare, threatened, or endangered if the habitat or the species is threatened. On a global scale there are no peripheral species. Species may have a large range and be very common, be rare over a large distribution, be rare with a small distribution, or be abundant but with a small distribution. There are many permutations of these definitions. Introduced and exotic species further complicate the definitions; but for the purposes of this chapter, we are concerned only with native species that occurred here before European settlement.

The evolution of *endemic* fish species requires long-term and relatively stable habitat, such as large river systems like the Missouri River; geologically old, large, deep lakes such as Lake Baikal in Russia; streams in isolated mountain ranges; or streams and lakes fed by perennial groundwater. Endemic species may evolve in such circumstances, especially if strong selection pressures are present. Large river systems may support several endemic species, such as the pallid sturgeon and the sicklefin chub, which occur only in the Missouri River drainage and nowhere else in the world.

Isolated mountain ranges may harbor endemic species—the Black Hills has several native species that are isolated and disjunct. Over time, if the environment remains stable and these populations remain viable, a new species could arise that is fit for the special conditions in the Black Hills. The lake chub and mountain sucker are two Black Hills native fish species that are disjunct from populations further west. Bailey and Allum (1962) speculated that the stock of the longnose dace in the Black Hills, based on size of its swim bladder, had its origins in the headwaters of the Missouri River. At one time in geologic history there were water connections that connected the Black Hills to these western populations, but these waterways have been lost due to stream capture events.

The Black Hills are now drained entirely to the east by the Cheyenne River, which flows to the Missouri River in central South Dakota. The geological process of stream capture is readily seen at the point where the Belle Fourche River (actually the northern fork of the Cheyenne River that encircles the northern Black Hills) changes sharply from a northeast flow to a southeast flow near Colony, Wyoming. This is the point where the rapidly eroding Belle Fourche River captured the headwaters of the Little Missouri River, perhaps allowing the fish fauna of the Little Missouri River to enter the Black Hills Region.

Perennial spring-fed streams of the sand hills country in south-central South Dakota are another area that provides habitat for several glacial relict fish species that were isolated there since the last ice sheet retreated more than 10,000 years ago. The northern redbelly dace and the pearl dace are two of these species.

Determination of "rarity" begins with the compilation of historic information. Unfortunately, early records of fish distribution in South Dakota are sparse and sometimes poorly documented, and few specimens exist in museums. One of the best sources of information is the report of Evermann and Cox (1896). In 1892, they began an investigation of Nebraska, South Dakota, North Dakota, and Wyoming to determine the best locations for two federal fish hatcheries. During the course of this project, they inventoried the fish species in many streams and lakes and compiled the available ichthyological reports for the region up to that date.

The next publication of importance was at one time the comprehensive list of South Dakota fishes by Bailey and Allum (1962). This publication provides detailed information on the distribution and status of fish species in South Dakota, on zoogeography, hybridization and on the scientific name changes for each species up to 1962. More recent fish surveys using a variety of fishing gears (Figures 12.1 and 12.2) at 600+ sites have been more comprehensive than previous surveys. Data from the recent surveys has been published (Hoagstrom and Berry 2007, also see Chapter 3 in this book).

In 1981, the South Dakota Natural Heritage Program was established within the South Dakota Department of Game, Fish and Parks. Natural heritage programs form a nationwide network under **NatureServe**, (*www.natureserve.org*, accessed on 12/12/06) a nonprofit conservation organization. All Natural Heritage programs use compatible databases and are charged with developing a list of rare, threatened, or endangered plants and animals for each jurisdiction, which they then monitor. Heritage programs develop the lists of rare

Figure 12.1 The Natural Heritage database uses information gained by stream surveyors using seines in low-velocity streams with little in-stream structure. (photo by C. Berry)

Figure 12.2 Electrofishing with backpack electro-shockers is done in streams with snags and boulders. Electrofishing is non-lethal and most fish are identified and returned to the stream. (photo by C. Berry)

Table 12.1 Rare, Threatened[1] and Endangered[2] fish in South Dakota (from South Dakota Natural Heritage Database, South Dakota Department of Game, Fish and Parks). Species, ranks, and status may change frequently. **LE = Listed endangered LT = Listed threatened SE = State Endangered ST = State Threatened.**

SCIENTIFIC NAME	COMMON NAME	FEDERAL STATUS	STATE STATUS	GLOBAL[3] RANK	STATE[3] RANK	DISTRIBUTION STATUS
Ichthyomyzon unicuspis	silver lamprey			G5	SA	P
Acipenser fulvescens	lake sturgeon			G3	SA	P
Scaphirhynchus albus	pallid sturgeon	LE	SE	G1/2	S1	E
Lepisosteus osseus	longnose gar			G5	S3	P
Anguilla rostrata	American eel			G5	S3?	P
Alosa Chrysochloris	skipjack herring			G5	S3	P
Hiodon tergisus	mooneye			G5	SU	P
Couesius plumbeus	lake chub			G5	S1	D
Nocomis biguttatus	hornyhead chub			G5	S3	P
Notropis blennius	river shiner			G5	S2?	P
Notropis heterolepis	blacknose shiner		SE	G5	S1	D
Notropis percobromus	Carmine shiner			G5	S2	P
Notropis shumardi	silverband shiner			G5	SX	P
Notropis topeka	Topeka shiner	LE		G3	S3	E
Phenacobius mirabilis	suckermouth minnow			G5	SH	P
Phoxinus eos	N. redbelly dace		ST	G5	S2	D
Phoxinus erythrogaster	S. redbelly dace			G5	S1	P
Phoxinus neogaeus	finescale dace		SE	G5	S1	D
Macrhybopsis gelida	sturgeon chub		ST	G2	S2	E
Macrhybopsis meeki	sicklefin chub		SE	G3	S1	E

Table 12.1 continued.

SCIENTIFIC NAME	COMMON NAME	FEDERAL STATUS	STATE STATUS	GLOBAL[3] RANK	STATE[3] RANK	DISTRIBUTION STATUS
Macrhybopsis storeriana	silver chub			G5	S2	P
Margariscus margarita	pearl dace		ST	G5	S2	D
Carpiodes cyprinus	quillback			G5	S3	P
Catostomus catostomus	longnose sucker		ST	G5	S1	D
Catostomus platyrhynchus	mountain sucker			G5	S3	D
Cycleptus elongatus	blue sucker			G3/4	S3	IE
Hypentelium nigricans	N. hog sucker			G5	SH	P
Ictiobus niger	black buffalo			G5	SU	P
Moxostoma erythrurum	golden redhorse			G5	SH	P
Fundulus diaphanus	banded killfish		SE	G5	S1	P
Fundulus sciadicus	plains topminnow			G4	S3	PE
Percina caprodes	logperch			G5	S3	P
Percina maculata	blackside darter			G5	S2	P
Percina phoxocephala	slenderhead darter			G5	SX	P

[1]A threatened species is a species likely to become endangered in the foreseeable future.

[2]An endangered species is a species in danger of extinction throughout all or a significant portion of its range (applied rangewide for federal status and statewide for state status).

[3]Ranking of G1 and S1= critically imperiled, extreme rarity (five or fewer occurrences or very few remaining individuals or acres) or especially vulnerable to extinction; G2 and S2 = imperiled, rare (6-20 occurrences or few remaining individuals or acres) or very vulnerable to extinction; G3 and S3 = rare (21-100 occurrences) over broad range, or found locally (even abundantly) in a restricted range, or vulnerable to extinction throughout its range; G4 and S4 = apparently secure, though it may be quite rare in parts of its range, especially at the periphery; a long-term concern; G5 and S5 = secure, though it may be quite rare in parts of its range, especially at the periphery; GU and SU = possibly in peril, but status uncertain; more information needed; GH and SH = historically known, may be rediscovered; GX and SX = believed extinct, historical records only. ? = rank is inexact (e.g. G2?); SA = accidental or casual in the state

[4]Distributional status is E = Endemic; D = Disjunct; I = Indigenous; P = Peripheral.

species by consulting with biologists, compiling all available literature, examining museum specimens, and conducting independent field surveys.

Currently, 34 of the fish species known to occur in South Dakota are on the South Dakota Natural Heritage Program list of monitored species (Table 12.1). Species are assigned state and global rarity rankings (Table 12.1). Two South Dakota species are listed under the federal Endangered Species Act of 1973. Nine species are listed under the South Dakota State Endangered Species Law (SDCL 34A-8). The list is dynamic and changes as new information becomes available. For example central mudminnow and trout perch were removed from the list in 2006 because surveys showed that they were more common than thought.

Central mudminnow

Central mudminnows, a common fish over their large range, occurs throughout the Great Lakes states and south into Missouri and southern Illinois. In South Dakota, this species occurs in Owen's Creek in Day County and possibly in other areas as well. Its population in eastern South Dakota was once thought to be an isolated and possibly disjunct population, but more recent information indicates this species is common across Minnesota. The 2–4 inch long, long-lived (7–9 years) mudminnows have unusual traits. They gulp air from the surface, feed under ice in winter, eat small fish, and spawn in the spring on flooded, streamside vegetation (Pflieger 1997). There have been no studies of the central mudminnow in South Dakota.

Trout-perch

The trout-perch has spiney-rayed fins like a perch and an adipose fin like a trout, thus it combines characteristics of both soft-rayed and spiny-rayed fishes. Adults grow to 5 inches long. Only two species are known, and one of them resides in South Dakota. The range of the trout-perch extends throughout most of Canada, the Great Lakes Region, upper Mississippi River basin, and eastern portion of the Missouri

River basin. In South Dakota the trout-perch occurs in the Big Sioux River and its tributaries, including lakes Kampeska and Pelican. These populations are peripheral to populations in Minnesota and Iowa where the species is more common. However it is not common or widely distributed, being found at only 3 of 33 sites sampled recently in the Big Sioux River (Hayer et al. 2006).

Species Accounts for State and Federally Listed Fish

The following species accounts discuss the status of each threatened or endangered fish species in South Dakota.

Pallid sturgeon

The pallid sturgeon is endemic to the Missouri River and the Mississippi River below the confluence with the Missouri River. It does not occur in tributaries or in the Mississippi River above the confluence with the Missouri River. In South Dakota, the pallid sturgeon was once common in the Missouri River, but large impoundments have destroyed most of its habitat and blocked the only migration route. Captive breeding and stocking may be the only conservation tool left to prevent extinction of the pallid sturgeon.

Blacknose shiner

The blacknose shiner is found from Maine and eastern Canada west throughout the Great Lakes Region (Backlund 1995). A disjunct population occurs in the Sand Hills Region of Nebraska and extreme southern South Dakota. In eastern South Dakota, the blacknose shiner historically was found in Prairie Creek (a James River tributary) and likely other similar East River streams. Today the blacknose shiner in South Dakota is found exclusively in a few groundwater-dominated streams in Todd and Tripp counties.

Topeka shiner

Topeka shiners (Figure 12.3) once ranged from Kansas and central Missouri north through Iowa and eastern Nebraska

Figure 12.3 Endangered Topeka shiners are more common in South Dakota than in the other five Great Plains states in its range. During spawning season, males develop bright red fins. (photo by C. Berry)

to southwestern Minnesota and eastern South Dakota. No other species of fish shares this historic range. Today the Topeka shiner is found in many tributaries of the James, Vermillion, and Big Sioux river basins in South Dakota. Outside of South Dakota and Minnesota, this species has declined to a few isolated populations. Stream channelization, impoundment of tributaries, and introduction of nonnative piscivorous fish are considered the primary reasons for the decline of Topeka shiners outside of South Dakota and Minnesota. The states of Kansas and Missouri currently have active propagation programs in place to reintroduce hatchery-reared Topeka shiners to historic locations.

Fish surveys and stream inventories in South Dakota since 1990 reveal that Topeka shiners are more abundant in the state than expected (Wall et al. 2004). In fact, the Topeka shiner story in South Dakota is currently a "good news" story, not only for the state, but also for the species in general. More Topeka shiners occur in South Dakota than in the other five states within their distributional range, primarily because of better water and grassland habitats within eastern South Dakota. Because of the good habitat base and because South Dakota was already working under a management plan designed for rare species, the state was excluded from critical habitat listing and has not had to comply with stricter Federal regulations. With increasing land use changes, largely driven by corn, soybean, and oil seed processing plants, the quantity and quality of remaining wetland and grassland habitats could decline, and with them the quality of habitat for all stream fishes.

Northern redbelly dace

The northern redbelly dace is fairly common throughout the upper Great Lakes Region east to the St. Lawrence and Atlantic drainages in New England. This species is less common in the upper Missouri River basin and the Mackenzie River Drainage of Canada. Similar to the blacknose shiner, several scattered populations remain in South Dakota. A glacial relict population occurs in several streams of the Sand Hills Region of Nebraska and in the Big Sioux River Drainage. This population of northern redbelly dace is disjunct from any other population of its species.

A population of northern redbelly dace was recently documented in Skunk Creek, a tributary of the Grand River (Morey and Berry 2004). This stream was characterized as having an abundance of aquatic macrophytes and a heavy groundwater influence that is typical of other northern redbelly dace streams in South Dakota. Because other Grand River tributaries share these characteristics, it is likely that more northern redbelly dace populations exist in this basin.

Finescale dace

The finescale dace exhibits a wide distribution from the St. Lawrence drainage of New England west through the upper Great Lakes Region north to the Mackenzie River Drainage and Arctic Circle. Disjunct populations have been documented in the upper Missouri River Drainage, the Sand Hills of Nebraska and South Dakota, and the Black Hills of South Dakota. Currently, populations of finesacle dace are in Cox and Mud lakes of the Black Hills (Shearer and Erickson 2005). Current management efforts are underway to reintroduce the finescale dace to other historic locations throughout the state. Finescale dace may cross (hybridize) with northern redbelly dace, which complicates species identification in some areas.

Sturgeon chub

The sturgeon chub is endemic to the Missouri River and its tributaries (Welker and Scarnecchia 2004, Galat et al. 2005), but no specimens were found in the flowing portions of the Missouri River during a recent comprehensive survey (Berry and Young 2004). A few reports have also documented the sturgeon chub in the Mississippi River between the Missouri and Ohio river confluences. This species prefers turbid waters with swift currents and has declined through much of its range due to dams built on the Missouri River. In South Dakota, viable populations still exist in the White and Cheyenne rivers (Fryda 2001). In 2007, the National Park Service's Inventory and Monitoring Program focused on the status of the sturgeon chub in Badlands National Park.

Sicklefin chub

Similar to the pallid sturgeon, the sicklefin chub is endemic to the Missouri River and Mississippi River below the Missouri River confluence (Dieterman and Galat 2004). Only two records occur of this species inhabiting tributaries of the Missouri River (both from the lower Kansas River). This species requires swift, turbid flows and has experienced declines due to dams on the Missouri River. The only recent records for the sicklefin chub in South Dakota are from the unchannelized portion of the Missouri River below Yankton and near the Niobrara River confluence. One specimen was found in South Dakota during recent surveys (Berry and Young 2004), but it is more common in other parts of the Missouri River (Galat et al. 2005). Sampling with a benthic (bottom) trawl has revealed that this species may be more common than once thought (Berry et al. 2004).

Pearl dace

The pearl dace shares a similar distribution to the finescale and northern redbelly daces, ranging from the St. Lawrence and Atlantic drainages throughout the upper Great Lakes Region and scattered across Canada. The pearl dace also exists in the upper Missouri River basin of Montana and North Dakota. Two disjunct populations occur, one in the tributaries to the Chesapeake Bay area of Maryland and Virginia, and the other in the Sand Hills Region of Nebraska and southern South Dakota. A handful of streams (including those in LaCreek National Wildlife Refuge) in Todd, Tripp, and Shannon counties contain the only known populations of pearl dace in South Dakota.

Longnose sucker

The longnose sucker is perhaps the most widespread sucker in the northern United States and Canada. Its range extends from the St. Lawrence and Atlantic drainages in New England west throughout the upper Great Lakes Region, throughout Canada and into Alaska. The longnose sucker inhabits a variety of clear, cold-water streams in the Rocky Mountain Region of the upper Missouri and Platte River basins. Vegetation makes up much of its diet (Baxter and Stone 1995).

In South Dakota, the longnose sucker exists in several streams in the northern Black Hills Region, where it is thought to be an isolated and disjunct population. The species is common in Wyoming streams, including Redwater and Sand creeks in the Black Hills (Baxter and Stone 1995). Historic records indicate this species occurred in French and Castle creeks. Land use changes and water quality problems have likely reduced its distribution in the Black Hills.

Banded killifish

The banded killifish occurs in drainages of the middle and upper Atlantic Coast states, and the Great Lakes Region west into eastern North and South Dakota. This species has only been reported in several eastern glacial lakes within South Dakota. While once thought to be an isolated population, more recent information indicates its distribution is peripheral and the banded killifish is much more common in Minnesota.

The black and white vertical bands on the body make this small (2-3 inches long) fish distinctive and give it its name. Killifishes are also called "topminnows" because they have a protruding lower jaw and a mouth that is well-developed for surface feeding on insects.

Conclusion

The future of the 34 fish species being monitored by the staff of the South Dakota Natural Heritage Program is uncertain at best. Except for the pallid sturgeon, the eight other fish species listed as threatened or endangered for the state are small or very small in size and none are economically important species. Because they are state or federally listed, their status in South Dakota will continue to be monitored, and, when possible, management plans to enhance these species will be activated.   ❖

Chapter 13

Exotic Species

Jeffrey S. Shearer

66 America is under siege by invasive species of plants and animals, and by diseases. The environmental, economic, and health-related costs of invasive species could exceed $138 billion per year... 99

— L. Ludke, U.S. Geological Survey, 2002

"Flying carp may pose danger to river" was the newspaper headline in the *Brookings Register* (April 2, 2005). The newspaper had previously reported that a Yankton fisherman fishing downstream from Gavins Point Dam was hit in the head by a 30-pound fish that fell into the boat and was later identified as an exotic Asian carp. "No bait needed for this 30-pound catch," read the headline. Exotic species are more than an interesting news story, they cost the U.S. about $138 billion a year in environmental, economic, and health-related costs (Pimentel et al. 2000, Pimentel 2002).

Wherever humans have explored and settled, their actions have left "footprints" on the landscape. The impacts left by these footprints can vary greatly. The case of exotic species in lakes and streams is a prime example. Most sources rank exotic species introductions second only to habitat degradation as the primary reason for loss of native fish diversity and abundance. Exotic fishes are said to have a "Frankenstein effect" (Moyle et al. 1986) on native fishes. Six general mechanisms enable the new species to succeed by displacing native fishes: competition, predation, inhibition of reproduction, environmental modification, transfer of new parasites or diseases, and hybridization. Introduced species are usually more successful in altered environments than in unaltered environments.

The vast majority of exotic species and their associated impacts go unnoticed or are poorly understood. Often only those that are the most ecologically or economically damaging (e.g., zebra mussels in the Great Lakes) receive publicity; however, every major watershed in the United States has been impacted to some extent by the introduction and spread of exotic species. And the number of invaders and their ranges have increased in recent years. In the late 1800s, only a handful of exotic fish could be found in this country's waters. The common carp is the most notable, first introduced to the United States in the 1830s. Today as many as 500 nonindigenous fish species inhabit our lakes and streams (Fuller et al. 1999).

The boom in exotic fish introductions is aided, in part, by the variety of distribution methods. One underlying factor remains the same—human actions, whether accidental or intentional, have always played a role in the introduction and spread of exotic species. Some species have been *unintentionally* introduced through bait bucket transport, the aquarium trade, boat or barge transport, and a variety of other methods but natural resources agencies share some of the blame.

Some fish species were *intentionally* introduced for recreation, such as the brown trout imported from Europe in the late 1800s. Others were stocked as a biological control of nuisance aquatic plants; examples are grass carp to control submersed plants and silver carp to control algae. Mosquitofish have been stocked to reduce disease-carrying insects. Recently, fishery biologists are being more careful, and have developed protocols for evaluating the potential dangers of introducing an exotic species (Courtenay and Stauffer 1984, see American Fisheries Policy Statement 15 at www.fisheries.org, accessed June 20, 2007).

As varied as modes of introduction are, the terminology used to describe exotic species can be just as diverse (and confusing). Fighting the invaders is such a common problem that biologists just use the term ANS without definition, because colleagues know that ANS means Aquatic Nuisance Species. The terms nonindigenous, nonnative, invasive, translocated, and introduced, are also used interchangeably to define with exotic species; however, each term has a slightly different meaning.

All trout in the Black Hills of South Dakota are nonnative, but few people would think of them in the same context as the common carp. Additionally, fish that are native to one river basin may be considered a nonindegenous species in an adjacent basin. For example, channel catfish have a long history in the Missouri River, providing food for the Lewis and Clark Expedition of 1804–06. The introduction of channel catfish to the Colorado River basin, however, has been detrimental to the endangered Colorado pikeminnow (*Ptychocheilus lucius*).

For the purposes of this chapter, exotic species are those species from waters outside the United States or Canada. South Dakota has several exotic aquatic species and the remainder of this chapter will examine selected species of exotic fish, invertebrates, and aquatic plants, and their impacts.

Bighead carp

Bighead carp (Figure 13.1) have received as much attention as any exotic species in South Dakota waters in recent years. This species was imported into the United States from China in the early 1970s by the aquaculture industry in the lower Mississippi River basin. Bighead carp were used to control zooplankton in aquaculture ponds and reared as a food source for markets in New York City and Chicago.

Figure 13.1 Bighead carp can weigh up to 50 pounds and are becoming more numerous in the Missouri River and lower segments of the James and Big Sioux rivers. (photo by J. Kral)

Flooding and accidental releases have allowed this species to escape from private ponds and spread throughout the Mississippi River basin. The bighead carp entered South Dakota via the Missouri River by the late 1990s. Today the bighead carp inhabits the Missouri River below Gavins Point Dam and lower portions of the James, Vermillion, and Big Sioux rivers. Gavins Point Dam and the Falls of the Big Sioux River provide barriers to further upstream movement.

A major concern regarding bighead carp is their potential impact on the food web in rivers and streams since they can reach high densities. Many commercial fishermen on the Mississippi River report bighead carp and silver carp as their primary catch in nets. Their high numbers coupled with their large size (current state record is just over 49 pounds) results in their consuming vast quantities of zooplankton and other fine organic material. Their diet has serious competition implications for early life stages of all fish species and perhaps more important is the potential competition with our native filter-feeding species such as the paddlefish. Other species decline when the carrying capacity of a body of water is dominated by the total weight of bighead carp. Carrying capacity is the total weight of all organisms or the supportable load.

Silver carp

Silver carp are among the most recent exotic species to invade South Dakota. The U.S. Fish and Wildlife Service documented the first silver carp at the confluence of the James and Missouri rivers during the summer of 2003. Silver carp were imported from China during the early 1970s by the aquaculture industry to control phytoplankton in rearing ponds. Flooding and accidental releases allowed silver carp to escape private facilities and spread throughout the Mississippi River basin. In South Dakota, the silver carp currently inhabits the lower Big Sioux, James, Vermillion, and Missouri rivers.

Silver carp are well adapted for life in large rivers such as the Missouri. Silver carp may move 200 miles over the course of one year. A spring rise in water levels is a spawning cue for these fish. Natural reproduction has not been documented in South Dakota; however, this is one aspect of silver carp life history currently being examined by researchers.

The impacts of silver carp are similar to those of bighead carp. Silver carp are filter feeders and directly compete with other species for food. They are also a threat to humans. When disturbed, silver carp display an unusual behavior by jumping out of the water. Because of their large size, this behavior is a serious threat to boaters who may be struck by a airborne fish. They may also be transported to new waters in bait buckets.

Young silver carp are similar in appearance to gizzard shad, a common baitfish in the Missouri River. An angler seining gizzard shad may unintentionally transfer silver carp to other waters. Current South Dakota regulations prohibit using bait collected in the Missouri River below Gavins Point Dam in other state waters.

Grass carp

Grass carp were first imported in the early 1960s from Asia and were used by natural resource agencies and aquaculturalists to control aquatic vegetation problems. Today triploid strains of grass carp are commercially available and permitted for use following genetic testing to verify sterility. Either accidentally or intentionally, fertile grass carp were released into river systems and have been documented in 45 states. In South Dakota, the grass carp has moved into the Missouri

River below Gavins Point Dam and upstream into the James and Big Sioux rivers.

As the name suggests, grass carp consume aquatic vegetation. When grass carp reach high densities, they can remove all vegetation from a lake, changing the ecology of that system. The turbid nature of our rivers, however, restricts the growth of aquatic plants, possibly explaining why grass carp numbers are low in South Dakota. Introduction of this species into glacial lakes could be of concern; however, reproduction is unlikely because grass carp generally need river habitats to reproduce.

Grass carp also could pose a serious threat to Sand Lake National Wildlife Refuge, an impoundment on the James River, where aquatic plants (e.g., sago pondweed) serve as an important food source to thousands of migrating waterfowl. A small water-control structure is the only barrier separating Sand Lake from the downstream James River with its grass carp populations.

Common carp

Common carp are so widespread and have inhabited U.S. waters for so long that most people forget they are an exotic species. In fact, the common carp may have been the first exotic fish species introduced into this country after being imported from Europe as a food source in the 1830s. Today the common carp inhabits just about every major watershed in the United States outside of Alaska. The common carp made its first appearance in South Dakota waters in the early 1880s from stockings by the U.S. Fish and Fisheries Commission. It now inhabits every major stream system and most lakes in South Dakota outside of the Black Hills.

Common carp have an omnivorous diet, are prolific spawners, and are tolerant of a wide range of environmental conditions. These characteristics enable them to reach a high abundance, inhabit a variety of aquatic systems, and out compete other fish. Resource managers often renovate lakes with rotenone (a common piscicide or fish poison), draw down lake water levels during winter, or permit commercial harvest to control common carp populations. LaCreek National Wildlife Refuge routinely renovates refuge ponds to prevent an overabundance of common carp from destroying aquatic vegetation, which is needed for waterfowl and shorebirds. Common carp are also a problem in Sand Lake National Wildlife Refuge and other wetlands managed for waterfowl (Clark et al. 1991).

The electrical barrier is an unusual tool used to block upstream migrations of fish such as common carp into lakes (Verrill and Berry 1995). An electrical barrier is a concrete lining of the bottom and sides of a stream channel. In the surface of the concrete are electrodes with 3-foot spacings that run perpendicular to the stream flow. The pulsed, direct current field is nonlethal, but does stun fish and cause them to float back downstream. Electrical barriers are used in some rivers of the United States to block the spread of Asian carp.

European rudd

European rudd currently inhabit nine lakes within South Dakota, but receive relatively little attention when compared to the carp species. Originally imported into the United States in the early 1900s as an ornamental fish, European rudd were later sold as baitfish in some states. Introduced through bait buckets and intentional stockings, European rudd have spread throughout many lakes and streams in the eastern and midwestern United States. Because of their similarity in appearance to golden shiners, a common baitfish, European rudd are sometimes advertised as "hybrid golden shiners." Bait buckets are the most likely way they were introduced into South Dakota waters.

Since predatory fish consume European rudd to a limited extent, biologists have found they can control populations in lakes where predatory game fish are abundant. One of the management objectives for Lake Alice in Deuel County is to maintain walleye populations at high levels to control the rudd population. Similar regulations are in place for walleyes and largemouth bass in Newell Dam, Butte County. European rudd, however, can grow to lengths that are too big for most predators (16-inch rudd were recently sampled in Lake Alice) and are capable of reaching high abundance when predation is low. Their diet is quite varied, ranging from zooplankton and macroinvertebrates to plant material, and it overlaps the diets of several other fish species, resulting in competition for food sources and habitat degradation.

Curlyleaf pondweed

Curlyleaf pondweed (*Potomogeton crispus*, Figure 13.2) is to the aquatic plant world what the common carp is to the fish world. This plant has been around so long and is so wide-spread that few people give it the attention that more recent exotic plant species receive. Curlyleaf pondweed is native to Europe and was introduced into the United States by aquarium hobbyists during the 1880s. Today curlyleaf pondweed can be found in 46 of the lower 48 states (excluding Maine and South Carolina). In South Dakota, curlyleaf pondweed was first documented in Burbank Lake in 1965. Since then it has spread to a handful of lakes including Oahe, Sharpe, and Lewis and Clark and one stream, Rapid Creek.

In large windswept reservoirs, such as Oahe and Sharpe, most species of aquatic vegetation do not become established to the same extent as they do in smaller protected waters. Where curlyleaf pondweed does present a problem is in marinas and protected bays. LaFramboise and Hipple lakes (side channel waterbodies on Lake Sharpe) are good examples. Curlyleaf pondweed starts growing in late winter and early spring (often under the ice), which gives it a competitive advantage over other aquatic plants. Thick stands of curlyleaf pondweed reach the water's surface by late spring, impeding boat travel and recreational use. By early July, the thick vegetative stands die and wash ashore, creating a nuisance along beach and boat ramp areas.

At moderate population levels, curlyleaf pondweed may actually be beneficial as a refuge area for young fish in Missouri River reservoirs where other aquatic vegetation is largely absent. Exotic plant species, however, rarely grow in moderation and in the glacial lakes of northeastern South

Figure 13.2 A carpet of curlyleaf pondweed (*Potomogeton crispus*) covers the surface of a lake (above). The pondweed is so dense that boating and fishing are impossible (below). (photo by J. Shearer)

Dakota (e.g., Roy and Enemy Swim lakes) this plant could choke out stands of native vegetation.

Eurasian watermilfoil

Information on the history of Eurasian watermilfoil (*Myriophyllum spicatum*) is scarce. Some reports indicate the first documented case found in the United States was in Washington D.C. in 1942. Other sources indicate Eurasian watermilfoil was introduced much earlier into the Chesapeake Bay. Currently Eurasian watermilfoil is found in all but a few of the lower 48 states. Eurasian watermilfoil reproduces through seeds and plant fragments, enabling this species to spread and quickly colonize new areas. In South Dakota, Lake Sharpe is the only waterbody to contain this plant; first discovered there in 1999 in bay areas at West Bend, Joe Creek, and along Hipple Lake.

Eurasian watermilfoil is often the focus on aquatic plant control programs; for example, Minnesota and Iowa Departments of Natural Resources, have entire programs devoted to the control of Eurasian watermilfoil. Lake associations spend countless dollars each year treating dock and waterfront areas with herbicides or attempting to harvest the plant. How could an aquatic plant draw so much attention and concern? A primary reason is the potential impacts Eurasian watermilfoil has on the ecosystem of a lake or a stream.

Eurasian watermilfoil can completely alter a small lake or pond. Where Eurasian watermilfoil is especially dense, a lake's surface may resemble a "lawn." This plant starts growing before most native aquatic plants, and quickly establishes thick stands by early summer. Aside from creating a nuisance for boaters, anglers, and swimmers, these thick stands shade most other vascular plants, plankton, and benthic algae growth in a lake, completely altering primary production within the trophic food web.

In late summer, Eurasian watermilfoil starts to die and the decomposition of plant material combined with warm water temperatures creates low dissolved oxygen levels in the water. Summer fish kills can occur following a large die-off of Eurasian watermilfoil.

Zebra mussels

Ask any angler or boater in the Midwest what comes to mind when they think of exotic species and chances are zebra mussels (*Dreissena polymorpha*) will be their answer. Since their introduction into the Great Lakes in the mid-1980s, zebra mussels have been the target of numerous prevention and public education programs by federal, state, and private entities. This black and white striped mussel, which is about the size of a dime, is indeed the poster child for exotic species in the Great Lakes and upper Mississippi River basin.

Zebra mussels were first discovered in Lake St. Clair in 1988 and were likely imported from Europe in the bilge or ballast water of transoceanic vessels. By 1999, zebra mussels had spread across 21 states in the Mississippi River and Hudson River basins. Today this bivalve can be found in 23 states and in the Provinces of Quebec and Ontario. During the summer of 2003, zebra mussel veligers (larval stage of the mussel) were found below Fort Randall and Gavins Point dams in South Dakota. While no adult zebra mussels were discovered, the presence of veligers suggests that adults have become established in the Missouri River.

Zebra mussels have the potential to completely alter an aquatic system, just as Eurasian watermilfoil does. These mussels are filter feeders that take plankton and other organic matter from the water column. In high densities (up to 200,000 per square meter in some lakes), zebra mussels can filter vast quantities of water, greatly increasing water clarity, which in turn encourages the growth of nuisance aquatic plants. Zebra mussels also directly compete with native mussel species and young fish that rely on plankton as a food source.

Zebra mussels can attach to any solid substrate, including living mussels and crayfish. So far the greatest impact to man is damage to water control structures, irrigation pipes and intake pipes. As a result, water treatment plants often have to implement costly control procedures to prevent or remove the build-up of zebra mussels. What impacts zebra mussels hold for the Missouri River and South Dakota remain to be seen, but their recent discovery has raised concerns.

Asian clam

For the most part, Asian clams (*Corbicula fluminea*) have taken a back seat to zebra mussels in terms of publicity and attention. Native to southeastern Asia, the first Asian clams in the United States were collected in Washington state in 1938. Today the Asian clam has spread throughout most coastal and midwestern states. In South Dakota, this species was first documented in 2003 near Vermillion in the Missouri River. Subsequent investigations have reported Asian clams upstream in the Yankton area as well. Asian clams have a typical clam shape, are usually brown and less than 2 inches in diameter.

The most significant impact associated with Asian clams has been their habit of attaching themselves to pipes and other hard surfaces (called biofouling) of water treatment and power plants. Similar to zebra mussels, Asian clams can clog water intake pipes and irrigation canals, thus increasing operating costs to treat and remove. Asian clams as well as zebra mussels reproduce through the dispersal of free-floating larvae (veligers). The free-floating larvae are microscopic and easily transported in bait buckets, livewells, and bilges by unsuspecting boaters.

Other Exotics

The exotic species discussed above are by no means unique to South Dakota, nor are these the only exotics that degrade the biodiversity of our lakes and streams. Plants such as purple loosestrife (*Lythrum salicaria*) can dominate riparian zones and wetlands, and tamarisk (*Tamarix* spp.) absorbs vast quantities of water through its root system, potentially reducing stream flows. At the watershed level, lakes and streams are directly tied to ecosystem functions in the uplands where crested wheatgrass (*Agropyron desertorum*) and smooth brome (*Bromus inermis*), two exotic grass species, have invaded.

Compared to the problems in other states, South Dakota's aquatic resources are in pretty good shape. Trout streams in the Black Hills have not been infected with whirling disease, a parasite that has devastated some trout fisheries in the Rocky Mountains. The round gobie (*Neogobius melanostomus*) and quagga mussel (*Dreissena bugensis*), species that have plagued the Great Lakes, have not invaded our lakes. The list of exotic species just outside the South Dakota border, however, is extensive and is justifiable reason for concern in future decades.

Conclusion and recommendations

As South Dakota attracts more outdoor enthusiasts the potential to introduce more exotic species increases. South Dakota offers great outdoor recreational opportunities. The Missouri River reservoirs generate millions of dollars in income from visiting campers, boaters, and anglers. The high water years of the 1990s turned prairie potholes into new lakes in northeastern South Dakota, and anglers were

13.1

Suggestions for Reducing the Spread of Exotic Pests

- CLEAN all plant material from boat and trailer.
- DRAIN your livewell and bilge before leaving the boat ramp.
- FLUSH your livewell and cooling system with hot water (140 F) between lakes.
- DRY boat and equipment for several days between lakes.
- DO NOT DUMP bait bucket contents into any body of water.
- DO NOT RELEASE fish where they were not caught.

quick to notice. Today highly mobile recreationists from all over the Midwest travel to South Dakota for fishing, hunting, and other outdoor opportunities. Exotic species could hitch a ride to South Dakota. We should be aggressive in educating the public about transporting exotic species to South Dakota (Box 13.1).

The annual fishing handbook states that, "A person may not release fish, reptiles, amphibians, mollusks or crustaceans not native to South Dakota into public or private waters within the state, other than an aquarium, without written authorization from Game, Fish and Parks."

All exotics are not ecologically and economically disastrous, but the unknown "Frankenstein effects" associated with these species are troubling and we should be conservative when investigating the value of stocking new species. Each introduction of an exotic species shifts the natural diversity of South Dakota's lakes and streams one step further from their former natural status.   ❖

Chapter 14

Fish Hatcheries and Stocking Practices: Past and Present

Michael E. Barnes

" The present day superintendent of a modern fisheries station should be a master of a number of trades, have a working knowledge of several professions, and be an all around office man in addition to having a thorough knowledge of fish culture. **"**

— D.C. Booth, circa 1930

The dust shot out from underneath the wheels of the buggy as it rambled through well-worn ruts in the prairie sod. The severe drought that brought an end to the Great Dakota Boom still gripped the plains of Dakota Territory in 1886. In addition to the hot and dirty pioneers riding on the hard wooden bench, this buggy also carried travelers who were enjoying clean, cold-water and plenty of space to move around in as they journeyed toward the Black Hills. The contented travelers were trout swimming in water-filled cream cans, on their way to be stocked in Black Hills streams.

Private citizens Richard Hughes and Samuel Scott had purchased their special cargo in Leadville, Colorado just a few days earlier, presumably from Dr. John Law. They carried ice to keep the fish cool during the two-week return trip to Rapid City. Amazingly, the brook trout survived this perilous journey and became the first fish stocked in the Black Hills. Scott, one of Rapid City's founders, and Hughes, a journalist after his gold mining days, had high hopes for these trout. They envisioned naturally-reproducing trout populations that would spur tourism in the Black Hills.

Just before the first trout entered the water at Cleghorn Springs outside Rapid City, a different fish stocking story unfolded on the eastern side of the Missouri River. The Territorial Fish Commissioner had put in a request for the U. S. Fish Commission (which became the U.S. Bureau of Fisheries in 1903, which in turn became the U. S. Fish and Wildlife Service in 1940) to stock another species. This species was expected to provide an inexpensive and sumptuous food source that could be eaten throughout the year. This fish was the common carp.

After traveling in a specially equipped railroad car, the carp were transported by real horse power and stocked into several eastern lakes in 1886. This was the first recorded fish stocking in the territory that would become South Dakota three years later. The stocked carp quickly illustrated that fish stocking was not a panacea for fisheries management, for the "prized" fish soon came to be despised by both anglers and government officials.

Federal Fish

Four years after private citizens stocked trout in the Black Hills, the federal government entered the trout stocking business in South Dakota. In 1890, the U.S. Fish Commission stocked brown trout in Spearfish Creek and the next year they stocked brook trout in a few other locations as well. Based on the success of these stockings, the U.S. Congress authorized construction of a fish hatchery at Spearfish in 1896. John S. Johnston, one of the first homesteaders in Spearfish, sold his property near the mouth of Spearfish Canyon to the federal government in 1898. Johnston had used his property to raise "troutlets" and had stocked thousands of pond-raised fish around the Spearfish area since 1895.

Originally called the Spearfish Fish Cultural Station, the Spearfish National Fish Hatchery (Figure 14.1) was completed in 1899 and the first fish from the hatchery were stocked in 1900. Over the next 10 years, the U.S. Bureau of Fisheries stocked over 15 million rainbow trout, brown trout, brook trout, cutthroat trout, lake trout, and Atlantic salmon in

Figure 14.1 The new Spearfish National Fish Hatchery in 1899 would 100 years later become the Nation's only fish culture museum (D. C. Booth Historic National Fish Hatchery) and a major tourist attraction in the Black Hills. (U.S. Fish and Wildlife Service photo)

South Dakota. Although it first served as a subsidiary to the federal hatchery at Leadville, Colorado, Spearfish National Fish Hatchery later became a center for federal fisheries operations in the western United States. In 1937, the Spearfish hatchery was one of 87 National fish hatcheries; a staff of nine was paid $15,000 in total wages (Earle 1937).

During the early years of statehood, the federal government either directly stocked fish, provided the state with fish to stock, or gave fish to private individuals to stock (Figure 14.2). Stocking records are not complete and might not be the most reliable. For example, six "goldfish" were recorded as stocked in 1890, but those fish are now believed to have been Tench, a Eurasian import. Black bass, crappies, rock bass, and "sunfish" were stocked in Lake Kampeska in 1890 or 1891. Several species of trout were stocked throughout the Black

Figure 14.2 Trout reared at the Spearfish hatchery were transported in creamery cans by horse and buggy for stocking creeks in the Black Hills. Labeling on the cans says "Department of Commerce and Labor, Spearfish So. Dakota, Bureau Fisheries. (U.S. Forest Service photo)

Hills in the 1890s. Brown trout were stocked in Spearfish Creek in 1890, which was just seven years after brown trout were first imported into the United States from Germany.

Trout were also stocked before 1900 on the eastern side of the state in some unlikely places: brown trout in Turkey Creek near Yankton; cutthroat trout in Collins Springs near Dell Rapids; brook trout in Turkey Creek near Wakonda; and lake trout in Lake Kampeska, Lake Hendricks (near White), and Big Stone Lake. Another unique stocking was 125,000 whitefish in a lake near White in 1900.

The U. S. Bureau of Fisheries was the primary source of fish for stocking in South Dakota after the turn of the century. In 1913, the Bureau provided the state with two million game fish fry (Figure 14.3). The records are somewhat ambiguous, but non-native fish (e.g., chain pickerel) were stocked in southeastern lakes, and common carp were stocked at least until 1914.

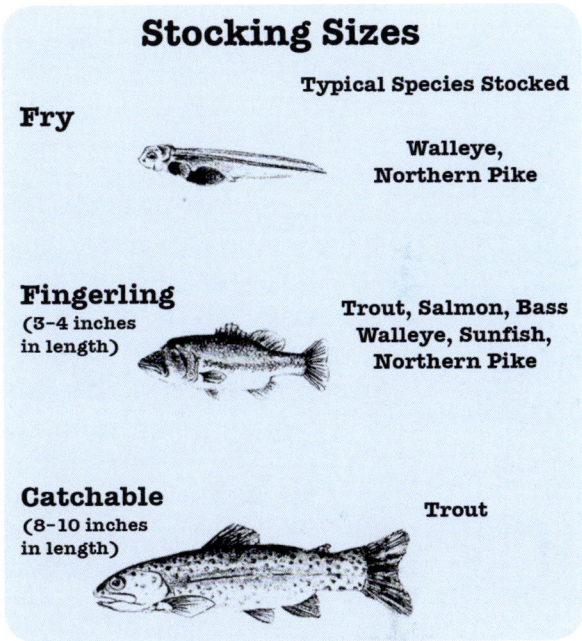

Stocking Sizes

Typical Species Stocked

Fry
Walleye, Northern Pike

Fingerling
(3-4 inches in length)
Trout, Salmon, Bass Walleye, Sunfish, Northern Pike

Catchable
(8-10 inches in length)
Trout

Figure 14.3 Fish are stocked as fry, fingerlings or catchables. Fry (sac fry) only have a few days of nutrition in the "yolk sac" as they learn how to find natural food. Fingerlings must be fed artificial diets while being reared in raceways or ponds at the fish hatchery, however, some fingerlings are harvested from the wild. Catchables are used for put-and-take fishing. Catchables are expensive because of the feed, maintenance, disease control and other care that is needed to rear large fish. Drawing by Anna Mehlhaff.

State Fish Rearing Begins

The state of South Dakota first appropriated funds for fish propagation in 1909. The newly established Game and Fish Commission was authorized to spend up to $200 for fish rearing and stocking (SDGFP 1959). These early expenditures were spent mostly on trapping fish from a lake and then transferring them to other waters. For example, in the 1910s, bullheads were netted from Lake Alice in Deuel County and then transferred around the state in cream cans carried on railcars and automobiles.

W. F. Bancroft, the first state game warden, lobbied for a state-owned fish hatchery around 1915. He wrote, "We heartily favor the construction of a state fish hatchery as soon as a competent man can be secured to operate it." That "competent man" was Frank Purcell, from Madison, Wisconsin, who was hired to run South Dakota's first state hatchery. Lake Kampeska State Fish Hatchery, near Watertown, was built in 1916 to incubate walleye and northern pike eggs so that newly hatched fry could be stocked into state waters.

The success of the Kampeska hatchery varied greatly from year to year. The hatchery used steam boilers starting in 1919 to control water temperature, but this tempering plant was not the most reliable or consistent. In 1922 the hatchery was shut down before any egg incubation even occurred. Ice jams and the unpredictable spring weather interfered with hatchery operations. In addition, spawning crews ventured onto wet and cold Lake Kampeska only to find that the walleyes captured by their nets had already released their eggs.

Another bad year in 1923 at the Kampeska hatchery led the state to look at alternative locations for a walleye egg hatching facility. Land was purchased in 1924 for $415.30 and a new "pike and pickerel" hatchery was completed on the shore of Pickerel Lake in 1929. Although the Kampeska hatchery was plagued with problems, it was still operated for a number of years after the Pickerel Lake hatchery started production, which proved to be fortunate from a fish production perspective. In 1933 the Pickerel Lake hatchery wasn't able to hatch any live walleye fry due to an "unknown green slime" coating everything in the incubation jars.

Although egg incubation likely didn't occur past the mid-1930s at the Kampeska hatchery, it wasn't removed from the state hatchery listing in the South Dakota Department of Game, Fish and Parks (SDGFP) annual reports until 1945. Shortly after the Kampeska hatchery closed, the Pickerel Lake hatchery upgraded its technology by installing propane water heaters for the egg incubation water. This improvement helped make the Pickerel Lake State Fish Hatchery the primary state-owned walleye and northern pike hatchery until 1983.

The 1920s were a time of cooperation with local sportsmen's clubs. Numerous nursery ponds that were built around the state and volunteers removed fish for stocking. Fish species that are not routinely stocked today were the focus of many of those early rearing efforts. Clyde B. Terrell of the South Dakota Game and Fish Commission described the results of some of these stockings, saying in 1923 that "places such as Red Lake and White Lake…which were stocked with bullheads a few years ago, now furnish the people of the locality with all the fish they wish to catch." Trout of all species continued to be stocked, along with channel catfish, largemouth bass, sunfish, "silver" (white) bass, crappies, walleyes, northern pike, and yellow perch. Some unusual stockings from this time period include lake trout in Orman Reservoir and a stock pond near Buffalo, and golden shiners in Whitewood Creek.

Even though it was early in the history of fish rearing in South Dakota, hatchery supervisors identified a problem that is voiced still by state hatchery staff. Complaints were sent in 1920 from the northeastern part of the state to administrators in Pierre that, "… men that have been employed by the Fish Department for some time left on account of low wages paid by the state, compared with those paid by other lines of business."

Rapid City Trout

While the federal government continued to produce trout at the Spearfish hatchery for stocking in the Black Hills and other locations in the western United States, the state of South Dakota decided to construct its own cold-water hatchery to rear trout from eggs to fingerlings (Figure 14.4). In 1918

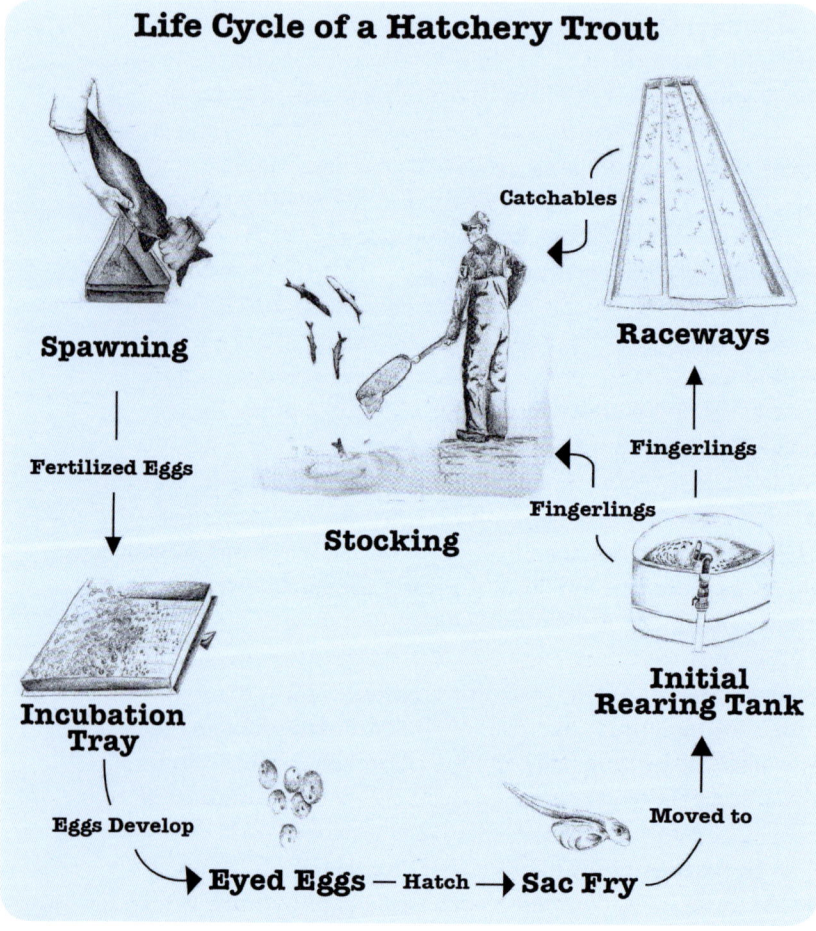

Figure 14.4 Life cycle of a hatchery trout. Eggs and sperm are "stripped" from wild or captive broodstock (upper left). The eggs develop in incubation trays submersed in running water. The developing embryo has large dark eyes visible in "eyed" eggs. After hatching, the fish are fed scientifically prepared commercial fish food. Fish are stocked as fry, fingerlings or catchables. Drawing by Anna Mehlhaff

a temporary trout hatchery began operating on East Main Street in Rapid City. Almost as soon as it started, the hatchery and two million brook trout eggs that had just started hatching were destroyed by a fire.

A replacement hatchery was erected with lumber harvested from the State Game Park (which became Custer State Park that same year). The hatchery building looked like a house and for good reason—if the hatchery didn't work out, the building could be resold as a residence with minimal financial loss.

Using city water, which did not contain chlorine at that time, the Rapid City hatchery was producing about two million small trout a year for stocking by 1923. However, the hatchery was plagued with water quality and supply problems. In 1922, 150,000 sickly looking brook trout eggs were relocated to the abundant and pure water of Cleghorn Springs west of Rapid City where they "recovered immediately."

When a new hotel started using Rapid City water in 1927, the hatchery water supply became more unpredictable and problematic. Finally in 1928, a trout hatchery was permanently established at Cleghorn Springs, near the site where Hughes and Scott had stocked the first trout. The older hatchery in the city proper was abandoned in 1930 after hatching its last group of brown trout eggs. It was subsequently sold and moved to a different location in Rapid City, where it continues as a residence today.

The springs at the mouth of Cleghorn Canyon were used to grow fish for many years prior to state operation. Dan Cleghorn, the original land patent holder, constructed ponds adjacent to the springs for common carp culture in the 1890s. This initial fish-rearing attempt likely failed because the cold 52º F spring water was more ideally suited to trout and salmon than to warmer water fish like carp. In the 1920s, George Wagner persuaded the Izaak Walton League of Rapid City to lease the Cleghorn Springs ponds for trout rearing. Due to the success of the League's program, the SDGFP began negotiating for the purchase of the "famous" Cleghorn Springs in 1927.

Cleghorn Springs State Fish Hatchery has changed through the years. Initially it consisted of a brick hatching house and 12 rectangular earthen ponds. Lester Ripple, the first hatchery supervisor, referred to the original hatchery in 1930 as, "…a place of beauty as such state property should be." A later renovation in the mid-1940s added 29 earth-bottom ponds with stonewalls, and seven earthen ponds.

Only one hatchery building remained standing after the 1972 flood of Rapid Creek but the loss of the hatchery manager's family due to the flood is sadly the lowest point in the history of hatcheries in South Dakota. In addition to demolishing the hatchery, the 1972 flood also destroyed the state-owned hatchery manager's residence. After the flood, the entire hatchery was rebuilt in only 14 months, complete with flood-proof buildings and concrete raceways.

During its tenure as the oldest operating state-owned hatchery in South Dakota, Cleghorn Springs State Fish Hatchery has grown rainbow trout, brown trout, brook trout, cutthroat trout, lake trout, and chinook salmon, primarily for lakes in the Black Hills and Missouri River reservoirs. Walleye eggs were also hatched at Cleghorn Springs until about 1959.

Bass Ponds Everywhere

South Dakota also began rearing warm-water fish such as largemouth bass, sunfish, and bullheads. Seven acres were purchased for a bass hatchery at Lake Andes in 1923. The construction of rearing ponds for bass spawning and subsequent fingerling production was completed in 1925. A small cottage was built adjacent to the ponds to house the hatchery manager, with plantings of shrubs on the grounds for both landscaping and to "provide food for the bass."

By 1926, the Lakeside Hatchery was operational on Cottonwood Lake near Redfield. Not only were warm-water fish species reared at the Lakeside Hatchery, this multipurpose state facility also produced game birds (waterfowl and pheasants) and farmed fur-bearers. Just a few years later, bass and sunfish from the Lake Byron Bass Hatchery north of

Huron were stocked in the vicinity of Faith. By 1939, the Twin Lakes Bass Hatchery near Woonsocket began production. These facilities were instrumental in producing fish for the numerous west river stock ponds and approximately 1,000 small impoundments that the Works Progress Administration constructed during the 1930s. In 1949, South Dakota had six state fish hatcheries named Lakeside, Cleghorn, Pickerel, Twin, Byron and Andes (Tunison et al. 1949).

From 1934 to 1973, ranchers in western South Dakota were able to get largemouth bass and panfish for stock ponds through the U.S. Fish and Wildlife Service farm pond program and Soil Conservation Service. Federal hatcheries often just mailed small fish in plastic bags, which ranchers retrieved at the post office and then stocked the fish themselves. After this program was discontinued, Conservation Officer Michael "Mick" Muck from Philip started what was referred to informally as the "Bad River Bass Hatchery." Officer Muck used a concept with largemouth bass that was first attempted with walleye natural rearing ponds on the eastern side of the state in the 1960s. During spring, he collected adult bass from stock ponds with ample bass populations and transferred a small number of them into shallow ponds. These newly stocked waters were then seined the next spring, producing yearling bass for stocking. This program provided most of the largemouth bass planted in West River ponds until 1989.

In the 1950s additional bass ponds were constructed at Drake Springs in Minnehaha County and near Big Stone Lake. Rearing of fish at all of these warm-water hatcheries and ponds, except one, continued until some time in the late 1950s or early 1960s. The Big Stone Lake pond is still used today, producing panfish species in a cooperative venture between the SDGFP and a local outdoor sports club.

The Depression and War Years

Fish stocking and hatchery production went through a relative lull during the years of severe drought prior to World War II. In 1933, fish were gone from half the lakes due to low water levels. By 1936, only six natural lakes retained game fish populations. The focus of fish stocking and fish production during the dry years was to concentrate stocking in the few lakes with water to "provide as good a fishing as possible." Major work was put into netting fish, primarily bullheads, from low-water lakes and moving them to other locations with more abundant water supplies. Fish salvage was an established fish management practice in the 1930s (Nielsen 1993).

Bullheads were also grown in rearing ponds, and stocking records from 1932 detail the release of "jumbo bullheads" into many water bodies. In 1935, more than 150 bodies of water were stocked with bullheads. In addition to the normal stockings of typical predatory and panfish species from both state and federal sources, stockings of "calico" bass (possibly Sacramento perch), "river" chubs (maybe creek chubs), green sunfish, orangespotted sunfish, and many "minnows" occurred during this time period.

Trout stockings continued throughout the state during the Dirty Thirties as well. In addition to the stockings in the Black Hills, it was standard practice to stock the display trout from the state fair into eastern waters. In 1940 water returned to the State, and the cool-water and warm-water fish hatched and grown at Pickerel Lake and Cleghorn Springs hatcheries played a big role in fish reintroductions. A then record 50,884,230 fish were hatched and distributed in South Dakota in 1940.

It was also during the 1930s that the Spearfish National Fish Hatchery water supply changed. Water flows from the springs originally supplying water to the hatchery were greatly decreasing (and eventually stopped completely). Because of the water supply problems in 1935, the U.S. Bureau of Fisheries paid $10 per year for an artesian well flow of 500 gallons per minute on the McGuigan property just to the north of Spearfish. The McGuigan substation was only used for one year to produce 10,000 fish and was demolished in 1936.

In 1936, a diversion channel was dug just downstream from the Homestake Mining Company hydroelectric plant to create a steady supply of surface water from Spearfish Creek for hatchery use. A stone building was built close to the channel that same year. This structure, now named the "Snappers Club," in the Spearfish City Campground, was only used for a few years for trout egg incubation. The diversion channel still flows through the campground and continues to funnel water from the aquaduct-supplied power plant to the Spearfish Hatchery.

Hatchery Expansion

Throughout the nation, the mission of fish hatcheries had changed from supplying food to supplying recreation (Wood 1953), and South Dakota followed the trend to attract tourists to the Black Hills.

McNenny National Fish Hatchery

The Spearfish National Fish Hatchery was plagued by water quality and supply problems, so the U.S. Fish and Wildlife Service entered into a lease agreement in 1946 with the newly created SDGFP for the construction of a hatchery near the Wyoming border. The state had purchased property in 1943 from Judge James McNenny, who had kept the property open as a public recreation area, raised trout in some of the spring-fed ponds, and run a guest ranch.

By 1952, artesian wells and trout rearing facilities were completed at McNenny National Fish Hatchery. Because the state owned the land, a percentage of the fish produced at this federal facility were stocked in South Dakota public fishing waters each year. McNenny was operated as a substation of the historic Spearfish hatchery and became part of the Spearfish Fisheries Complex.

The staff at the McNenny National Fish Hatchery was instrumental in the development of the dry, pelleted food that has been fed to trout and salmon since the 1960s. As an example of what was fed to trout before dry diets became commercially available, the Spearfish hatchery purchased 2,897 pounds of inedible beef livers and 15,123 pounds of pork "melts" in 1953. Sheep liver, beef lungs, beef melts, salmon eggs, and "plucks" from condemned cattle were all fed to trout prior to 1960 at the Spearfish Fisheries Complex.

A Unique Trailer House Hatchery

A unique state hatchery that operated only a few years began operation in May of 1965 when an old trailer house was converted into a mobile hatchery and temporarily placed on state land adjoining the then Big Sioux Conifer Nursery outside of Watertown. The trailer house hatchery incubated 25 quarts of walleye eggs in 1965 and also hatched 150,000 rainbow trout eggs. The plan was to use the hatchery to produce trout for eastern South Dakota lakes, perform walleye egg incubation experiments, and test the water from the newly drilled well for full-scale fish production. The mobile hatchery, along with possible development of a permanent hatchery at the Big Sioux Nursery site, was abandoned due to lack of funding and political support. In 1958 or 1959, the SDGFP established a temporary walleye egg hatchery with 20 egg jars in Estelline. This hatchery was probably used for only a year or two.

D.C. Booth Historic Fish Hatchery

In 1983, federal budget cuts forced closure of the Spearfish Fisheries Complex. The U.S. Fish and Wildlife Service, still owned the Spearfish National Fish Hatchery but allowed the city of Spearfish to use the hatchery as a tourist facility. The city opened a gift shop and maintained the old hatchery building, which had been converted into a museum by the U.S. Fish and Wildlife Service in 1982 (Smith 1987). In a unique agreement, the state used the ponds and raceways for fish production during the summer, avoiding most of the water supply problems that had plagued the hatchery in the past. In 1989, the federal government resumed control of the now-named D. C. Booth Historic National Fish Hatchery and renovated the facility. The SDGFP continues to use the rearing units at the hatchery for trout production seasonally.

McNenny State Fish Hatchery

The state of South Dakota assumed control of the McNenny Federal Fish Hatchery on July 1, 1983 and the SDGFP immediately spent over $500,000 in renovations to improve water supply, incubation, and indoor rearing tank systems. Now under state operation, McNenny State Fish Hatchery stocks all of its production in South Dakota public fishing waters. McNenny primarily raises trout for the Black Hills and Missouri River reservoirs, in addition to producing most of the chinook salmon for Lake Oahe. In something of a reversal from the past, McNenny uses the D. C. Booth (Spearfish) hatchery as a substation for fish production during the summer months. The McNenny and Cleghorn hatcheries now produce all of the trout released in the Black Hills.

Blue Dog Lake State Fish Hatchery

Blue Dog Lake State Fish Hatchery started operations in the spring of 1983 (Figure 14.5). Situated on the northwest

Figure 14.5   Blue Dog Lake State Fish Hatchery has 700 egg incubation jars and 30 rearing tanks inside the hatchery building and 53 acres of ponds. Blue Dog is a walleye hatchery although 13 other species have been reared there from time to time (SD Game, Fish and Parks photo).

corner of Blue Dog Lake in Day County, this hatchery is by far the largest of the three state-owned hatcheries. With 700 incubation jars for hatching walleye and northern pike eggs, the number of walleye eggs incubated often exceeded 150 million in the 1980s and 1990s. Unlike the Pickerel Lake hatchery, where all the fry had to be stocked or transferred some distance to remote rearing ponds, Blue Dog Lake State Fish Hatchery has over 53 surface acres of ponds on the hatchery grounds.

Fingerling walleyes and northern pike became a routine part of production and provided fisheries managers with fish numbers and sizes that were inconceivable prior to the construction of the Blue Dog Hatchery. The 30 indoor rearing tanks were used for a number of fish species, including chinook salmon and rainbow trout. At least 13 fish species, including paddlefish, muskellunge, largemouth bass, smallmouth bass, yellow perch, northern pike, and bluegills, have been reared at Blue Dog, although propagation has always concentrated on walleye fry and fingerlings. Blue Dog personnel also assumed responsibility for the natural rearing pond program in the northeast in the late 1990s.

"Natural" Rearing Ponds

The SDGFP began using northeastern natural rearing ponds for walleye production in 1963. The idea was to stock hatchery-reared fry in small, very productive, non-fish-containing wetland ponds in the spring, let the fish grow over the summer, and then harvest the fingerlings in the fall (Kinnunen 1996). These ponds also usually winterkilled, thus creating another fish-free rearing pond the next spring.

Jim Sprague, a fisheries manager in the northeastern part of the state, was the first to use this technique to harvest a few large walleye fingerlings in 1963 and 1964. The program was discontinued in 1965 because the abundant vegetation in the wetlands made seining difficult and unproductive. Nearly 20 years later, however, the program was restarted. The first few years were good learning experiences (i.e., not too successful), but in 1984, Fisheries Manager Ron Meester finally harvested a number of large fingerling walleyes from a Marshall County pond. Meester used frame nets, instead of seines, to harvest the fingerling walleyes in the fall, thereby eliminating the vegetation problems that had plagued earlier efforts.

Natural rearing pond use quickly expanded and by 1987, 11 ponds produced over 270,000 walleyes (6-8 inches) for fall stocking. In the early 1990s, the state entered into agreements with the Sisseton-Wahpeton Indian Tribe, private individuals in South Dakota, and a Minnesota company to obtain even more large walleye fingerlings for fall stocking. Yellow perch and black crappie fingerlings also were grown in natural rearing ponds starting in 1993, and the program spread throughout the entire eastern part of South Dakota. Typically 13 or 14 natural ponds are now used each year for fish production.

Gavins Point National Fish Hatchery

Four years after the completion of Gavins Point Dam, the Gavins Point National Fish Hatchery produced its first fish in 1961. Congress authorized the hatchery with the Flood Control Act of 1944 that granted the Department of Interior a permit for "constructing, operating, and maintaining a fish hatchery and other fish cultural improvements" on land owned by the U.S. Army Corps of Engineers. Two Congressional acts provided subsequent funding for the hatchery on June 13, 1956, and construction started in 1958 just west of Yankton. The ultimate operation of the hatchery was designated as mitigation for the construction of the Missouri River dams by stocking game fish species in the newly created reservoirs.

Gavins Point National Fish Hatchery became the largest fish hatchery in South Dakota with 26 rearing ponds and eight raceways. Two cooperative agreements with the state of South Dakota arranged for some of the fish reared at Gavins Point to be stocked in state waters. As a federal facility, however, only part of its production was destined for South Dakota's lakes and rivers. In addition to the Missouri River reservoirs, fish from Gavins Point have been stocked in waters of the national grasslands, Indian reservations, military bases, and national wildlife refuges. Fish have also been stocked in state-managed lands outside of South Dakota and certain species, like paddlefish, have been shipped internationally to Russia and other countries. More than five billion fish, representing at least 20 different species, have been stocked from the hatchery since it started operation.

Gavins Point National Fish Hatchery has grown since 1961. In 1983, the state paid to construct 10 rearing ponds, each 1.3 acres, at the hatchery. The U.S. Army Corps of Engineers subsequently reimbursed the state for the cost of construction. With the listing of the pallid sturgeon as an endangered species in 1990, additional facilities were added to assist with sturgeon recovery. An advanced rearing and broodstock holding facility was constructed in 2004 to complement the relatively new sturgeon building and the endangered species culture building. Although it continues to rear the typical sport fish usually associated with fish hatcheries, production at Gavins Point has evolved over time to focus on native river fish species that have been adversely affected by the construction of the six Missouri River mainstem dams. In addition to pallid sturgeon, the hatchery has been the primary producer of paddlefish stocked in South Dakota and has also reared other native river fish like blue catfish, flathead catfish, blue suckers, sturgeon chubs, and flathead chubs.

Stocking Missouri River Waters

The closure of Oahe Dam in 1959 and its completion in 1963 created the fourth largest reservoir in the United States. Gone was the flowing muddy Missouri River. In its place was a clear water reservoir stretching over 230 miles with a maximum depth greater than 200 feet. With permanent cold-water habitat, Lake Oahe soon became the site of a number of relatively unusual fish stockings.

Kokanee and coho salmon were introduced in 1970; followed by lake trout, rainbow trout, and opossum shrimp (*Mysis velicta*) in 1972; Bonneville cisco and spottail shiners in 1973; lake whitefish in 1977; brown trout in 1981; steelhead (migratory rainbows) in 1982; cutthroat trout and lake herring in 1984. Few of these stockings were successful, especially compared to the rapid colonization of Lake Oahe by rainbow smelt. Smelt were first found in Oahe in 1973, having been discharged from upstream Lake Sakakawea in North Dakota.

Another escapee from North Dakota proved well suited for the cold water of Oahe and as a predator on the abundant smelt. Chinook salmon were first detected in Oahe in 1976 and became numerous enough for state fisheries crews to spawn some fish in 1981. Eggs spawned from Lake Oahe salmon and eggs imported from Lake Michigan were used during the 1980s to sustain the salmon fishery on Lake Oahe.

Whitlocks Spawning and Imprint Station

In 1984, Whitlocks Spawning and Imprint Station was built on the shores of Whitlocks Bay, near Gettysburg. The spawning station was used during fall to collect fish for spawning (including rainbow and brown trout for the few years that those fish were stocked) and to increase numbers of salmon for anglers to harvest. The last stockings of Lake Michigan source fish occurred in 1988. Since that time, the salmon population in Lake Oahe has been maintained almost entirely with progeny of Lake Oahe fish.

Lake Oahe had been used as a northern pike brood source for a number of years, starting almost immediately after the reservoir began filling in 1966. As the reservoir aged and walleye began to dominate the fish community, Lake Oahe became the focus of state walleye spawning efforts each spring. Several facilities on the reservoir were developed to facilitate walleye spawning.

Spawning Stations

The Foster Bay Spawning Station was located in the lower part of the reservoir where the Cheyenne River emptied, whereas the Grand River Spawning Station was built upstream where the Grand River joined Lake Oahe near Mobridge. Each spawning station had a boat ramp and equipment to assist with spawning walleyes and northern pike. These stations were used heavily during the later 1980s and 1990s when Lake Oahe was the primary, and often only, source of walleye eggs for fish production in South Dakota. Siltation in the reservoir and the advent of the "spawntoon" boat (pontoon boat modified as a spawning platform) led to closure of these spawning stations in the late 1990s. Although fish were also spawned on the Moreau River arm of Lake Oahe, a permanent spawning station was never constructed at that site.

American Creek Spawning and Imprint Station

In 1985, the American Creek Spawning and Imprint Station at Chamberlain started operations. This facility was originally designed to be multi-purpose—an imprinting station for trout and salmon; a holding facility for lake run brown trout, smallmouth bass, and sauger; a spawning building for walleyes and paddlefish; and a headquarters for fisheries management crews working on Lake Francis Case and Lewis and Clark Lake. Chinook salmon, cutthroat trout, and brown trout were released from the station from 1986 to 1988, but Lake Francis Case lacked the necessary habitat for these fish to thrive. After discontinuing trout and salmon stockings in Lake Francis Case, the facility was renamed the American Creek Fisheries Station.

The primary fish-rearing focus of the American Creek Station has been paddlefish spawning and restoration since the late 1980s. After adult paddlefish are netted from Lake Francis Case, they are transported and put into 1,400-gallon circular tanks at American Creek. The eggs are surgically removed from the females, fertilized, and incubated for a few days before being shipped to Gavins Point National Fish Hatchery in Yankton (Graham et al. 1986). Each year since 1987, between 500,000 and 1.5 million paddlefish eggs are usually collected and incubated at American Creek.

Subimpoundment rearing ponds

For a time, the Missouri River reservoirs were stocked with fish propagated in small subimpoundments adjacent to the reservoirs. These rearing ponds were filled during spring, stocked with fish, and then subsequently drained into the reservoir. The Snake Creek subimpoundment, built in 1968 next to Lake Francis Case, was the first of its kind. The successful rearing of northern pike fingerlings was spotty at best and the pond was not used after 1986. The Platte Creek subimpoundment, built in 1983, also had unpredictable rearing success and was converted to a kid's fishing pond in 1990. Other subimpoundments, such as Blue Blanket, Oahe, and Spring Creek, were constructed in 1983 and 1984 on the upstream reservoirs, used for a few years, and then abandoned.

Fish for Spawning

The first fish stocked in South Dakota came from sources outside the state. After the early hatcheries were built, brood fish supplying the eggs and sperm needed for reproduction came from a variety of sources.

Trout and Salmon

Trout eggs initially came entirely from out-of-state broodstocks until Cleghorn Springs State Fish Hatchery and the Spearfish Fish Cultural Station began to hold older fish for spawning. Subsequently, eggs came from South Dakota hatchery broodstocks, as well as from fish in hatcheries (federal, state, and commercial) or lakes elsewhere. At one time, two different strains (genetic types) of rainbow trout and one strain of brown trout broodstock were held at Cleghorn Springs. This has changed over time, so that in 2004 no trout broodstock were held at any hatchery in South Dakota. All trout eggs now either come from federal broodstock hatcheries or from sources in Wyoming.

Chinook salmon eggs are spawned from wild fish that ascend the fish ladder at Whitlocks Spawning Station in October. Eggs from North Dakota salmon have also been used in the past and Montana fish may become a future source of salmon eggs.

Walleye

Most of the initial state walleye spawning efforts occurred on Lake Kampeska. During the late 1930s and early 1940s, spawning operations shifted to Pickerel Lake and other lakes in the northeastern part of the state. Spawning efforts then shifted again to Lake Oahe from the 1960s through the mid-1990s, particularly with the building of spawning stations. In the late 1990s, walleye spawning became more opportunistic and broodstocks were used for spawning wherever they could be most easily obtained.

Relatively unknown lakes in the northeastern part of the state that filled with water in the early 1990s have provided most walleye eggs since 2000. For a number of years in the late 1990s, Park's Pond supplied nearly all of the eggs used for walleye stocking in South Dakota. For some unknown reason, the walleyes swimming in this former natural rearing pond were all females. While this made egg collection relatively easy, males had to be transported from other locations to fertilize the eggs. Walleye eggs from Minnesota, North Dakota, and Nebraska broodstocks have also been stocked in South Dakota.

Warmwater Species

Warm-water broodfish, such as largemouth bass, have come from a variety of places. Initial stockings by the U.S. Bureau of Fisheries came from out of state. When South Dakota began rearing largemouth bass, the first broodfish came from places such as Lake Andes or Wylie Lake near Aberdeen. Both largemouth and smallmouth bass broodfish have been held at the Blue Dog Lake State Fish Hatchery, where fish from wild sources supplement the hatchery broodstocks. Bluegills, crappies, and other sunfish came from any number of unrecorded sources.

Hybrids

Artificial spawning and propagation has allowed fish hatchery personnel to tinker with Mother Nature. At least four hybrid fish have been produced and stocked in South Dakota. The first recorded hybrid stocking was that of a bluegill and pumpkinseed cross from the Lake Andes hatchery in 1950.

In the early 1980s, tiger muskies (muskellunge x northern pike) were first stocked into Lake Mitchell, and over the next few years in Lake Sharpe, Orman Reservoir, Lake Kampeska, and Lake Poinsett. Since 1998, only Orman Reservoir has received these fish.

Splake, produced by fertilizing the eggs from a lake trout with sperm from a brook trout, were first stocked into Deerfield Reservoir in the Black Hills in 1975. No stockings occurred again until 1984, after which splake were stocked into Pactola Reservoir until 1991 and Deerfield Reservoir until 1999.

The most recent hybrid to start swimming in South Dakota waters was created by crossing saugers with walleye. These "saugeyes" also occur naturally in South Dakota, particularly in the Missouri River reservoirs and tailwaters (VanZee et al. 1996). Hatchery-produced saugeyes were first stocked at LaCreek National Wildlife Refuge in 1986, and in the Little White River Project and Wall Lake in 1990, with subsequent stockings throughout the state in places like Richmond Lake, Elm Lake, Mina Lake, and numerous smaller water bodies. Most saugeye stockings occurred from 1995 to 1999, with only a few waters receiving any of these fish after 2000.

Introduced species

Introductions of fish continued in the latter half of the 20th century. Muskellunge were first put into Amsden Dam in 1975 and then stocked in several other water bodies during the next 36 years.

Smallmouth bass were first stocked in 1964 in the Redwater River, a tributary to the Belle Fourche just north of the Black Hills, and other western streams and creeks were stocked with smallmouths in the late 1960s. After Angostura Reservoir received these fish in 1971, stockings became more widespread. Subsequent introductions into the Ft. Randall tailwaters (1972), Lake Sharpe (1980), the northeastern lakes (1982), Lake Oahe (1983), and Lake Francis Case (1985) proved to be extremely successful. Many of these populations are now maintained entirely by natural reproduction.

Spottail shiners were stocked into the Missouri River reservoirs in 1973. Most large reservoirs and even some natural lakes subsequently received stockings of these fish, with Pactola Reservoir getting the last stocking in 1992.

Gizzard shad stockings started with Lake Oahe in 1982, followed by Orman Reservoir and Lake Marindahl in 1983, Lake Thompson in 1989, Angostora Reservoir in 1990, and Shadehill Reservoir in 1999.

Golden shiners were stocked in Ottumwa Dam in 1966, the first recorded golden shiner stocking since the 1920s. A few other west river dams were stocked in the 1960s, and then in the latter half of the 1970s, golden shiners were stocked in waters across the state.

Emerald shiners were stocked into Angostora Reservoir in 1990 and Pactola Reservoir in 1992.

Private Hatcheries

For some time prior to the 1950s, fish hatcheries in South Dakota could only be operated by governmental agencies. After legislation in 1953 made it legal to grow fish commercially in South Dakota, private individuals began growing fish for stocking in privately-owned waters or to sell for food.

Tom Simpson started Trout Haven in the Black Hills in 1953 with catchable-sized trout that he purchased from commercial hatcheries in Idaho, but he quickly shifted to growing his own fish. Trout Haven became popular with tourists who could pay by the inch for the trout they caught from the well stocked private ponds—and no fishing license was required.

When Trout Haven Ranch began shipping fish, its first exports went to Colorado in 1966. Over the years, the hatchery has provided fish to 16 states and three Canadian provinces. South Dakota purchased 40,000 pounds of catchable rainbow trout from Trout Haven Ranch after the Rapid

City flood of 1972 to fill the void until production resumed at the soon-to-be-rebuilt Cleghorn hatchery.

Tom's son Steve took over operation of the hatchery in 1974 and, in cooperation with his wife, opened the Eden Valley Trout Farm nearby. This is the largest cold-water hatchery complex in the state, producing over 300,000 pounds of rainbow trout each year. Brown trout and brook trout have also been reared at this complex, but were only grown for a few years because "There was a lot more work than money in them."

Bill and Babe Cummins opened the Black Hills Rainbow Trout Farm just north of Spearfish in 1954. They grew fish primarily for restaurants and grocery stores, but some fish were also sold to private landowners in South Dakota. Allen Dittman purchased the farm in 1969 and added a fee-fishing pond on the hatchery grounds. After a flood in 1972, the hatchery was immediately rebuilt. The Great Flood of 1976 ended the business until Dittman's sons resurrected the Black Hills Rainbow Trout trademark in the 1990s. Since 2003, this hatchery has been leased, with most of its production sold in Colorado.

The rainbow trout production from the Linstad Trout Farm generally ends up released into private stock ponds in western South Dakota. Wayne Linstad started the business in the 1960s just a few miles north of Spearfish and continues to sell mostly fingerling trout to ranchers and other private pond owners.

Many other trout hatcheries, such as Howard Morrison's ponds on Little Spearfish Creek, have come and gone in the Black Hills area. Private hatcheries have also grown tilapia at Philip and catfish in the southeastern part of the state, but these species have been grown for market.

The "Volunteer" Stocking Crews

No recounting of fish stocking in South Dakota would be complete without mentioning that the public has intentionally or unintentionally stocked fish. State regulations say, "A person may not transplant or introduce live fish or fish eggs into public waters." However, lakes in the Black Hills are replete with examples of novel fish seemingly appearing out of nowhere. Fish like largemouth bass, northern pike, green sunfish, rock bass, yellow perch, and countless other species that were not stocked by the SDGFP abound in waters more suited for trout. Where did these fish come from?

No one knows for sure. For example, unnatural fish species were found in Deerfield Reservoir soon after all fish were intentionally removed by lowering water levels and applying fish poisons. Illegal fish stocking also became apparent during the high-water years of the 1990s when new lakes were created in the northeastern part of the state. Many of these species were probably stocked inadvertently when anglers illegally dumped bait buckets. This may explain the widespread distribution of most bait species, as well as the occurrence of nonnative fish, such as European rudd in places like Newell and Sheridan lakes.

Swimming Forward

Fisheries management in South Dakota will almost always entail at least some stocking of hatchery-reared fish. South Dakota's unpredictable precipitation, widely-fluctuating temperatures, and other climatic vagaries often don't allow Mother Nature to provide suitable numbers of fish to satisfy the angling public. Recovery of endangered fish, like the pallid sturgeon, is an important mission for fish culturists in the future. To meet future fish production needs, all three state-owned hatcheries will be renovated by 2010.

Despite the continued demand for hatchery-reared fish, the days of stocking willy-nilly are gone (Stroud 1986). Greater understanding of the negative impacts of hatchery fish on wild fish behavior, genetics, and numbers will lead to even more justification for where and when fish are stocked. Instead of just recording how many fish are stocked, fisheries personnel will focus more on how those fish fare after they leave the hatchery. (Do they survive? Are they ever caught?) As hatchery staffs incorporate novel rearing techniques that maximize the survival of stocked fish, the numbers of stocked fish may well be reduced in the future.

Acknowledgments

There is no way this chapter could have been written without the help of many people. I am indebted to Randi Smith of the D. C. Booth National Historic Fish Hatchery, Dennis Unkenholz of the SDGFP, and retired South Dakota Fisheries Chief Robert L. Hanten. I also appreciated the contributions of Brian Blackwell, Herb Bollig, Steve Brimm, John Carriero, Rick Cordes, Jack Erickson, Robert P. Hanten, Ron Koth, Keith McGilvray, Ron Meester, Michael Muck, Will Sayler, Craig Soupir, Gerry Wickstrom, and Keith Wintersteen.

Information for this Chapter was primarily obtained from U.S. Fish Commission Reports, such as the Report of the U.S. Fish Commission Investigation Team, Annual Fisheries Reports (Conservation Highlights) of the South Dakota Department of Game, Fish and Parks, Conservation Digest articles, hatchery brochures, the South Dakota Department of Game, Fish and Parks stocking database, personal interviews, and material (particularly records and correspondence from Les Ripple) archived at the DC Booth Historic National Fish Hatchery. ❖

Chapter 15

Commercial Fisheries and the Baitfish Industry

Robert L. Hanten and Robert P. Hanten

> " The ubiquitous carp and brown trout
> Came from Europe with sponsors devout.
> *Salmo trutta* won fame, but who is to blame
> For the cyprinid we should'a kept out? "
>
> — Anonymous

People have exploited South Dakota's fisheries resources for subsistence, barter, recreation, and profit for many years. Archeological evidence from Indian villages indicated that buffalo fish and catfish were common fare for tribes along the James River and the state's eastern waters. Bone fish hooks, willow fish traps, and fish spears are common artifacts from most village sites from the Missouri River eastward. It's doubtful that the native people relied on fish as a commercial commodity, but they were certainly used for subsistence and some local bartering.

Pioneer immigrants fished with hook and line, seines, nets, traps, and spears taking all species by the wagon load during the 1860s and 1870s before game and fish laws were enacted. A few made a business of selling fish to augment their finances. Commercial fishing was not as big an industry as commercial harvest of big game, prairie chicken, and waterfowl but a variety of fish did find a place in the market. The fish industry continues with ups and downs and it is also a tool of fish management.

The eastern portion of the state supported most of the commercial fishing until 1960. Common native game fish were pike, perch, and sunfish whereas the common commercial fish were suckers, minnows, drum, bigmouth buffalo (Figure 15.1) and common carp (Box 15.1).

In the Missouri River, catfish, sturgeon, and paddlefish are by law game fish, but were considered commercial fish under special circumstances. The Missouri's unique turbid river ecosystem was changed in the 1950s and 1960s into large, deep, cool-water lakes and a clear-water river system. The fish community also changed and, for a while the Missouri River supported a substantial commercial fishery.

Figure 15.1 Bigmouth buffalo (*Ictiobus cyprinellus*) was one of the top commercial fishes in South Dakota. (U.S. Fish and Wildlife Service photo)

The western half of the state, with the exception of the Black Hills has several turbid warm-water rivers and streams, just a few shallow warm-water lakes, and three large irrigation reservoirs. Catfish, suckers, and minnows are the major fishes in the rivers, but commercial fishing never developed in the western part of South Dakota.

The Early Years

The Dakota Territory consisted of both present-day states of North and South Dakota. When the more conservation-

15.1
All Of A Carp Is Used

Most of the common carp and buffalo fish are sold to Stoller Fisheries, a wholesale broker in Spirit Lake, Iowa. Stoller ships some fish on ice to wholesalers in Chicago, New York City, and Memphis; some are shipped live to Eastern markets for religious holidays. Larry Stoller, President of Stoller Fisheries, takes pride in "excellent rabbinic control" during the processing of kosher fish products for Jewish markets. Carp and buffalo may also be processed into fillets, fish sticks, and patties for mid-western markets where specialty fish food restaurants feature deep fat-fried carp and buffalo fish sandwiches. Some carp are smoked. Some carp meat is minced for dehydrated soup bases and fish stocks. Carp caviar, a Greek speciality food called *tamama*, is made from carp roe (eggs). The roe is also sold for use in sauces, creamed soups, dips and garnishes.

Stoller's could process more of the carp than just the flesh. Carp skin can be tanned into fish leather because the skin has a thick collagen layer as does the dermis of mammal skin. Carp pituitary glands, which are the source of reproductive hormones, can be harvested and sold ($285/oz in 2006) to fishery biologists to inject into other fishes to force spawning. Carp heads (non-split and split) are sold for commercial trapping baits, and guts can be sliced into 6-inch sections and sold as catfish bait. Air bladders may be dried and ground into powdery isinglass, which is used mainly for the clarification of wine and beer. The skeleton, known as the "rack," may be ground to bone meal. The remaining offal may become fish meal, formulated fish food, or liquid fish fertilizer, a low-pH plant food used on flowers such as geraniums.

minded residents became concerned about dwindling fish populations, the territorial legislature in 1883 passed the first laws establishing fishing seasons and methods. The laws did not sit well with people who depended upon game and fish for their livelihood. All sorts of methods were used to capture fish, but new laws stated that anglers were allowed to take fish only by hook and line and with only two hooks per line. In 1889, South Dakota became a state and required a license to hunt and fish. It was legal to own a fish trap, gill net, or seine, but not to use it for taking food (game) fish. Illegal fishing, however, was difficult to prove unless the user was caught in the act. In 1899, spearing became legal for buffalo fish and suckers, but all other species had to be taken by hook and line.

A fish commissioner was appointed by the governor to protect and restore the over-exploited fisheries. His duties were to stock fish received from the U.S. Commissioner of Fisheries and to open and close stocked waters. County sheriffs, deputies, and constables enforced the laws and justices of the peace in the counties tried the violators. Illegal netting, dynamiting, and selling of fish continued with the attitude that fish and game belonged to the public and could be used as they (the people) wanted regardless of long-term consequences.

The fish commissioner arranged for fish stocks from the U.S. Fish Commission. Unfortunately, one of the species selected to replace the "inferior native species" was the common carp. Common carp fry were stocked in present-day South Dakota waters from 1886 until 1911. There is no guarantee that South Dakota waters would be free of common carp today if they had not been stocked then. Common carp were being introduced throughout the nation at that time and since most of South Dakota's waters are connected to the Mississippi and Missouri river drainages, within a short time they would have immigrated. As early as 1914, the state was contracting with commercial fishermen to remove the excessively abundant populations of common carp, along with bullheads, buffalo fish, suckers, and freshwater drum.

In 1893, fish wardens were hired to enforce the new laws and stop indiscriminant netting and other violations. Up to six fish wardens could be appointed by each county board to serve at their pleasure with no compensation except the fines that they collected. They did a big business. The enforcement incentive was high because half of the fine was split between the warden and local justices of the peace and the other half was allotted to the state. The attitude of some people, however, remained that fish and game resources were their rightful due, and that no one could stop them from taking what they wanted.

In one instance, a person that the warden had arrested for seining assaulted him with a shovel handle, and the warden was forced to shoot the violator to save his own life. Wardens' comments at the time indicated that their most serious problem was "…a certain class of foreigners who infest the fields and woods at all seasons intent only upon the slaughter of all kinds of birds (game and fish), and who set at defiance of game and trespass law…"

In 1909, the South Dakota Department of Game and Fish and the Game and Fish Commission were created by the state legislature. A state warden was appointed to enforce the laws and regulations and manage the day-to-day operations of the department. A law was also passed placing the seining of rough fish under the jurisdiction of the Department of Game and Fish. The first record of department-contracted rough fish removal was that same year when 566,235 pounds of non-game fish were seined from the state's lakes and streams.

The new law provided a mechanism to contract with commercial fishermen to remove rough fish, and a percentage of the harvest value was paid to the state for supervision and administration. Second class wardens were hired by the day to ensure game fish were handled correctly and returned to the water. The new law also allowed the commission to permit private groups to seine rough fish without compensation to the state, provided they paid for the warden supervisor. The deputy state wardens determined which waters could be commercially fished by private groups. Required reports of these operations included the numbers and pounds of each species taken, and the amount paid to the state.

The original objective of the rough fish commercial program was to reduce competition with the more desirable game fish by removing nongame buffalo fish, bullheads, suckers, and common carp. The program would also provide a source of low-cost food to the state's citizens during economically difficult times. The cost of the program was paid for by the fees charged to commercial fishermen either on a flat rate or on a percentage of their catch.

The Fishermen and Harvest Techniques

Browns Valley, a small community on the Minnesota-South Dakota border, was the focal point for the commercial fishermen who did the contract seining for rough fish. Both Minnesota and South Dakota enlisted their services. The commercial fishermen were numerous in the region because of the numerous lakes in the area where legal and illegal commercial fishing was common before state regulations.

Family names associated with commercial seiners over the years from Browns Valley included the Randalls, Peichowskis, Reeds, Raws, Millers, Ewalds, Pages, and Jarkas to list a few. Other commercial fishermen from South Dakota included Bill Williams and Tracy Muchler from Watertown and Jack Raw from Lake Norden. Contract commercial fishermen on Missouri River reservoirs in the 1960s and 1970s included Richard Holcom from Chamberlain and Ed Walker, Melvin "Mutt" Brewer and Mark Jackman from Mobridge.

Commercial fishermen were a colorful bunch of characters to say the least. Their difficult profession caused them to work hard, talk rough, and play hard when they could. When things were not going well on the lake, the air was "blue" and not because of the cold temperature. Usually the boss directed the seining from a vantage point where he could see the whole operation. In the days before wireless communication devices were common, the roar of the seining boss echoed across the winter ice as they cussed some worker for not properly tending the net.

These fishermen were the icons of the trade. Their skillfulness and expertise was, and still is, paramount to the successful business of rough fish and bullhead removal from the state's waters. The reservoir netters had the talent and tenacity to fish the large open-water lakes in all kinds of weather, hauling heavy loads of netting gear and fish in small open boats. They knew where to place the nets for the best catches. These fishermen were, and some still are, masters of their trade.

The winter seiners often exhibited a sixth sense in locating concentrations of rough fish under three feet of ice, and they possessed a special talent for laying out a seine that covered nearly a square mile of lake area. They were usually able to avoid snags that could hang up or tear big holes in the net and allow the catch to escape.

It is a spectacle to see fishermen lay out a 5,000-foot net for a winter seine haul. They start by cutting a large hole in the ice where the net starts. Then they cut numerous holes in the ice (originally chopped by hand and now drilled by power augers) that are 30 feet apart in the huge circle that the seine will enclose. A running board is a long board that is pushed from hole to hole under the ice to deploy a towrope. Engines pull the towrope and the net along in the predetermined circle, eventually reaching the take-out hole where the net and the rough fish catch are removed (Figure 15.2).

The take-out hole size and placement are extremely important. The hole needs to be in water deep enough to accommodate the net and the catch so the rough fish can be removed and the game fish returned to the lake without

Figure 15.2 Dipping for buffalo fish from a seine hole; wooden boxes await filling with fish for transport by the truck parked on the ice in the background (SDGFP photo)

losses from suffocation or over-crowding. And, the ice around the take-out hole must be accessible for large trucks that will haul the rough fish to market. Catches of rough fish in one seine haul can exceed 100,000 pounds. These operations take all day, starting before dawn and often ending after dark. It is strenuous and dangerous work on ice and in cold water and cold weather.

While winter seining under the ice is the most common method of removing rough fish, commercial fishermen also seine and trap in open water where fish concentrate. Bullhead thinning operations with bullhead pocket nets are done both during ice cover and in open water. Pocket nets have two hoop nets at either end of a 30-foot lead or block net.

Local sportsmen's groups often cooperate in removing common carp from waters, either because of their special interest in improving sport fishing or the lack of interest by commercial operators. With nets often borrowed from the state, these groups seine open water lakes or trap rivers and streams during the common carp spawn when concentrations of fish generally provide suitable numbers for their effort to be worthwhile. Harvested common carp are sold or given to the local people or farmers for hog food.

The Eastern Lakes

In the 100+ years that rough fish removal has been practiced in eastern South Dakota waters, the harvest has varied annually between 3.8 thousand and 3.5 million pounds (Figure 15.3). The average catch for the 55 years for which data are available was just over 2 million pounds per year. Common carp, buffalo fish, bullheads, and freshwater drum comprise the bulk of the catch. Other species such as suckers, and in some rare cases white bass, are caught and also removed, but they commonly make up less than two percent of the catch.

Many of the larger lakes like Poinsett, Kampeska, Madison, Brant, Pelican, Big Stone, and Traverse need almost annual attention from commercial fishing. Medium and smaller waters may be fished several years in a row and, after the rough fish population has been reduced, can be left alone until

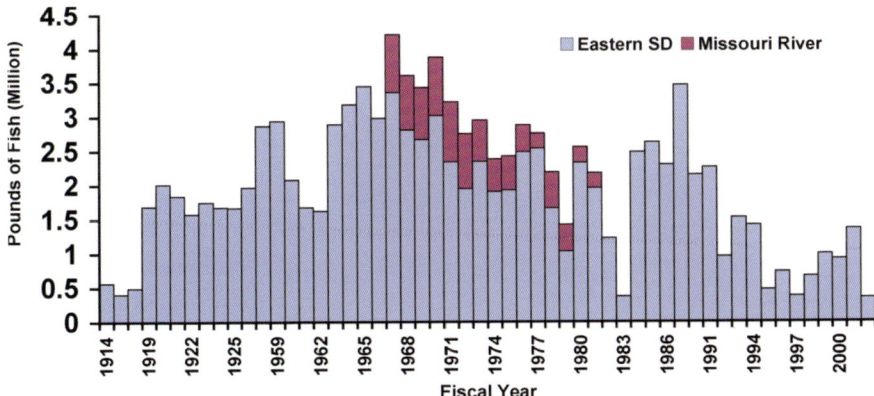

Figure 15.3 Pounds of commercial fish harvested in South Dakota from 1914 to 1925, from 1957 to 1986, and from 1989 to 2004. The figure was constructed by synthesizing annual reports of the Department of Game, Fish and Parks and published articles (1861–2004) about the fish resources of South Dakota.

the populations become a nuisance again. There are about 100,000 acres of water in 100 lakes east of the Missouri River that often need the rough fish removed. While some of these may just need bullhead populations thinned (Figure 15.4), most of the waters also have over-abundant populations of common carp and buffalo fish.

The opportunities for commercial fishermen wax and wane with the weather, much the same as for the dry land farmer. Unlike the farmer, however, there is no crop insurance or

Figure 15.4 Bullhead scoop; commercial fishermen are sometimes hired to harvest bullheads where bullhead populations are a nuisance. After the population is reduced, the remaining bullheads and other fish usually grow larger (SDGFP photo)

price support. Commercial fishermen are on their own when drought and winterkill occur. During the hundred years of commercial fishing history in South Dakota, there have been two major and several minor droughts. After the exceptionally high water years of the early 1900s, the prolonged dry periods of the late 1920s and 1930s caused devastatingly low water levels in South Dakota. In 1932, Madison, Oakwood and Poinsett lakes winterkilled, and by 1933 all fish were gone from half of the state's natural waters. By 1936, only six lakes had enough water to support fish.

In 1926, the state passed a new law prohibiting commercial fishing licenses and repealed the old commercial seining laws, however, contract rough fish removal was still allowed on some lakes. Commercial fishermen seined fish in lakes that were sure to winterkill, and rescued and moved game fish to better waters and tried to market the rough fish. The state also had crews doing rough fish removal and fish rescue.

The South Dakota Superintendent of Fisheries, R. L. Ripple, commented, "Even the beastly carp have suffered from the drought." Reports in 1931 showed that 1.3 million pounds of common carp and other rough fish were seined, but they could not be sold in eastern markets and were used mostly as hog feed. In 1932, an estimated 650,000 pounds of common carp were distributed to relief organizations, and in 1933, an estimated 2.5 million pounds of common carp from Lake Andes were given to the needy. In 1934, only Lake Kampeska provided commercial fishing receipts ($196) from contract operations, compared to six lakes that produced about $5,000 in receipts the year before.

Things got even worse for commercial fishermen. In 1935, the state decided to use state crews to remove rough fish. Again, only Lake Kampeska was seined and 275,000 pounds of common carp and buffalo fish were taken, which the state sold to local people for 3–4 cents per pound. The state netted $3,750 after buying out a commercial fisherman's equipment and paying the 15-man crew to seine fish.

Water conditions improved in the late 1930s and by 1940 three waters were fished for rough fish. The water conditions in 1942 were the best in years, and many lake levels were restored. Four waters that year were netted for rough fish, grossing the state about $9,400. The same trend continued through 1944, with four different waters fished each year. The catch and receipts increased as more lakes came back to full production. By 1945, the state was again contracting for rough fish removal, and received about $22,300 that year.

Sport fishing was good over the next 10 years and commercial removal of rough fish also increased with receipts ranging from $15,500 to $56,200 per year. Pounds caught and total catch records were not kept from 1926 until 1956. From the late 1950s to 1961 there was another but lesser drought. Catches averaged 2.3 million pounds per year in the late 1950s, but declined to 1.6 million pounds in 1960.

Water levels recovered through the 1960s and catches went up to three million pounds per year, peaking at over four million pounds by 1967 with the Missouri River impoundments adding to the total catch (Figure 15.3). Commercial fishing in the Missouri River actually started in 1957, but the available records did not differentiate catch location and Missouri River catches were included with those from the eastern South Dakota lakes until 1967. State biologists Don Monroe and Don Warnick attempted to summarize the activities of the commercial fish industry in periodic reports (see references). Much of the information presented here is taken from these unpublished reports.

The 1970s were plagued with another major drought. Waters like Poinsett, Oakwood, and Madison winterkilled in 1976. By 1978, over 200 fishing lakes were out of production. By 1983, catches had declined to a low point of only 380,439 pounds because eastern waters were still recovering and the Missouri River buffalo fish populations had collapsed.

When improved water levels again restored the fisheries in the eastern lakes during the late 1980s and early 1990s, commercial catches went up, peaking at nearly 3.5 million pounds in 1989. Annual catches were good until the mid-1990s when abundant rains flooded sloughs and marginal lakes to depths that had never been witnessed. The fish populations followed the waters and became diluted in the thousands of acres of new lakes. Former duck marshes like Lakes Thompson, Whitewood, Preston, Waubay, Bitter, and many others that had been only a few feet deep became 20-30 feet deep, capable of sustaining fish life throughout the year.

Commercial catches declined, but populations increased because the whole aquatic ecosystem in eastern South Dakota changed and tripled or quadrupled in size. Traditional netting areas, except in the more stable lakes, were gone. The need for rough fish removal was low because nature had expanded water volumes and decreased population densities. By 1997, commercial catches were down to 375,597 pounds. They slowly returned to a million pounds per year by 1999 and remained there into 2001. With all the new waters, there will be a future need to harvest rough fish when populations become excessive.

Missouri River and Reservoirs

Since the first settlement along the Missouri River, a small group of fishermen trapped, hoop netted, and set lined catfish and rough fish for their own use and for the local market. These fishermen have persisted to the present with their small but locally important commercial fisheries. By 1937, they were required to have an annual license. Regulations now restrict the areas fished, the net mesh size, fish size limits, and where, when, and how they can fish.

Attempts were made to document the size of their annual catch and commercial value, but viable figures could not be obtained because of the variability of participants from year to year and the reliability of the information. Hoop netters have at times provided the state management crews with a source of large fingerling channel catfish, also known as fiddlers, for stocking other waters. Today's hoop netters can truly be categorized as modern-day "River Rats." The people of Yankton still celebrate Riverboat Days as a part of their town's heritage and affectionately call selected people "River Rats."

In 1957, contract commercial fishermen were recruited to remove the booming populations of buffalo fish and common carp developing in the new Missouri River impoundments (Figure 15.5). As the reservoirs filled, flooded uplands created abundant new feeding, spawning, and nursery areas for expanding populations of both game and rough fish. The large, highly fluctuating reservoirs added nearly 400,000 acres of commercial fishing waters to the state and presented challenges not encountered on the eastern natural lakes and smaller impoundments.

Fishermen who were experienced in netting large reservoirs in Oklahoma and Texas came to South Dakota to remove rough fish. The reservoir fishermen used larger trap nets than had ever been seen in South Dakota. They also fished large mesh gill nets that caught big buffalo fish and

Figure 15.5 Commercial fisherman, R. Holcom, netting buffalo fish on the Missouri River. Flooded timber in the background tells of the 1970s when commercial fishing was booming in reservoirs. Holcom used 4-inch mesh nets that caught only the largest (8 to 15 pound) common carp and buffalo (photo by K. Wooster)

common carp but allowed smaller game fish to pass through unharmed. The fishermen only fished during open water periods beginning in the spring after the game fish had completed their spawning activities. Cold storage facilities were constructed and maintained at Mobridge and Chamberlain to keep the catches fresh until the fish could be transported to eastern and southern markets. To protect the new fisherman's investment, limited entry contracts were granted by the state.

The expenses of reservoir commercial fishermen were much higher than the expenses of eastern lakes fishermen; however, the high quality of their catches and the timing of fish harvest activities kept them competitive. Their techniques worked well and they enjoyed a relatively good fishery for about 15 years. During that time, approximately 500,000 pounds of buffalo fish per year were marketed from Lake Oahe and Lake Francis Case. Common carp, catfish, freshwater drum, suckers, and goldeyes were among other species marketed.

The U.S. Bureau of Commercial Fisheries became interested in reservoir commercial fishing, and the agency set up a Lake Oahe field station in 1963 at Mobridge, SD. Their goal was to investigate the potential of large inland reservoirs for commercial fish production. They focused on common carp, buffalo fish, goldeyes, white bass, and other less used species. They were also interested in the longevity and stability of the commercial fisheries. The field station was closed in 1970 when funding was assigned to other regions. By then it was clear that the commercial fishing potential of the northern reservoirs was limited and catches declined greatly.

In 1983, commercial removal of rough fish from the Missouri River impoundments was suspended. The abundance of common carp and buffalo fish had followed the typical boom and bust pattern of newly created waters and the economics of the reservoir commercial fisheries had become severely marginal for contracting fishermen. The fish management benefits were also questionable. A moratorium on commercial fishing was placed on Missouri River reservoirs to determine if the buffalo fish commercial fisheries (the economic backbone of the industry) would recover. After five years, it was evident that buffalo fish populations would not regain the levels needed to support a sustained commercial fishery.

Fisheries managers and anglers also questioned commercial netting of other low volume, high value species like goldeye and white bass and the potential conflicts with the nationally important walleye sport fisheries and the developing salmon fisheries. Contract commercial fishing on Missouri River impoundments was suspended indefinitely except for a special salmon fishery at the Whitlock's Bay Spawning Station on Lake Oahe.

Chinook salmon adults caught by the state's fisheries crews during their egg-taking operations each fall on Lake Oahe are marketed through a contract with commercial game processors. This began in 1985 and continues each year. The fish are processed locally and sold fresh or smoked. The volume is quite low and has varied from 3,200–20,000 pounds per year. This operation uses the salmon that die after spawning and makes them available as a quality food product.

Commercial Fishing Industry Problems

Fisheries managers and sportsmen have debated the benefits of commercial fishing for as long as the program has existed. Controversy over the real benefits to fisheries management, contracting procedures, detrimental effects on the in-lake environments, and accusations of fraud have tended to keep the program in continual flux. Low market prices for rough fish also have been a nagging problem with the industry for many years and threaten its future.

Recreational anglers accuse commercial anglers of illegally marketing sport fish, but with the supervision required at any removal operation, it is doubtful much illegal marketing has occurred. Some objections to the use of a public resource for private profit always arise. The situation could be compared to mining and irrigation on public lands. Commercial fishing benefits to public waters probably outweigh detriments because the state receives a service at no cost to the taxpayer or sportsman.

It is difficult to document the long-term benefits of rough fish removal to the sport fisheries. Results are generally overstated and promoted without good data. Large quantities of rough fish are certainly removed, but whether it really helps sport fishing has never been determined because of the many variations in fish abundance and sportsmen's harvest with or without the program.

Game fish survival after being returned to the lake by commercial operators has not been thoroughly investigated, but losses have occurred due to *immediate* netting mortality and *delayed* stress-related factors. The immediate loss of game fish because of injury does not appear to be excessively high, but delayed mortality has never been measured for winter-seined game fish.

Dragging a large net across acres of lake bottom in the winter under ice cover definitely has some negative effects on lake habitat. However, the detriments, which have been investigated, are assumed to be about the same as the feeding and rooting effects of the large populations of common carp that were to be removed.

A few out-of-state fish brokers have controlled the wholesale market of rough fish over the years. These buyers have been able to manipulate opportunity and prices to their advantage outside the range of normal supply and demand principles of the competitive system. Consequently, local commercial fishermen sometimes cannot get a fair market value for their fish. That controlled market, coupled with the lack of normal price increases, has put South Dakota's commercial fish industry in serious trouble.

Development of the aquaculture industry has also hurt commercial fish markets. Catfish and trout are now raised in captivity and are increasingly replacing wild fish at fresh and frozen food markets. Aquaculture of other species like tilapia and salmon are taking up additional market shares as people have become concerned with the origin and possible contamination of wild fish stocks that have an unknown history compared to pond and pen-reared fish.

The value of the fish marketed commercially has not materially increased over the past 90 years. Common carp that were selling for 4 cents a pound in 1915 are only worth 5–7 cents a pound in 2006. The prices for other species have also remained low while operational costs, including labor, equipment, and transportation, have risen dramatically. A commercial fisherman who owns his equipment can stay a bit ahead and make a modest profit, but any new contractor trying to start up is in a losing situation because of interest on loans and the high cost of new equipment.

Baitfish Industry

A baitfish is defined as any fish of the minnow family (except grass, common, silver and bighead carp, and goldfish), fish of the sucker family (except buffalo and carpsucker), and fish of the stickleback family. Most bait vendors primarily sell fathead minnows and white sucker "chubs," but many also sell golden shiners and creek chubs. Anglers may seine 12 dozen baitfish of any combination but not including state and federally listed, rare, and endangered species.

The baitfish industry in South Dakota has evolved into a well-established business. In the early 1900s, however, the collection of bait was considered an individual right and had no organized commerce. Retail bait dealers have changed from "Mom and Pop" bait shops that harvested and sold their own bait to modern-day one-stop retailers. Wholesale and export bait dealers have also modernized equipment and techniques that reduce losses while hauling and holding large quantities of bait.

As the demand for bait increased over time, so did the need for additional regulations and monitoring to protect the resource. Two items have remained constant in this industry—the harvest of baitfish is a labor-intensive, dangerous job and the South Dakota Department of Game, Fish and Parks (SDGFP) still struggles with getting accurate harvest and sales estimates of baitfish. The baitfish industry is filled with many colorful characters and unusual business ventures.

The largely unregulated taking of bait from public and private waters occurred until 1939 when the South Dakota legislature enacted laws requiring resident and nonresident retail and resident wholesale bait dealers to be licensed. A retail license allowed a person to raise, trap, seine, buy, and sell baitfish to the public. A wholesale bait license permitted the licensee to raise, trap, seine, buy, and sell baitfish to any licensed retail or wholesale bait dealer. These laws were implemented to protect and perpetuate the bait resources of the state and are still in place today.

The baitfish industry is based primarily in eastern South Dakota where productive, marginal lakes and wetlands provide ideal habitat for native fathead minnows. Leasing bait-fishing rights is a second source of income for some landowners. Wetlands have many values and most are difficult to quantify, but aquatic bait species is one wetland product that has a definite value. The wholesale value of fathead minnows as bait averaged $94 per wetland acre (range = $0–$186 per acre, 1988 values) in six wetlands.

Fathead minnows can survive the South Dakota winters because of their tolerance of low dissolved-oxygen levels. Fathead minnows thrive in marginal lakes and wetlands where zooplankton are abundant and fish predators are few. The fathead minnow's multiple spawning capabilities enables populations to annually increase from several hundred fish per acre to hundreds of thousands per acre. In 2001, harvest estimates of minnows from public waters ranged from about 1 pound to about 220 pounds per acre.

White sucker chubs are taken from private waters or are imported. In private waters managed for chub production, bait dealers collect white sucker spawn, hatch the eggs, and stock resulting fry for later harvest.

The early years of the baitfish harvests and sales were primarily "Mom and Pop" retailers (resident retail dealers). It was not uncommon to purchase minnows or chubs from a retailer that operated a small, seasonal bait shop. They typically trapped their own bait from nearby ponds and creeks and sold the catches out of garages or outbuildings.

Many a northeast South Dakota angler can recall buying a quarter's worth of bait from Edwin "Ed" and Emma Pries. Ed sold bait out of his garage in Watertown from 1940 until near his passing in 1977. He ran one of the few places where bait was sold year round. Ed seined and trapped bait within a 50-mile radius of Watertown. He kept three groups of baitfish in his garage: fathead minnows, shiners, and chubs and dace. He also sold frogs in late summer and fall. In the 1940s and 1950s, most anglers thought fathead minnows were only good for catching crappies and yellow perch. Shiners, chubs, and frogs were the bait of choice for walleyes, pike, and bass. Nightcrawlers were a novelty bait.

Before World War II, most people fished during the spring and summer only. In the fall, a few old timers might sit on pails and shore fish around Pelican Lake but that was it. Fall was hunting season and ice fishing was almost non-existent due to thick ice and lack of power augers. With the advent of power augers (built from military plane motors that operated the bomb bay doors) after World War II, ice fishing became popular and demand for bait in the winter increased.

Don Monroe and in later years Don Warnick were the main state commercial fish supervisors who oversaw the bait industry. They monitored and reported on the harvest and economics associated with the commercial baitfish industry. The accuracy of baitfish harvest and sales data has always been subject to doubt. The doubt stemmed from high non-reporting rates of baitfish dealers of their catches and sales, the agency's use of projected values for non-reporters, accuracy of baitfish volumes, and discrepancy in reporting format (i.e., small chubs reported as minnows). In general, "interesting" accounting, *extreme* protectiveness of baitfish harvest locations, and questionable business practices were the norm in the wholesale bait business.

Ongoing rivalries raged between bait dealers over who had the right to trap which sloughs. In an attempt to beat their competition, bait dealers would lease the baitfish harvest rights from farmers who had productive sloughs. If a bait dealer caught someone harvesting bait in a slough he had leased, there was "war." And rivalries intensified during years when baitfish were scarce.

Bait harvest is a tough business requiring hard work during inclement weather and sometimes dangerous conditions (cold water, poor ice conditions, etc.). Bait dealers harvest baitfish year-round using a wide range of gear, including box traps with leads, cloverleaf traps (Figure 15.6), and seines. A proficient bait harvester uses a wide variety of techniques. Bait harvest for personal use is allowed with legal minnow seines, traps and throw nets.

A unique technique is used to harvest bait during the middle of winter as dissolved oxygen levels drop. A hole is cut in the ice and an oxygen tank is placed on the ice. A tube and air stone delivers air to the bottom of the slough. Box traps are stacked over the rising column of air bubbles. A sheet of plywood is placed over the hole and snow is piled on top to prevent the hole from freezing. Fathead minnows, attracted to the oxygen-rich water, are caught in the box traps. The traps are then lifted to the surface and the bait fish removed.

Figure 15.6 **Wholesalers harvest fathead minnows with several kinds of traps, including the cloverleaf trap (photo by Anderson Bait).**

Export of baitfish from South Dakota boomed from less than 13% of total sales (2,313 gallons of minnows and 500 gallons of chubs) in 1970 to over 63% of total sales (33,021 gallons of minnows and 1,650 gallons of chubs) in 1985 (Welsh 1986). The number of minnows harvested from South Dakota marginal lakes and wetlands also shot up during this time period (Figure 15.7). In addition, both known and unknown transportation of threatened or endangered

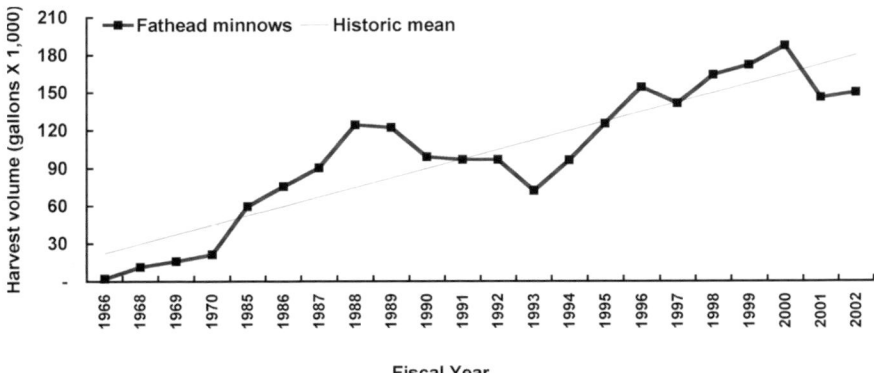

Figure 15.7 **Volume (gallons) of fathead minnows harvested in South Dakota, and historic mean from 1966 to 1970, excluding 1967, and from 1985 to 2003.**

species across state lines was a serious problem in the baitfish business. The species of concern included central mudminnows, blacknose shiner, northern redbelly dace, finescale dace, banded killifish, and game fish which were illegal to posses or sell as baitfish.

In the mid-1980s, surrounding states began to tighten their baitfish import and export laws. With the massive increase in exportation of bait out of South Dakota, the legislature enacted laws to protect and sustain the baitfish resources of the state, including a law that authorized licenses for nonresident wholesale baitfish dealers and export baitfish dealers.

The retail bait business was changing, and by 1985, South Dakota retail bait dealers were becoming one-stop bait-and-tackle shops and most retailers were purchasing their baitfish from wholesalers. According to 1986 state records, retailers purchased approximately 85% of the minnows and chubs from resident wholesalers. The days of small garage bait shops were gone.

The retail baitfish business of today is more modern and is usually connected with other retail services. It also can be quite lucrative. It is not uncommon to stop at one store and purchase baitfish, *and* gas, snacks, sporting goods, groceries, and beer. At most shops, baitfish of several species and sizes can be found year-round. Today's retail dealers typically have tanks with controlled water temperature, filters, and agitators, all of which greatly reduce the loss of baitfish. Some retailers can even inject bags of live baitfish with oxygen, and if the water is kept cool, the fish will live for days. The 2001 retail business estimated gross value of fathead minnows sold to anglers was approximately $1.4 million.

The wholesale and export bait businesses are also more modern but still remain a labor-intensive profession that can be profitable with the technological advancements in transporting and tanks for holding large numbers of baitfish. This has become possible with the use of truck-mounted, insulated holding tanks that are equipped with oxygen, air stones, and agitators. Chemicals added to the hauling tanks can reduce stress and disease during transport.

South Dakota baitfish are shipped to retail outlets, private hatcheries, and public hatcheries in surrounding states. Many South Dakota export bait dealers have established large baitfish distribution routes that cover thousands of miles and crisscross the central United States. In 2001, approximately 157,000 gallons of minnows and 1,150 gallons of chubs were exported to the following states, listed in order of decreasing amounts: Minnesota, Wisconsin, Tennessee, Illinois, Nebraska, Kansas, Iowa, North Dakota, Ohio, and Missouri. In 2001, approximately two-thirds of the baitfish exported from the state was destined for Minnesota and Wisconsin. The estimated gross value of the 2001 wholesale fathead minnow harvest was approximately $2.5 million.

Unusual Bait Business Ventures

Folks that harvested frogs were called frog dealers and they mostly operated in northeastern lakes of the state. The frogs used by dealers in South Dakota were leopard frogs that were very abundant in wetlands in the late summer and fall. Before the 1980s, young frogs were widely used as walleye bait. Bait frogs (frogs less than 3 inches in body length) were sold at an average price of $1 per dozen in the late 1960s and 1970s.

Live frogs (frogs longer than 3 inches) were also sold to biological supply companies at an average price of 50 cents per pound. Many school dissection frogs came from northeastern South Dakota.

The number of frog dealers fluctuated from 15 in 1967 to six in 1970. The abundance of frogs in different years and the availability of markets usually reflected the number of dealers. The estimated economic value of this market ranged from a high of $15,019 in 1967 to a low of $2,932 in 1970.

In later years, the leopard frog population collapsed from a combination of disease and other die-offs. The frog population never fully recovered to historic numbers, and the commercial frog business dwindled. In 1988, the South Dakota legislature repealed the resident frog dealer license to protect the remainder of the frog population. Residents can harvest a variety of reptiles and amphibians for personal use as bait, including the leopard frog.

Carl Lowrance was the founder of Lowrance Electronics Company, known for its "Green Box," which was the first depth finder available to recreational anglers. He also started a business venture based on tiger salamanders that were abundant in some wetlands. He harvested larval tiger salamanders, known as "sallies" by people in the business. Sallies were collected from large permanent sloughs in northeastern South Dakota in the late 1960s and early 1970s. Sallies were marketed to bass fishermen in the southern United States.

Lowrance developed a piece of equipment that mounted on a boat and operated something like a rotary screw trap used to collect migrating fish on large rivers and streams. When run through the sloughs, it collected larval salamanders. However, the short larval life cycle and the logistics of such a business made it a short-lived venture.

Future of Commercial Fishing

Commercial contract rough fish removal may end because few new fishermen are recruited and the business has a very low profit margin. If SDGFP determines that commercial rough fish removal has benefits to sport fishing, then the agency may have to develop a system of incentives and subsidies for fishermen in the future. On the other hand, the future of the baitfish industry seems brighter. There will be a demand for baitfish as long as there are walleyes and perch and northern pike to catch.

Acknowledgements

The authors would like to thank those who provided insight, information, and photographs for this chapter, specifically Mike Barnes, Jerry Broughton, Deb Burtts, Dale Gates, Carroll Hanten, Todd Kaufman, Kyle Potter, Roger Pries, Randi Smith, Craig Soupir, and Don Warnick. A special thanks to the staff at the South Dakota State Library in Pierre who assisted with locating archived reports and other information. The authors would like to thank Kevin Wooster, the *Mitchell Daily Republic*, Richard Holcom, and Chris and Matt Anderson for allowing use of reprints and articles, and for providing photos. Reports used to write this chapter can be found in the references section at the end of the book, particularly those by D. Warnick, D. Monroe, and R. Welsh. ❖

Chapter 16

Evolution of Modern Sport Fishing in South Dakota

Tony Dean

"The small boy starting forth with a sapling, a cotton string and a bent pin and returning with a few minnows has probably experienced all the joys of a millionaire angler with his thousand dollar outfit."

— D.C. Booth, 1930

During the summer of 1968, I remember driving out to the Oahe Tailrace, casting a Doll Fly from shore and nearly always catching a few walleyes. I'd just moved to Pierre from eastern Iowa and I was in awe of the great fishing I found here. Some days I'd spend my noon hour munching on a burger while casting spoons into the LaFramboise embayment on the south side of Pierre, usually landing several pike before heading back to work. I'd gaze longingly out at the water and see dozens of boats. Few South Dakotans owned boats then…and most we saw were Iowa or Minnesota licenses.

Thankfully, I didn't need a boat to take advantage of my next South Dakota fishing discovery. There was a stock pond over every hill west of the Missouri River and nearly all of them held largemouth bass. I didn't find out about them until the following summer, but once discovered, I went every spare moment. I'd fish a different dam every time out, and the fishing was fantastic. Most of these small ponds were loaded with bass so unsophisticated they viewed a Hula Popper as a brand new lure. It reminded me of that famous scene in *Field of Dreams*. I'd ask, "Is this heaven?" And of course, the answer would be, "No, it's South Dakota."

Over the past 30 years, I've realized when it comes to quality fishing, this is the promised land, but I've also seen significant changes in our sport. I saw great fisheries like Lake Oahe boom, then bust. I witnessed the biggest explosion in angling knowledge the sport has ever seen, and saw how it transformed South Dakotans from shore anglers with little fishing knowledge to boat owners and skilled anglers. They began buying boats as dealerships sprung up in every town that could support one, and by the early 1970s, boat sales were booming. The premier fishing boat was the 16-foot Lund Pro Angler powered by a tiller-steering, 4-cylinder Mercury 50 HP outboard. It was a big time fishing boat and was usually equipped with a Lowrance 2330 flasher and a Minn Kota electric motor. It sported splashguards on the transom and the boat operator always wore a rainsuit. You could buy one brand new with all the toys for around $7,500.

The Missouri River reservoirs teemed with fish; walleyes in Lake Sharpe and Francis Case and northern pike…really big pike…in Oahe. On three consecutive sunny spring days in April, 1981, Al Lindner and I caught and released 57 northern pike between 10 and 29 pounds. Most ran 15 to 17 pounds and Al told me it was some of the best northern pike fishing he'd ever experienced.

Then North Dakota officials stocked rainbow smelt in Lake Sakakawea; the silvery baitfish quickly spread down the system, and Oahe's walleye population grew and prospered. But so did our equipment and level of angling expertise.

We traded fiberglass rods for graphite, Doll Flies for Fuzz-E-Grubs, bell sinkers for bottom bouncers, River Runts for Rapalas (Figure 16.1), and any angler worth their salt subscribed to *Fishing Facts* and *In-Fisherman*. We learned about "structure" and began to fish what we thought it was, even if we didn't always understand why. However, the wisest among anglers had already been doing so, following the advice of longtime *Sports Afield* fishing editor, Jason Lucas, who decades earlier, advised anglers to fish "edges."

Figure 16.1   **An explosion of new types of gear and the knowledge of how to use them has been the story of recreational fishing in the last 30 years (photo by C. Berry).**

At the same time, the Black Hills provided outstanding stream trout fishing; the stock ponds of central and western South Dakota offered largemouths and panfish, and the Glacial Lakes of northeastern South Dakota yielded walleyes, northern pike, bluegills, yellow perch, and some crappies. But the Missouri River reservoirs were providing such fantastic fishing that they overshadowed the rest. The big reservoirs garnered regional and then national attention. Even so, the biggest walleyes usually came from Big Stone, Roy and Enemy Swim lakes, and you didn't need a boat because most were caught while wading and casting minnow baits after dark. Many of the Black Hills streams were producing wild browns with Spearfish Creek also boasting rainbows. Each fall, a handful of anglers were catching giant browns that migrated out of Pactola Reservoir into Rapid Creek. Yes, this was fishing country.

But the sport was changing and the next two decades would see a knowledge explosion that would turn ordinary anglers into efficient predators capable of quickly reducing fish populations, even in large bodies of water.

The first walleye tournament ever, the South Dakota Governor's Cup, was held on Lake Sharpe in 1973, and it too had long-lasting impacts on fishing here and elsewhere. From a local event that brought together a hundred or so of soon-to-be fishing icons, tournaments grew and impacted fishing way beyond anyone's imagination, and as a participant, I saw those changes from a front row seat. Many who would later become fishing "stars," fished that initial event, including Al and Ron Lindner, Babe Winkelman, and others. Over half of the entrants ended up working in the fishing and marine industry.

The success spawned others, and for the next decade, I fished a tournament nearly every weekend until I finally grew out of it. I began doing fishing seminars, writing for *In-Fisherman*, and participating in all the things that would also impact fishing, and in retrospect, not always in a positive way.

Tournaments unveiled new techniques and locations to local anglers with the inevitable heavy exploitation following each event. The rash of in-depth knowledge in fishing magazines and books, television shows, and seminars, was spread over a 20-year period during which more fishing information was dispensed than during any comparative time span in fishing history. Finally, the fishing guide profession grew rapidly, enabling those who had neither boats nor knowledge, to become as successful as the experts.

Seminars drew huge crowds. My first, at Stich's Marine in Sioux Falls back in 1979, drew over 2,000 fishermen, and while most crowds were smaller, it wasn't unusual to have over 400 people pay from $4 to $10 each to listen to a fishing pro detail and demonstrate new location keys and techniques (Box 16.1).

16.1
Legendary South Dakota Anglers[1]

The title "legendary angler" is often the parent or grandparent who introduced us to fishing and had an impact on our lives. Some anglers become experts at fishing, and influence many people on a regional or national level. They have an impact that benefits freshwater sportfishing, and are recognized as legendary anglers by The National Fresh Water Fishing Hall of Fame and Museum (http://www.fresh-water-fishing.org/). Visitors to the Hall of Fame always remember being inside the huge musky (photo courtesy of The Hall).

The Hall of Fame was founded in 1960 as a tourist attraction and a museum to display the artifacts of the sport of fresh water angling. The Hall of Fame also publishes annual record fresh water fish catches (NFWFHF 2006) for various kinds of fishing tackle. For example, the *all-tackle* world record goldeye (*Hiodon alosoides*, 3 lbs, 13 oz) was caught in the Oahe Tailwater, as was the 10-lb *fly fishing* division record (1 lb 9 oz).

Legendary anglers from South Dakota are:

Mike McClelland, who wins walleye tourneys and writes fishing books (see references)

Bob Probst, a walleye angler who can lay claim to many walleye fishing innovations

Tony Dean, a writer and regional TV outdoors show host (*http://www.tonydean.com/*) who advocates natural resource conservation and has contributed greatly to the "Fishing Has No Boundries" program for less privileged and handicapped anglers.

---

[1] Information supplied by book editors.

The Lindner brothers and Babe Winkelman came up with television shows and I soon followed suit. We held nothing back, offering knowledge as though it would be gone tomorrow. South Dakota anglers now had boats…well equipped ones, and were armed with the knowledge that would turn them into efficient anglers (Figure 16.2). By the early 2000s, nearly every boat on South Dakota water was piloted by an angler who had a good understanding of how to find and catch fish.

Figure 16.2   Anglers travel in big, fast sonar-equipped boats guided by Global Positioning Systems that can put them right on top of a hot spot in Lake Oahe (photo by C. Berry).

Today, I reflect and wonder, what is this thing we created? Sometimes it resembles a monster that cannibalizes its young. We taught people to fish, but we didn't do a very good job of teaching them to fish responsibly. And we all learned the hard way that a summer harvest can exceed the supply of a spring spawn in even the largest of lakes.

Ironically, I remember a warm, February day back in the early 1970s when Bob Hanten, Jack Merwin, and I fished walleyes in the stilling basin below Oahe Dam. Bob was South Dakota's fisheries chief and Jack would later head the Game, Fish and Parks Department.

Pointing to the dam above us, Bob said, "We're lucky these lakes like Oahe are so big because their sheer size alone means fishing pressure will never be a problem." Today, the retired fisheries chief sings a different tune, and like me, I think if he could, he'd turn the clock back.

The fishing pressure of today has a big impact on fisheries. It happened on Lake Oahe in the mid-1990s, more recently on Lake Francis Case, and a few years earlier, following a record runoff in eastern South Dakota, winter anglers needed just six weeks to harvest more than 250,000 big yellow perch from Cattail-Kettle Lake. Similar yellow perch harvests were repeated at many of South Dakota's newly created lakes during the same period. And that proved that efficiency among fishermen wasn't limited to those in boats during the open water season.

Just as summertime anglers travel in big, fast sonar-equipped boats guided by Global Positioning Systems that can put them right on top of a hot spot provided by a fishing buddy who fished it a few days earlier, winter fishermen have their own high tech toys. Sonars work even better for ice fishing, especially flashers, devices that operate in real time. Three-color flashers show fish on the outside edge of the signal as green marks that turn yellow as they move toward the center and bright red when the fish is immediately below the angler. The fisherman can watch his bait and presentation in relation to the fish, see how the fish is reacting and respond accordingly. And if that isn't enough, he can drop a real time camera down the hole and watch the action in living color.

Portable ice fishing shacks that set up in seconds and tow easily along with lightweight, modern augers that quickly bore through thick ice, enable winter anglers to operate with nearly the same degree of mobility and comfort as their summertime counterparts. It changed ice fishing too, a sport that was once characterized by drinking beer and dealing poker hands.

When top anglers like Gary Allen and former Nebraskan Bob Propst began guiding, it started another trend that's showed no signs of slowing. In Pierre alone, there are at least a hundred full and part time fishing guides. If you don't have a boat or fishing knowledge, you can hire a guide whose skills and equipment offer a shortcut to success.

I've long believed guides could do much to help improve fishing, but too often, they adopt adversarial roles with fisheries managers, and because they deal with many people, can influence them against status-quo management. If anything, fishing guides and biologists should be allies because guides who are on the water every day are often the first to spot trends, and when biologists work closely with guides, they can provide them with important data that turns them into partners. Fisheries managers need to reach out to the influencers, and when they do, fishing and fisheries resources win.

Working together isn't a new idea. In states that are more heavily fished and have seen the consequence of unregulated pressure, guides have come to realize the wisdom behind specific regulations. And once they do, they most often become champions of management.

I remember a day back in the early 1990s when I fished Montana's Big Horn River, and my guide was the legendary Frank Johnson. He explained his rules. First, he said, we'd be keeping no fish. That was fine with me. A couple days later, I asked Frank how he'd handle it if a client insisted on keeping fish.

"I'd tell him he'd have to find another guide," he chuckled. "He'd probably have to go a long ways because there isn't a guide around here who doesn't understand that with the pressure we have, we could wipe this fishery out in a month or two."

The high tech tools used by anglers aren't the only things that have affected fishing. Add the Internet and a cell phone in every pocket and you begin to realize how word of a hot bite

Figure 16. 3 With the Internet and a cell phone in every pocket, word of a hot bite can spread quickly resulting in hundreds of anglers descending on a single body of water, which can be over-fished (South Dakota Tourism photo)

can spread so quickly, resulting in hundreds, even thousands, of anglers descending on a single body of water (Figure 16.3). I maintain an outdoor website that carries fishing reports and I always have mixed emotions about it.

Add it up and you can't help but realize how difficult it is to manage fisheries these days, and it isn't likely to become easier. And because everything happens so quickly in this information age, I am not surprised that fisheries managers find themselves behind the curve more often than not.

Can we sustain the kind of walleye fishing we've known? Most anglers think so, and a decade ago, I would have agreed. Now, I'm not so sure, and certainly not unless we can figure some innovative ways to reduce harvest. Other species, notably bass and trout, have faced heavy pressure in other regions and have managed to survive and even prosper. But that's due mostly to catch and release practices.

The first talk of the importance of catch and release came from the noted trout angler, the late Lee Wulff, who said, "a fish is too valuable to be caught just once," a quote that was repeated often enough that most trout fishermen practice catch and release these days. Ray Scott of the Bass Anglers Sportsman Society borrowed that concept from Wulff and was so successful at instilling it in the minds of southern, rural bass anglers that in many areas, peer pressure results in most bass being released.

But trout or bass don't taste as good as a walleye, a fish that tastes too good for its own good. If you set a limit on walleyes, most will fish until they reach it, and many have no qualms about filling live wells or freezers. I've heard some biologists say they believe that encouraging catch and release will eventually turn things around. I don't think so and the more I see, the more I am convinced that the only catch and release program that works with walleyes is one that is mandated.

So, what more can we do? Set the limit at two fish? One? That's already been done in some Wisconsin and Minnesota waters. Screw slot limits down so tightly that it is difficult to keep a fish? Minnesota did that on Mille Lacs Lake and it wasn't popular with many anglers, but regulations like this become exactly what I alluded to earlier with mandated catch and release. Whether anglers like them or not, they work, but only because they force compliance. And yet, such regulations may not be the answer because they could drive anglers from the sport and then there won't be enough license monies to manage the fisheries.

I suspect complex regulations will be required because each body of water is different, has different predator-prey ratios, food sources, and water quality. But angler complaints about complexity may not hold much water. Generally, it amounts to little more than knowing how to read a tape measure or yardstick, something any angler capable of running a GPS or reading a sonar ought to be able to handle.

The alternative might be regulations that are even more unpopular. Example: Ban live bait for walleyes. That wouldn't be popular with bait dealers, though it would be effective. Eliminate weekend fishing? Ban fishing on four days during each week? Cut night fishing on clear water lakes? How about instituting a lottery system where only so many boats are allowed on a specific lake at any given time? I've experienced that in California while fishing for bass in San Diego's reservoirs. Close the season for lengthy periods, especially when harvest is high such as during June? What would you do if you were appointed the fisheries manager?

While fishing pressure has become a major problem in some South Dakota waters, it isn't the most serious threat to our fisheries. To enjoy good fishing, we need good water quality, and the best way to assure it is to maintain our present wetlands base and keep grass on the land. Unfortunately, too many fishermen do not see or understand the basic connection between good fishing and clean water. If we could make it either illegal or unpopular to drain wetlands or plow native prairie, we'd go a long way toward ensuring good water quality and good fishing. We are blessed with ample water resources that have historically produced good fish populations and fishing opportunities that are good enough to be considered some of the best in the nation. But if it is to continue, we need to make wise choices.   ❖

Chapter 17

Fishing With Artificial Flies in South Dakota

Alan Davis

> **"**There is no greater fan of fly fishing than the worm.**"**
>
> — Patrick McManus

At the infamous Munich conference of 1938, British Prime Minister Neville Chamberlain attempted to engage Adolph Hitler in a discussion about the joys of fly fishing. Hitler replied, "I know no weekends, and I don't fish!"

It is one of history's ironies that General Dwight Eisenhower, an avid fly fisher, accepted the surrender of Nazi forces six years later. During his two terms as President, Eisenhower took frequent vacations to go fly fishing.

Ike's 1953 visit to South Dakota was mainly to dedicate the Ellsworth Air Force Base, but he also did some fly fishing in the Black Hills. He spent two nights in the Game Lodge at Custer State Park and also took time on June 12 to catch some brook trout in French Creek. Did Ike know about President Calvin Coolidge's political problems when he fished for trout in Black Hills streams (Box 17.1)?

Ike was often criticized for taking time away from his public duties, but he never wavered in his devotion to fishing. I think I see it Ike's way. The pressures of contemporary life are formidable, but from time to time they can be escaped. It helps to remember what Bertrand Russell once said, "One of the symptoms of an approaching nervous breakdown is the belief that one's work is terribly important." As an antidote to stress, fly fishing is unsurpassed.

Fly fishing is enjoying growing popularity in America. It has, since its early beginnings in the British Isles, been associated with trout. Authors of a 1923 book on fishes wrote, "Perhaps the thorough-going angler will be disposed to scorn live bait and use only the artificial fly. It is not all of fishing to fish" (Jordan and Evermann 1969).

17.1

Black Hills Trout and Presidential Politics

The State Game Lodge in Custer State Park received national attention in 1927 when President Calvin Coolidge and staff occupied it for three months. It was Coolidge's Summer White House. President Dwight Eisenhower stayed at the lodge for several days in 1953.

Coolidge liked to fish but was unprepared for the uproar about his fishing techniques and the image of Republicans using worms instead of fly-fishing for trout. The cartoon on the next page is one of several that cartoonist and outdoorsman J. Darling drew about the uproar. The cartoon is titled "Agriculture information isn't the only thing Cal is getting in South Dakota." (Courtesy of the "Ding" Darling Wildlife Society)

Another important reason for the growth of the sport has been the recent interest in applying fly fishing methods to a wider array of sport fish, such as bass, pike, and panfish. Taken together, these factors would make South Dakota an ideal location for the aspiring fly fisher.

Although we cannot be sure who the first person was to try fly fishing in South Dakota, we do know fly fishing has

been popular for many decades with male and female anglers alike. Two women, who were ardent and successful fly fishers in the Black Hills were Helen Feuner (now deceased) and Ella Luverne, who still resides in Rapid City. Ella described how she and Helen tied their own flies and often fished the same waters together (though never sharing all their favorite fishing spots or secrets). Helen described ventures in which they took the train a mile or more up Rapid Creek, then fished their way back to Rapid City.

Through the years, much has been written about fly fishing in South Dakota, primarily in the Black Hills area. Literally hundreds of anglers have enjoyed the sport and several clubs or organizations have been created with fly fishing as the focus of their mission. The earliest one was called the Black Hills Fly fishing Club. Today the club is called the Black Hills Fly Fishers. A club with an interest in Gary Creek and other fly fishing sites in eastern South Dakota is the South Dakota Lakes & Streams Association headquartered in South Shore.

Generally, fly fishing is thought of synonymously with clear, cool-flowing streams and trout. Before 1886, the Black Hills of South Dakota had hundreds of miles of cold-water streams but no trout. In 1886, that changed when trout were stocked, and hatchery stocking of trout has continued through the years. D. C. Booth, Superintendent of the first fish hatchery in the Black Hills wrote (circa 1930), "Whenever a man plans a vacation he usually thinks of fishing and when he thinks of fishing he means game trout fishing." Booth went on to give more advice writing, "There is very little politics or religion in game trout, it is always a safe topic for discussion." Today trout are found in most of the streams and reservoirs throughout the Black Hills. Brook, brown, and rainbow trout have all been stocked in these waters.

The Black Hills Trout Management Area contains nearly 800 miles of primary cool-water streams for trout fishing plus 22 reservoirs, most of which are located on public lands and open to fishing year-round (Figure 17.1). Outside the Black Hills area there are few streams with water temperatures cool enough to sustain trout. Gary Creek in the northeastern corner of the state (Deuel County) is a unique trout stream in the eastern part of South Dakota. It supports a small population of brown trout. Fly fishing small streams is a unique fishing challenge because trees and brush often require special casts (Hughes 2006).

A growing cadre of fly fishers now enjoy catching many warm-water species of fish. The list of warm-water and cool-water fishes in South Dakota that can be caught with fly fishing gear is extensive. At the least, it includes bluegills, white bass, smallmouth bass, largemouth bass, northern pike, walleyes, crappie, and even common carp. There probably is no better fish for "training" young fly fishers than the bluegill. A young

Figure 17.1 A fly fisherman stands back from the bank of a small Back Hills stream and uses a "flip cast" to present the fly. Most of the creeks in the Black Hills are small, but some widen to 25 or 30 feet (e.g., Rapid, Spring, Boxelder, and Spearfish creeks). (Photo courtesy of Dr. Lester Flake)

angler can make a cast that ends up with the fly line in a tangle, a cork "popper" can land in the middle of that fly line, and a bluegill will still rise to take the popper! A forgiving species on which to learn, bluegills can be found on many of the farm ponds in central and western South Dakota and in several of the eastern glacial lakes that have submergent aquatic plant growth.

The tail-water release areas just below the large dams on the Missouri River provide cool-water releases and a mixed fishery for fly fishermen, including trout, white bass, smallmouth, and goldeye. Although few anglers actually use fly fishing gear in most of these locations, the fighting ability of a three-pound rainbow trout will really be noticed when using a fly rod.

Fly Fishing Equipment

Fly fishing derives its singular technical and aesthetic qualities from the characteristics of the equipment. The fly rod is the most important tool for the fisherman, and requires careful selection. The most immediately recognizable features of a fly rod are its long reach and the location of the reel seat at the end of the handle. The length is necessary to add distance to casts and control the fly line, which is heavier than standard fishing line. Most fly rods are between 7 to 9 feet long, depending on whether one intends to fish in open areas requiring long casts or in places more confined by surrounding trees and brush.

Rods are designated by weight, referring to the weight of the fly line they are designed to accommodate. Weights range from 3 to 15, with the majority of fishing situations requiring a rod between 5 and 8. Fly rods vary in their firmness, with the more firm rods producing longer casts. Almost all contemporary fly rods are made of graphite composite material because of its favorable strength to weight ratio. The prospective fly fisher is advised to shop around, trying the feel of many rods and handles before making a final choice.

Fly lines are larger than ordinary fishing line and are filled with tiny bubbles that enable them to float. Lighter fly lines have less impact on the water and are less likely to frighten fish. On the other hand a heavier line can cast for greater distance with heavy flies or in windy conditions. The ability of fly lines to unfurl smoothly is aided be the fact that they are not of uniform width, but are tapered on both ends.

Specialty fly lines exist, including lines for deep lake fishing that are designed to sink. Some lines are weighted more in the forward end to produce longer "shooting" casts. Of course, it is important to select a line that is appropriate for the size of your rod, flies, and the conditions in which you will fish. Many fishermen use a longer rod (9 ft.) with a relatively lighter line (5 wt.) as a compromise.

Most fly reels are made of aluminum, have a 1:1 ratio of handle to spool turns, and need not be expensive to perform well. The line is attached by cotton backing to the fly reel.

Fly fishing situations almost always require wading, so fly fishermen need to possess a good pair of waders or hip-boots. Waders are made of rubber, Gore-tex, or neoprene and can ordinarily be repaired. Even the most experienced fly fisherman will occasionally fall in the stream and waders can fill with water creating a safety hazard. A wading belt, worn tightly outside the waders above the waist, should be considered a standard piece of safety equipment.

Slippery stream bottoms make wading boots an absolute requirement for the serious fly fisherman. Crossing a distance of moss covered river rocks can feel like walking on greased bowling balls. Most wading boots have felt soles to provide stability, with some even incorporating steel cleats.

To fish lakes or large ponds, many fishermen prefer to use a float tube that surrounds the body and provides buoyancy. Anglers who are considering use of a float tube should seek professional instruction so that all safety precautions can be observed.

Reading Water

Fly fishing can be adapted to many types of water, with characteristics that can differ significantly from one to another. When fishing any body of water, the chances of catching fish are greatly improved by some basic knowledge of fish habitats. It helps to keep some general principles in mind. Fish cannot swim continuously against a current. In moving water they must rest to avoid exhaustion. Also, fish must try to avoid being seen by predators, such as osprey, cranes, and fly fishermen. They will seek overhead cover if it is available. Finally, for fish in moving water the margin between the nutritional benefit of food and the effort expended in feeding is especially narrow. Fish are more likely to meet their nutritional needs if they stay in areas that are relatively rich in food.

Hunger is a powerful motivation, sometimes strong enough to outweigh the protection and predation considerations. These three factors are important at all times, and are reliable indicators of where fish might be located. Fly fishermen look for water seams, which are points where one type of water condition meets another. Seams can occur at the junction of slow and fast water, or shallow and deep. More generally, the term seam can refer to any place where two environments meet, such as water and land, or shadow and light. Seams are high probability holding areas because of the combination of food forms from the two environments.

Cold water streams offer trout cover, rest areas, and a continuous supply of food. Fish in a stream always face the current when holding, so the standard approach for the fly fisher is to start down water and move upstream. Remember, fish can see shapes, shadows, and movement, and react defensively to these images. The most successful fly fishers approach a stream and enter it cautiously, keeping in mind that fish may be holding close to the banks.

The basic water conditions of a stream are referred to as flats, pools, runs, and riffles. Flats are sections of smooth water varying in depth from one to five feet. Long sections of a stream may consist of flats. As a rule, flats are not particularly productive areas because of the lack of deep holding spots and clear overhead visibility. Fish may be found along flats if undercut banks are present offering some protection from predators and a supply of terrestrial insects such as grasshoppers and beetles.

Upstream from flats one may find pools that are sections of smooth deep water. These sections typically contain the largest fish in the stream because of the favorable combination of factors. The depth makes them invisible to predators, feeding opportunities are usually adequate, and rocks and low areas in the stream bed provide comfortable holding areas.

Runs are areas of moderately smooth water from two to six feet in depth. Runs often contain boulders and other sub-surface features that permit trout to hold comfortably and select food moving through the area. As with pools, the combination of current and depth means it is necessary to present flies in a manner capable of getting deep in the water.

Riffles are choppy sections of stream in which the water runs quickly across rough stream bed at a very shallow depth. Riffles are rich in oxygen and therefore contain much insect life. Hungry fish can be found anywhere in riffles, but prefer to hold behind rocks at the bottom of a riffle section to see what will wash out. The seam where a riffle meets a run offers a strong likelihood of finding fish.

When fishing lakes and ponds look for weed beds and shelves along the edges where fish may be hiding. The edges of weed beds will tend to be the most productive (and snag free) places to fish. If a lake has no significant weed growth, try fishing the leeward side. Fish will gather near the largest accumulation of insects, which will be down wind. Again, look for seams of shadow and light and natural cover such as fallen trees.

Fly Casting

Casting is the method of presenting the fly to the fish, and the most distinctive feature of fly fishing. Practice is necessary, because you cast the line, not the lure. The would-be fly fisherman should spend some hours in an open grassy field getting the feel of casting before trying it over water. While one could spend a lifetime perfecting fly-casting, the essentials are not difficult, and a couple basic casts will serve you well in a large variety of situations. There are plenty of "how-to" books on fly fishing (Reynolds 2004, Rounds 2006).

The most commonly used cast in fly fishing is known as the overhead cast. This is executed by first letting the fly line out to the intended length. While fishing, the drifting current will perform this service conveniently. The rod is then drawn swiftly into a back cast, creating a bend in the rod that supplies energy to whip the fly line behind you. You should hesitate long enough to let the line unroll almost completely before beginning the forward cast. If the forward cast is begun too soon it will create a problem referred to as "wind knots," small knots in the leader and tipit (short length of mono filament at the end of the leader) that will break when a fish is hooked. You then bring the rod forward in the direction of the target, slowly at first but with increasing speed, so that the spring tension in the rod will propel the fly line forward with sufficient force to extend the line completely.

Figure 17.2 Sketch shows the line positions of an "overhead cast" with rod moving between the 10 o-clock and 2 o-clock position. (Drawing by S. Kahara)

The aim of the cast is not to hit the water, but to stop in the air just above the target, and fall lightly to the surface. Remember, it is not necessary to lash the rod like a buggy whip. If a clock face surrounded the fisherman, the rod tip would move back and forth between the 10:00 and 2:00 o'clock points, letting the spring tension of the rod supply the required force (Figure 17.2). It is common for fly fishermen to perform this back and forth casting movement one or two times before letting the fly fall to the water. This is known as "false casting," and is often done while simultaneously stripping line from the reel with the left hand to obtain the desired length of line.

In many fishing situations adequate space for an overhead back cast is not available because of brush or tree branches near the streamside. The most common solution to this problem is known as the roll cast (Figure 17.3). A roll cast is begun by letting the current carry the line out to the desired length. The rod tip is then brought slowly up to the 10:00 o'clock point behind the fisherman, with the line still extending down stream. The rod tip is then brought forward smartly in the direction of the target. The weight of the fly line will cause the line to unfurl in the form of a hoop, at last flipping the fly over to land upon the surface. Great casting distance is not possible when roll casting, but this should not be discouraging given that most fish are hooked relatively close to the fisherman.

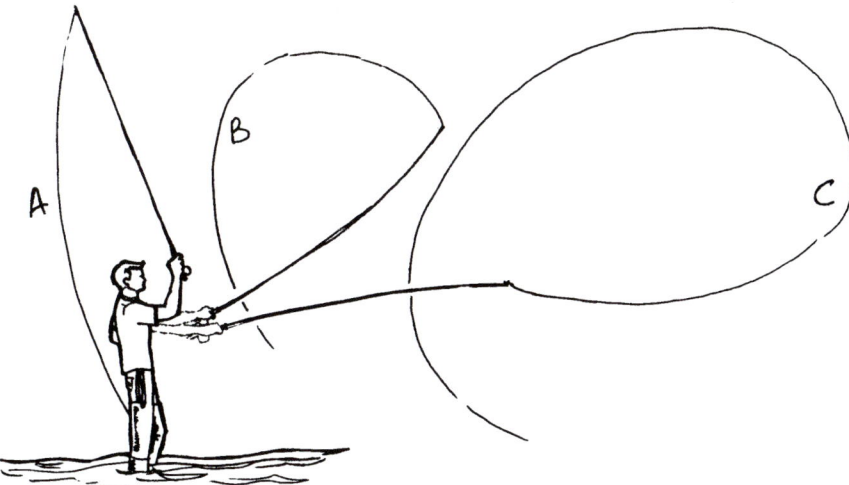

Figure 17.3 Sketch shows the line positions of a roll cast used when nearby trees preclude using the overhead cast. (Drawing by S. Kahara)

The most common casting mistakes are over-casting, and trying for too much distance. Many fly fishermen perform a large number of unnecessary casts out of boredom or impatience. Unnecessary motion is inefficient, and creates vibrations and shadows that can be detected by fish. When fly fishing, one should strive to keep movements deliberate, precise, and to a minimum. Also, avoid the temptation to perform big, booming casts that carry for a great distance. While impressive to watch, they are also usually not necessary and tend to hit the water with a splash. When this happens you will scare away any fish that might have been near. Think about how lightly an insect lands on the water and you will realize the importance of delicacy in fly casting.

Artificial Flies and Related Techniques

Artificial flies take many forms but most imitate insects. Stream and lake insects are called "benthic macroinvertebrates. Benthic because they live on the bottom or on submerged vegetation and wood, and macroinvertebrate because they can be seen with the naked eye and do not have a backbone. Benthic macroinvertebrates are important to the stream ecology, to the biologist, and to the fly fisher. The Izaak Walton League of America has a guide to aquatic insects that is a handy resource for anglers, students, biologists, and river monitoring teams (IWLA 2006)

Benthic invertebrates have a wide variety of life styles and modes of existence. The causal observer can see the largest insects flying above the water, skating on the surface and diving and swimming among submerged vegetation. Biologists classify aquatic insects as skaters, divers, swimmers, clingers, sprawlers, climbers, burrowers, and floaters. In lakes they live in the profundal zone (deep sediments), the littoral zone (shallow shore area sediments and vegetation) and the beach zone. In streams they live on bottom substrates such as mud, sand, rock, vegetation, and woody material. A special stream community is called the *drift*. Insects drift to migrate to new food patches, or because they are dislodged by the current.

Insects are important to stream and lake ecology because they are the main group that transports energy from decaying organic material such as leaves, and from primary producers such as free floating plankton or attached periphyton to higher trophic levels such as fish. This transportation duty gives insects an ecological label known as *primary consumers*. The assemblage of insects transports energy by collecting it with a wide array of methods. There are scrapers, net-building collectors, shredders, piercers, predators and parasites. Energy transportation from primary producers to secondary consumers (the fish) is complete when the insect is eaten. This is where the fly fisher tries to intervene.

Artificial flies vary considerably in type, size, and appearance, so great care should be taken in their selection. In general, flies should resemble as closely as possible the aquatic food forms recognizable in the locality being fished (Table 17.1). Many insects metamorphose from aquatic stages to winged adults. A large part of the diet of trout consists of mayflies, caddisflies, and stoneflies. These insects are eaten by trout in all of the stages of their life cycle; which include nymph, emerger, dunn, and spinner.

Fly patterns and designs resembling each of these insect stages catch fish (Figure 17.4). Considerable variation in size can exist from one locality to another, so it is always wise to seek advice from local shops, guides, and fellow anglers about what pattern and size of artificial fly performs best.

The oldest type of fly is known as a dry fly, which resembles mature dunns and spinners. Characterized by prominent tails and wings, these flies are fished on the surface and can produce the most dramatic rises and strikes in fly fishing. Dry flies are most effective when they are presented during a hatch. Fish that have ignored your best efforts for hours can go into a feeding frenzy when a hatch occurs.

When a mayfly or caddis fly hatch is at its peak, a dry fly of the right pattern dropped in the midst of the hatch can produce a strike on every cast. When the hatch ends, the fish will lose interest in your flies just as suddenly. It is, of course, vital to present a fly that closely resembles the hatching flies in size, shape, and color. Most fly fishermen spend a significant amount of time observing the water they plan to fish, to identify the insect types present and to record the precise times of hatches.

The standard approach to presenting dry flies is to cast the fly upstream and allow the current to carry the fly past pools, banks, or seams in the water. To

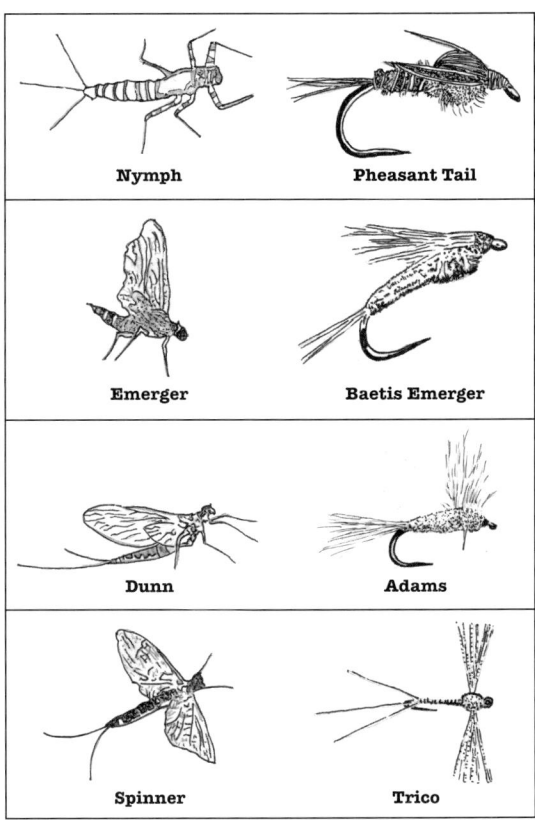

Figure 17.4 Life cycle stages of a mayfly (left) and the artificial flies that mimic them (right).

Table 17.1 Black Hills fly patterns suggested by Erickson and Koth (2002).

INSECT	PATTERN	SIZE	DATE USEFUL
Mayfly	Hare's Ear Nymph	14–20	Year-round
	Adams	16–20	Year-round
Caddis	Little Black Caddis	16–20	April–July
	Elk Hair Caddis	12–16	April–Sept.
Midge	Brassie	16–20	Year-round
	Griffith's Gnat	18–24	Year-round
Terrestrial	Grasshopper	8–14	May–Sept.
	Ant	16–20	May–Sept.
Stonefly	Hare's Ear Nymph	16–14	Year-round
	Bitch Creek	4–12	Year-round
Attractor	Wooly Bugger	6–14	Year-round
	Royal Wulff	14–20	March–Oct.

be visible to the fisherman and attract fish, the fly must set high on the surface and float in a manner that looks natural. The friction of the water current on the fly line tends to cause the line to drag the fly faster than the current, creating an unnatural appearance. Fishermen counteract water drag with a technique called line mending. Line mending involves using quick movements of the rod tip to flip a loop of line upstream when necessary so that it does not drag the fly too fast (Figure 17.5).

Dry flies have a tendency to absorb water with use and lose buoyancy. When a dry fly can no longer float it becomes ineffective. False casting can help to dry out a water-logged fly and extend its period of usefulness. Some anglers like to apply floatants to dry flies in spray or jelly form.

Fish rise quickly to strike the fly. Strikes are exciting, sometimes resembling small explosions in the water. Be alert for strikes and be prepared to quickly raise the rod tip to set the hook. What follows, of course, is the thrill of fighting the fish. A hooked fish will run, sometimes breaking the leader and

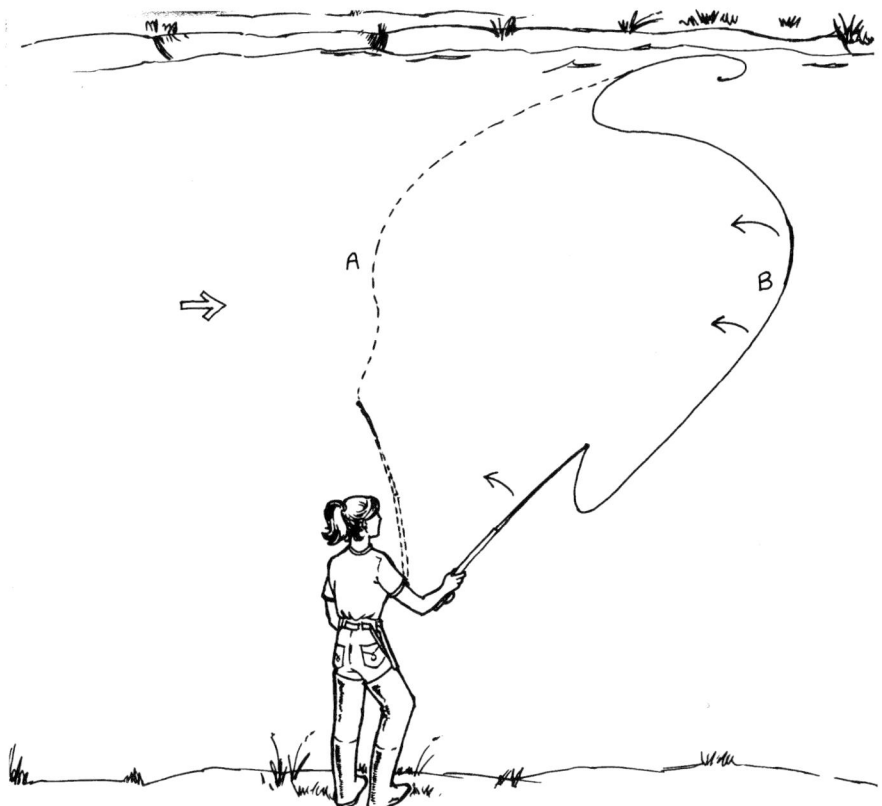

Figure 17.5 Sketch shows the line positions for mending the fly line to avoid rapid downstream movement by the fly. The river current carries the line to point B; the angler flips the line upstream to point A without casting. (Drawing by S. Kahara)

escaping. Playing the fish successfully requires keeping a tight line, but not so tight that it will break when the fish makes sudden jerks or runs.

When the fish has tired it can be reeled in close to the fisherman. At this point most fly fishermen use a dip net to pick up the fish. Barbless hooks are a popular choice, and on some streams they are the only type allowed. The advantage of a barbless hook is that it allows the fisherman to remove the hook easily and permit the fish to be released uninjured.

To release a fish without harming it requires great care and practice. Wet hands help to avoid damaging the eyes, gills, and mucous coat that protects the fish from parasites. Gently holding the fish prior to release and allowing water to flow over its gills for several moments gives the fish time to revive, increasing the likelihood that it will survive when released. The sport of fly fishing offers no greater satisfaction than releasing a fish back into the stream.

The adult phase of any aquatic insect is brief compared to the time spent in formative stages. For this reason trout are most accustomed to seeing their favorite insects underneath the water. To take advantage of this fact a relatively new and popular form of fly fishing has developed called *nymphing* (Osthoff 2006). During the majority of time, when a hatch is not occurring, fish are alert to the wide variety of food forms drifting beneath the surface. For this reason nymphing can be very effective.

Nymph flies are cast upstream and allowed to sink and drift past fish holding in pools and behind rocks. It is important for nymphs to sink quickly, even in fast current. Many are weighted with wire and some have a shiny metal bead next to the eye of the hook. In fast moving water, anglers often attach a small weight where the tipit connects to the leader as a way of bringing the fly quickly to the bottom of the stream. Because water at the bottom of a stream moves more slowly than water on the surface, line mending (Figure 17.5) is critical when nymphing. Nymph fishermen ordinarily make a large mend of the line immediately after the upstream cast has been made.

The chief difficulty with nymphing is the large number of missed strikes that occur because the angler cannot see the fly. The most widely employed solution to this problem is use of strike indicators. A strike indicator is a small highly visible object placed about six to eight feet up the leader. Strike indicators are made of materials such as cork that float. When the indicator suddenly dips below the surface, raise the rod tip to set the hook.

When fishing in unfamiliar water, one may not have the advantage of knowing the local insect forms or hatches. In such circumstances fly patterns resembling terrestrial insects can be a good choice. Ubiquitous insects such as ants, bees, and beetles are just as recognizable to fish as they are to the average picnicker. An especially popular terrestrial insect in the late weeks of summer is the grasshopper. In the hot days of August, trout lying close to stream banks can literally go wild for "hoppers." Many effective grasshopper patterns are available, and should be represented in any tackle box.

In recent years, there has emerged a new class of flies called attractors. These flies prove an exception to the rule of local resemblance, in that they do not really look like any food form the fish should recognize, but are effective at taking fish never the less. Attractors are popular with fly fishermen who fish in many different locations. Most attractor flies are fished in the same manner as dry flies.

An essential type of fly for any serious angler is known as a *streamer*. Trout and bass will often take streamers when they seem indifferent to other types of flies. Larger than most flies, streamers are designed to resemble a minnow or leach and are fished well beneath the surface. Streamers are ideal for large bodies of water, such as lakes or rivers.

The standard approach to streamer fishing on a creek or river is to cast upstream at an angle and allow the current to bring the streamer around in a sweep. It is important for the streamer to get down quickly to fish holding in deep water, so the faster the water the more weight the streamer must have.

In still water, streamers are cast out and brought back to the fisherman through a technique called line stripping, in which the fly line is guided through the index finger of the right hand and retrieved by a series of small tugs with the left hand (about 12 inches each) until the fly is too close to the fisherman to draw a strike. Line stripping can also be an effective technique for nymph fishing on lakes or reservoirs, since the jerking motion of the fly resembles the movement of an aquatic insect.

Large brightly colored streamers fished well below the surface are the preferred method of fly fishing for northern pike. A premier game fish, northern pike inhabit lakes and tend to be found in deeper water, often feeding within large weed

17.2

Dr. Lester Flake

Dr. Lester Flake, retired Distinguished Professor of Ornithology at South Dakota University was well known for his hunting and fishing adventures. He wrote from Utah "I am already thinking how I will miss my favorite spots for smallmouth, white bass, northern pike and other fish in South Dakota. It was a shame to live there for almost 20 years before really discovering the possibilities for eastern South Dakota fly fishing."

"In the mid-1980s I visited Lake Louise near Miller with my son Ryan. My son caught bluegills on the leeches but my black wooly bugger, a leach imitation fly, worked just as well. In 1992 I began offering a fly tying class in the adult education program at Brookings. I taught that class about 8 times with 7-12 people in the class each time."

Dr Flake gave these fly fishing tips:

• For white bass, I normally use a 6-weight rod and fish a small Clauser's below the surface. My favorite color was chartreuse green on the back with a white belly and a little sparkle.

• If you fish northern pike and they keep breaking your line, I recommend you add a tippet of hard monofilament...a 15-pound hard mono will keep most northerns on the line.

• One evening at Lake Thompson, I caught 8 walleye from 14-18 inches in about two hours wading near the shore tossing a Clauser's minnow trailed by a wooly bugger. Most walleye hit the trailing fly.

• Carp hit best on small nymphs like the prince nymphs or even crayfish imitations. My last carp was from Lake Thomson, a 15-pounder taken on a small black and white Clauser's minnow. The fish was one of the few that reached into the backing on my fly line.

• Fish for smallmouth with a Clauser's or wooly bugger; they will also take grasshopper imitations and small surface bugs. Sorry, my favorite smallmouth spots need to remain unknown but try...

• Enemy Swim and Clear Lake—good smallmouth bass fly fishing

• Lake Poinsett around Stone Bridge for white bass in May and early June

• Lake Norden Dam for white bass if they are releasing water

• Bitter Lake—northern pike and walleye

• Lake Cochrane—float tube or shore fish for largemouth bass and bluegill.

flats. For pike, fly fishermen prefer a longer rod of either a 9 or 10 weight. Large (4–6 inch) highly visible streamers with wire weed guards are recommended for pike. These are fished deep, usually with a weighted fly line. Because pike have very sharp teeth, a wire tippet is considered necessary. Special pike leaders are commercially available.

When fishing for northern pike, the streamer is cast long and retrieved in strips of 1.5 to 2 foot lengths. Northern pike are legendary for their size and fighting ability and may make several long runs before tiring. Because pike have teeth and gill rakers that are very sharp, a net, a fish-handling glove, and a hook removal tool are essential equipment.

When all else fails, there is even a fly that resembles an angleworm. The "San Juan Worm" consists of a short length of red or purple chenille tied in the middle to a hook and burned on each end to create a worm-like appearance. While the San Juan Worm may be an anathema to fly fishing purists, I can attest to the fact that they attract fish.

In moving water, the San Juan Worm should be fished in the same manner as nymph flies, casting upstream of the fish and letting the current drag the worm in front of the fish. It is important so get the fly well beneath the surface, so line mending and a weighted tipit are recommended. A split shot sinker about 18 inches above the hook on the tipit should be adequate to produce the desired result.

South Dakota offers great fly fishing opportunities (Box 17.2). I recommend that you experiment by going after a variety of game fish and try many locations. You'll have your best results after a combination of reading, conversation with fellow fly fishers, and time spent on the water. The time you spend studying fish, learning their location, feeding habits, and preferences will be rewarded. Even when fishing is slow, the water bodies where fish live are enjoyable to visit. And remember, "there is more to fishing than the fish." ❖

Chapter 18

Law Enforcement for Protection of Fisheries Resources

Bob Brown

"We ask a simple question
And that is all we wish:
Are fishermen all liars?
Or do only liars fish?"

— W. Fox in *Silken Lines and Silver Hooks*

Wildlife law enforcement is often misunderstood and almost always involves emotions and differing opinions – opinions about the type of training needed by Conservation Officers, and about the kinds of laws that can be enforced other than those related to hunting and fishing. Should Conservation Officers have other duties in addition to law enforcement (Box 18.1), and should they be allowed to extend their authority to private property to check anglers?

18.1
Law Enforcement Duties Of A Conservation Officer

Conservation officers enforce state laws and department regulations to ensure compliance, encourage wise and equitable use of the state's wildlife and fisheries resources, and promote public safety. Major duties include:

- Provide public education information regarding laws and regulations.
- Investigate information provided by private citizens regarding wildlife violations.
- Assist citizens in solving problems related to game, fish, and public lands and waters.
- Provide public information on fish and wildlife management activities.
- Conduct enforcement and surveillance activities; take appropriate enforcement action.
- Conduct investigations, collect evidence, interview witnesses and suspects, testify.
- Prepare affidavits, obtain and serve search warrants, arrest warrants, and subpoenas.
- Check licenses and safety equipment of hunters, anglers, boaters, and trappers.
- Monitor related businesses for compliance with laws and regulations.
- Provide training to co-workers and other law enforcement personnel.
- Plan and assist in search and rescue operations.
- Coordinate and teach Hunt Safe classes. Recruit and train volunteer instructors.
- Recommend changes to laws, regulations, department policies, and procedures.
- Coordinate with other agencies to enforce laws across jurisdictional boundaries.
- Enforce criminal laws other than fish and game laws when necessary.

In 2002, the Department of Game, Fish and Parks wrote a comprehensive report to answer these questions and many others (SDGFP 2002). Included in the report was a reprinting of my published article about the history of wildlife law enforcement in South Dakota (Brown 1993) and another good summary (Moum 1993).

I can summarize my 30-plus years associated with fisheries law enforcement in one word—*change*, mostly change in the angler. The number of anglers and their desire to catch "more" and "bigger" fish has been a powerful force. The result has been continuing changes in regulations, fishing technology, fish management, and enforcement procedures.

The purpose of this chapter is to review historical events in law enforcement with an emphasis on fishing and boating, and describe the change from the "fish warden" of the past to the Conservation Officer of today.

Dakota Territory

The Dakota Territory Legislature of 1883 marked the beginnings of fisheries law enforcement in the future state of South Dakota. The territorial governor appointed a fish commissioner whose duties were to stock fish, and to close and open stocked waters to fishing. Constables, sheriffs and deputies did the enforcement. Their main duty was "compliance checks," which is law enforcement jargon meaning *checking* anglers to determine whether they were in *compliance* with the law. The 1887 Legislature empowered judges to appoint attorneys to prosecute fish cases at a fee of no more than $10 per case.

Fish and wildlife populations in the early 1900s had been reduced by overharvest, which was labeled by one author as "unbelievable slaughter" (Hipschman 1959). For example, an article in the *Custer County Chronicle*, dated July 5, 1913, goes like this (Sundestrom 1994): "In 1912, state law already prohibited taking more than 25 trout from any stream by any person from April 1 to October 31. The law was supposed to keep people from dynamiting fish wherein trout thieves "plying their nefarious trade" in Hills streams fill a heavy glass bottle with lime, place a few drops of water in the lime, tightly cork the bottle and throw it into the trout holes. The lime, in slaking, develops tremendous force and explodes the bottle killing many trout which rise to the surface and are taken by thieves."

Statehood in 1889

Statehood became a reality on November 2, 1889, as South Dakota became the 40th state. The fish and game in the state became a public trust resource, owned in common by all the citizens of South Dakota, whether that wildlife was found on public *or* private land. The new Legislature was the trustee and they needed game wardens to protect the public interest in the wildlife resources. The first state legislative session to address fisheries needs was in 1893 when lawmakers instructed county boards to appoint fish wardens in each county.

Fish wardens received no salary, but split fine money with the state. It was said that a good fish warden and a favorable justice of the peace could do a land office business. A well-known fish warden was Lou Hawley, who operated a barber shop in the Cataract Hotel in Sioux Falls. He was named a fish warden in 1897 and served for 30 years, retiring at age 78 (Qualset 1945). Lou recalled that Governor Byrne offered him any state job he wanted. Lou said that he wanted a job as a game warden on one condition—that he be permitted to exercise just plain common sense in handling the game laws. He later wrote, "The fines I collected often comprised my salary."

In 1899, fish wardens were renamed "game wardens" and were given statewide authority to enforce laws that prohibited the sale of game. By 1901, the deposits of half of the fine money from game and fish convictions had grown into a substantial sum, so the legislature transferred the fine monies into the School and Interest Fund. Today, no money is kept by Conservation Officers. Present-day fine monies also go to schools, according to the South Dakota Constitution (Article 8-3, "proceeds of all fines collected from violations of state laws shall be…distributed by the county treasurer among and between all of the several public schools").

In 1903, wardens were appointed by the governor upon the petition of 10 citizens in a county. The "political appointee" wardens received a salary of $50 a month during the hunting and fishing seasons. Wardens were themselves guilty of a misdemeanor if they did not work during the various open seasons.

Game and Fish Department Created

The Game and Fish Department was created in 1909, and W. F. Bancroft was appointed as the first state game warden (chief). He was paid $1,500 a year. Bancroft proposed that game wardens be appointed on their qualifications and not on their political affiliation, and by 1913 his proposal was made into law. The first eight "qualified" wardens were former farmers, well drillers, and merchants. However, without political meddling, the eight wardens made 65 arrests in one year compared to only 46 arrests made by the *58 political appointee* wardens the year before.

Wardens were authorized to wear uniforms in June 1935— breeches, boots, tunic coat and campaign hat. An incident at the Sioux Falls Federal Courthouse prompted the authorization to wear uniforms. Two wardens were holding two poachers in an anti-room of the courthouse. A clerk mistook the wardens for poachers and set a court appearance time and bond for the wardens! Yes, wardens needed a uniform.

World War II brought change for sportsmen and wardens. Guns, ammunition, fishing tackle, artificial lures, and vacuum bottles were virtually nonexistent. Rationing gas to 3 gallons each week limited everyone's travel. Wardens had fewer hunters and anglers to check. Game populations increased. Soon the wardens were seeing high-ranking military officers, movie stars, and sports stars visiting South Dakota for the pheasant hunting. South Dakota was not a destination for anglers except for the trout fishing in the Black Hills. However, the new dams on the Missouri River would soon change recreational fishing and boating in South Dakota. Wardens would be dealing with new fishing regulations and boating safety.

Post-War Years Bring Change

In the late 1940s, college graduates were available in the job market, and a small cadre of biologists joined the staff of the Game and Fish Department. Law enforcement *and* biology were now being used to enhance and protect fish populations. The image and the duties of game wardens changed. The warden was no longer seen as a long-mustached individual who wore a badge and snooped around the country trying to arrest someone. He was a public relationist, educator, and fish and game manager combined (Peterson 1942).

The history of law enforcement, and all fish and game management in South Dakota, changed in 1957 with "the Gabrielson Report." Dr. Ira Gabrielson, a well-known national figure and former head of the U. S. Fish and Wildlife Service, was asked to examine the Department and to make recommendations. He recommended sweeping changes that included more scientific management of fish and game populations. He predicted more recreational fishing in South Dakota's future and recommended that the agency have a fish division that included more than just fish hatcheries and fish stocking.

New biologists and fisheries managers assumed much of the wardens' work, thus enabling the wardens to concentrate more on law enforcement. However, the changes were unsettling to the wardens who were older individuals and "kings of their counties." The records of the Department do not reveal much about the "power struggle" between the biologists and the wardens. The power struggle between the practical experience of the wardens and the science of the biologists solved itself in time. Working together was best for the Department and for fish and wildlife resources.

The 1958–59 Annual Report of the Department (Hipschman 1959) listed problems and advantages for fishing in South Dakota. The chief advantage was the recognition that "South Dakota has once been at dead zero and knows it can come back." The report stated, "The day will never come again when market hunters can kill wantonly…when people will depend solely on wild game and fish to provide for their table…the use of our resources will be largely for recreation." Within the list of problems was one problem that foreshadowed a major issue for law enforcement in the future. The report stated, "There is a question of obtaining hunting and fishing privileges on private lands for the sportsman."

From Warden to Conservation Officer

In 1970 the law enforcement position was renamed "Conservation Officer," a term that the public quickly shortened to "CO." Most of the new COs were coming on the job in their early 20s and many of them had university degrees in wildlife and fisheries sciences. Changes were numerous but one thing remained constant—the local CO was usually the first person who came to mind when a citizen evaluated the Game, Fish and Parks Department or had a question about fishing. It is that way today.

Fishing and boating violations vary from minor to extreme infractions (Box 18.2). COs have the latitude to use "officer discretion," especially when dealing with minor infractions. COs usually issue more verbal and written warnings than citations because all violations of the same regulation do not merit identical treatment. As an example, a youthful angler who misidentifies a fish and unintentionally exceeds the limit is far different than an experienced angler who is attempting to catch a second limit on the same day. Citations certainly play a role in law enforcement activities, but officer discretion plays an equally important role in achieving public compliance with natural resource regulations.

In 1972, fisheries regulations were relatively simple compared to today. The early regulations were usually "statewide" regulations. Daily limits and possession limits for a species were generally the same for all inland waters. Today, regulations are designed for a particular water body or a specific class of waters, and length limits are more complicated.

New equipment has dramatically influenced fishing success and changed enforcement situations for the CO. The open, 14-foot aluminum boat with a 20-horsepower motor has given way to a 17- to 20-foot (or larger) boat with a 200-horsepower motor. Larger boats sometimes make compliance

18.2
Examples of Fishing And Boating Violations

Minor Fishing and Boating Violations
Too many fishing lines
No life jackets
Fish length-limit violations
Fish house labeling
No boat registration
Boating after dark without proper lights
No observers while water skiing or tubing

Moderate Fishing and Boating Violations
Over limits of fish
Careless boat operation
Illegal fishing tackle (dip nets, snagging)
Unauthorized stocking of public waters
No fishing license

Extreme Fishing and Boating Violations
Illegal sale of fish
Illegal use of set lines and nets
Over-bags of game fish
Fishing with explosives
Fishing with shocking devices
Illegal baitfish transfer
Boating under the influence
Dumping toxics or sewage in water
Reckless boat operation
Lacy Act (interstate transport)
Trespass fishing

checks difficult and sometimes an officer must board the boat to personally inspect a live well and verify length limits.

During the early days of my career, the only tools that I needed to check anglers were binoculars and a copy of the small fishing handbook that provided me with the daily and possession limits for each fish species. The binoculars allowed me to determine angler activities before I "introduced" myself. I could check anglers both on shore and in a boat by simply asking to see their fishing license and looking in their cooler or bucket, or checking their stringer.

Individuals who violated regulations usually appeared in front of a Justice of the Peace, who determined the amount of the fine if the angler was found guilty. Today, COs carry a bigger fishing handbook and a bond schedule. The bond schedule is from the circuit judge who has set the penalty. A power of attorney provision allows a person to choose to either

18.3
Fisheries Management Duties Of a Conservation Officer

Conservation officers manage fisheries resources to conserve, perpetuate, and stock species; maintain and develop habitat; and enhance recreational opportunities. Typical duties include:

- Assist with lake management plans on public and private waters.
- Recommend stocking and habitat development.
- Design and implement habitat development projects.
- Recommend harvest numbers, size limits, and other harvest regulations.
- Operate and maintain fish-rearing ponds and trap and transfer fish.
- Monitor commercial fishing and authorize payments to commercial fishermen.
- Work with landowners to develop public fishing opportunities on private waters.
- Investigate fish kills and evaluate potential impacts on fish populations.
- Monitor fishing tournaments.

send fine money through the mail or set a date to appear in court.

Years have gone by and as biology and public desires changed, so did the law enforcement duties (Box 18.3). During the early years, fish warden duties were mainly focused on preventing illegal harvest, illegal marketing, or violation of state laws. COs today enforce laws and regulations but they also help implement Department programs for habitat, wildlife, and people. They represent the Department to the public, provide outdoor recreation education, and promote public safety. Change will surely continue in future years. I'm glad and proud that I was able to play a small part in the history of South Dakota law enforcement and fishing.

Tough Aspects of Law Enforcement

Wildlife law enforcement has always been a dangerous and demanding job. Laws alone are pointless unless someone is willing to enforce those laws. COs are fully certified South Dakota Peace Officers, and the training and equipment they receive is some of the best in the state. COs patrol our public waterways for reckless or intoxicated boaters, and investigate illegal activities of people carrying guns.

Anyone who has ever met a South Dakota CO has probably formed an opinion about that officer (Tilberg 1993). He might be the helpful guy who stocked fish in the farm pond, or the jerk who ticketed a relative for having too many yellow perch. CO's must *not* take things personally.

Conservation Officer Bill Antonides has some lessons for the prospective CO (Antonides 1993). His lesson Number One is—Everyone knows the area CO and many folks don't mind calling him or her at any hour of any day or night. Lesson Number Two is—A CO has roughly 200 different jobs and they are all top priority. Lesson Number Three is—There are a lot of people out there who really care about our environment and the wild creatures.

Some job responsibilities present unique challenges and opportunities. No one likes to suffer a financial loss. Landowners sometimes report instances of damage by wildlife to property, crops or livestock. Responding to these reports usually falls on the shoulders of a trapper or CO. However, during winters with extreme weather conditions, any Wildlife Division employee may be called to assist. Remedies that reduce or eliminate ongoing wildlife damage are much more numerous and varied than a few years ago. Whether the issue is Canada geese in soybean fields or deer on haystacks, the Wildlife Division is ready to assist.

It is tough on a CO when the public is disrespectful to their families when emotions run high. It is tough to read letters to the editor that would not be characterized as civil discourse. It is tough to accept that people are too complacent to call the TIPS (Turn In Poachers) hotline when they see violations. There are interesting exceptions however, such as the "TIP from the toilet" incident (Box 18.4).

There have been organized movements against the agency and attacks sometimes become personal because of Antonides' Lesson Number One—Everyone knows the local CO. The organized movements sometimes threaten the ability of the officers to make compliance checks, which are critical to wildlife conservation. Critical because the informal contact with anglers in the field provides the public with assurance that laws and regulations established by the South Dakota Legislature and Game, Fish and Parks Commission are being followed.

18.4

The Toilet TIP Incident

A poacher and another man were in a toilet at a state park and a woman on the woman's side overheard them discussing their over limit of walleyes. They bragged that they'd never get stopped going home because they were not pulling a boat. The eavesdropper called the TIPS hotline and the poacher was soon facing a Conservation Officer with lots of questions. They were fined $76.50 and lost all of their fish.

COs are allowed to conduct license and bag checks on private land under the "Open Fields Doctrine," a series of judicial rulings. Open fields means that COs can go onto private land to do compliance checks if they observe hunting or fishing activity. Challenges to the Open Fields Doctrine usually relate to hunting, however, another private land issue relating to fishing has been just as contentious recently because of high water levels in some northeastern lakes. The water and the fish are public trust resources but the high water levels have covered many acres of private land. Can anglers fish over private land? This situation has presented new challenges not only for COs, but also for legislators, judges, and policy makers, and the landowners.

Tomorrow's Challenges

Maintaining compliance with fish and wildlife laws so that fish populations will remain healthy will continue to be a challenge for the COs of tomorrow (Keyser 1993). Laws and regulations should be kept as simple as possible, understandable, and reasonable so that compliance won't be difficult. COs will have to work smarter because of new duties and challenges. Smarter may mean doing more public education to cultivate voluntary compliance.

Sportsmen and sportswomen, farmers, ranchers, city folks, and country folks all play a role in conservation. Although citations, fines and jail sentences provide a deterrent to would-be violators, it is the general public's attitude about conservation of South Dakota's resources that will do the most to preserve the resources for the next generation.

The COs of the future will continue to develop innovative tools and techniques to catch those who would steal the people's fish resources (Figure 18.1). New COs should listen to David McCrea, retired Law Enforcement Program Administrator who says, "Poachers, for obvious reasons, will never like game wardens. I learned that no matter what the circumstances, I was dealing with human beings, and each and every one deserved to be treated with dignity and respect,

Figure 18.1 Two men were arrested with 205 trout, which is 185 fish over their limit. They were fined a total of $3,000, lost their fishing privileges for five years and paid $18,500 in civil damages to the Department of Game, Fish and Parks. Total cost for the fishing trip was $21,500, and they did not get to keep their fish. (SDGFP photo)

even when I was issuing them a ticket or taking them into custody."

Antonides had a fourth lesson for new officers. Lesson Number Four is—There are few jobs with such stress, but there are also few jobs in which one can claim a little credit for making sure there are fish for kids to catch today and tomorrow. ❖

Chapter 19

Fisheries Management Practices: Past and Present

Dennis Unkenholz

> **"** Maintaining a good
> fishing pond is not easy. **"**
>
> — Robert L. Hanten

The philosophy concerning land 150 years ago was that everything was free. Land was free, game and fish were free, and the supply was thought to be inexhaustible. Everyone had his or her own idea of how many fish to take when an important daily issue was finding enough food. There were no stock dams, Missouri River reservoirs, or man-made lakes, but streams and lakes in Dakota Territory were very productive. The idea of "managing" any of these resources was unheard of for most of the 1800s.

As in the rest of the country, the fisheries of Dakota Territory were badly depleted by the end of the 19th Century.

Most historians mark the beginning of the North American conservation movement with the Governor's Conference of 1908 when President Roosevelt strongly advocated wise use of natural resources (Pinchot 1947, Nielsen 1993). South Dakota's Governor Coe Crawford attended the meeting and was himself a Progressive like Roosevelt. Crawford returned to South Dakota and encouraged the South Dakota legislature to work on this new idea called "conservation."

Crawford and the two governors who succeeded him (Governors Vassey and Byrne) focused on timber, grass, soil and water resource conservation (Crawford 1907). The Department of Game and Fish was created during Governor Vassey's administration in 1909. The Department functioned under a commission consisting of the Governor, Attorney General, and State Game Warden. The Department was incorporated into the Department of Agriculture in 1925 and in 1945 was renamed Department of Game, Fish and Parks and granted agency status. Even though Vassey formed the Department, Governor Norbeck (1916-1920) could be called the first "fish and wildlife conservation" Governor of South Dakota because of his interest in out of doors activities (Meyer 1975).

Management programs for fishes, aquatic habitats, and anglers have been used historically in South Dakota, with emphasis placed on one or more programs at any given time. Recently, fisheries managers develop objectives for a specific body of water, and then employ several approaches to managing fishes, habitats, and anglers. This chapter summarizes how fisheries management philosophy and approaches have evolved in South Dakota.

Three Parts to Fisheries Management

Fisheries management can be categorized into one of three approaches: 1) managing the biota, mostly the fishes but sometimes other aquatic animals and plants, 2) managing the habitat, which includes physical habitat and water quality, and 3) managing the human users, both recreational anglers and other outdoor enthusiasts.

Biota management first involves inventorying the kinds and numbers of fishes present, as the early observers did. Other types of fisheries management include stocking fish, trapping and transferring fish, introducing new fish species, removing rough fish, monitoring populations, and controlling fish with piscicides (fish poisons). These types of management are designed to enhance a species or group of species.

Habitat management includes improving water quality, reclaiming habitat, developing new habitat, impounding water, or diverting water. The expectations are that fish, and usually a preferred species or group of fish, will benefit from the habitat change. With improved spawning habitat, food sources, and cover, we expect fish survival to increase. On the other hand, habitat projects may simply provide underwater structure that will attract fish so anglers can increase catch rates.

Fisheries regulations such as creel limits, size restrictions, season limits, and a broad array of rules that govern angler conduct while fishing are examples of human management. These rules are generally aimed at managing the number or size of fish harvested. Such rules maintain or increase catch rates by anglers, improve fish size, and hopefully spread out the harvest among anglers.

Management of Biota (Mostly Fishes)

The science of fisheries management began with identifying the fishes in the waters of Dakota Territory. Surveys by the Federal Government produced some of the first fish lists (Evermann and Cox 1896). Contrast these incomplete (and sometimes inaccurate) lists with today's research that uses stable isotopes to help understand food webs or electronic signals to count fish or measure habitat. Just knowing what fish are present no longer provides fisheries managers with enough information to make proper management recommendations. Today's fisheries managers require data regarding fish health and condition, population structure, age and growth statistics, species interactions, annual mortality rates, and even angler use statistics.

Most fisheries workers in the late 1800s believed that fish culture was the key to the future of fisheries management. The U.S. Fish Commission distributed fish throughout the country. The duties of the Territorial Fish Commissioner included stocking fish fry received from the U.S. Commissioner of Fisheries, and regulating fishing seasons after fish were stocked (Game, Fish and Parks 1959).

Fish stocking

Fish stocking *was* fish management in those days and both the state and U.S. government were actively stocking fish. The first trout stocked in South Dakota came from the U. S. Fish Commission's hatchery in Colorado. This was the first example of successful biota management, and the results of that early trout stocking in the Black Hills are the basis for the fisheries that exist there today.

Initially fish were stocked to restore depleted fish populations or to add species that were desirable as food. Many native species did not meet expectations and were not favored in management. Today's managers stock fish when natural reproduction is inadequate (Figure 19.1). Trout fisheries in many Black Hills streams are maintained by routine stocking because natural reproduction cannot support the high demand for trout fishing, however wild trout reproduction supports some stream fisheries.

Stocking and hatchery fish have had, and still have, an important role in fisheries management in South Dakota. Construction of the Missouri River reservoirs created deep, large bodies of water

Figure 19.1   Stocking trout by "flipping," a technique used to avoid contaminating the net (and later hatchery water) with fish pathogens that might be in the lake or stream. (Photo by M. Barnes)

with cold-water habitats that were not suitable for the native species of the Missouri River. Studies conducted in South Dakota determined that Chinook salmon were better suited than other salmonids to the deep, cold water of Lake Oahe if there were cold-water prey fishes. Several species of prey fishes were stocked in South Dakota, but it was North Dakota that achieved success when they first stocked rainbow smelt in 1971 in Lake Sakakawea.

Rainbow smelt spread to all of the Missouri River reservoirs in South Dakota. Chinook salmon were stocked in South Dakota in 1982 and the fishery grew to a harvest of approximately 33,000 salmon in 1996 (Johnson et al. 1997). The goals of the salmon stocking program were to diversify the fishery, provide additional fishing opportunity, and provide a unique trophy salmon fishery on the prairie. The salmon program was a success.

Soon after the South Dakota Department of Game and Fish was established in 1909, emphasis was placed on fisheries restoration by stocking. The Department planned a state fish hatchery to augment the fish being stocked by the U.S. Bureau of Fisheries. The management philosophy of the time was to enhance or maintain fisheries by stocking game fish and controlling rough fish (Box 19.1). Little was done at that time to control the conduct of anglers, to regulate angler harvest, or to improve habitats.

Frank Purcell was hired in 1916 as the first Superintendent of Fisheries, and during that year work was completed on the Kampeska Fish Hatchery. The next 10 years saw a proliferation of hatcheries as the Kampeska, Rapid City, Cleghorn Springs, Lake Andes, and Pickerel Lake hatcheries were developed (see Chapter 14).

The drought of the 1930s took its toll on fisheries and by 1933 no fishes existed in many lakes in South Dakota. By 1936, the only lakes with fish were Enemy Swim, Pickerel, Blue Dog, Kampeska, Punished Woman, and Cochrane. As a result, the major goal of fish stocking following the 1930s was to restore fisheries in most state waters. As water conditions improved during the 1940s, fish stocking was the

19.1
Rough Fish Removal

Some ichthyologists don't like the term "rough fish" because all fish are important to the ecology of a lake or stream. Never the less, some nongame species such as common carp, bullheads, and buffalo fishes can be a nuisance when high populations "crowd out" game species that are the focus of most fisheries management. These nuisance species were called rough fish, and there has been a rough fish removal program almost annually in South Dakota since 1913. Commercial or contract fishing continues today by commercial fishermen who market the fish, mostly for human food, but for other products also. Supplies and markets are variable, the industry is small, and the benefit to sport fisheries is largely undefined. See Chapter 15 for more details on commercial fish management.

management choice and the "new water" proved to be very productive for fisheries. In 1945, the Department was reorganized and became the Department of Game, Fish and Parks (SDGFP). Its successful stocking program was an early example of effective biota management that used stocked fish to create, manage and restore fisheries.

Missouri River reservoirs

The SDGFP began investigating reservoirs in the early 1950s as Missouri River reservoirs were being filled by the U.S. Army Corps of Engineers. The fisheries investigations began in 1953 with the completion of Fort Randall Reservoir, now called Lake Francis Case. Fisheries surveys were also accomplished on Lewis and Clark Reservoir. The SDGFP continued reservoir fisheries work until 1965, when the U.S. Fish and Wildlife Service (USFWS) did most of the work.

Any historical account of fisheries management in South Dakota would not be complete unless it contained information on the North Central Reservoirs Investigation (NCRI). Tasked with studying the fisheries of the new reservoirs, the NCRI was established in 1961 with Federal biologists, offices, and laboratories in Yankton and Pierre. The NCRI scientists documented changes in fish populations as the dams were closed and reservoirs filled.

During the early years of filling, largemouth bass, bluegills, crappies, northern pike, and yellow perch thrived because of the newly flooded terrestrial vegetation that provided excellent habitat for these species. As the reservoirs filled and aged, the fish assemblage shifted to walleye, channel catfish, white bass, and included introduced species such as rainbow smelt, spottail shiners, lake herring, and smallmouth bass. Dominant fishes today are suited to the habitat available in the large deep Missouri River reservoirs.

The NCRI closed its doors in 1977. Their work was largely documentary in nature. For example, Walburg (1971) documented the loss of fish and plankton through the dams. Sixteen species of small fishes (< 1 inch long) were flushed downstream at rates of millions per day during peak flushing flows (e.g., a peak of 10 million freshwater drum per 24-hour period). These observations and others provided the basis for Missouri River reservoir research and development conducted by the SDGFP reservoir fisheries staff since that time.

Human management

The first formal regulation of fishing began in 1883 with Territorial legislation. Leaders realized that restoring fish populations required laws to govern fishing and enacted laws that restricted methods and seasons for fishing. For example, fish could only be taken by hook and line and no line could contain more than two hooks. In addition, yellow perch, northern pike, and muskellunge were protected from February 1 to May 1 in all waters except the Red and Missouri rivers. In 1893 the first fish wardens were hired to enforce fish statutes (see Chapter 18).

The new Department of Game and Fish in 1909 was charged with protection and conservation of fisheries resources and could implement rules according to legislative mandates. A specific *fishing license* was required beginning in 1925 instead of the traditional hunting and fishing combination license required since 1909.

The first creel *limit* was imposed in 1919 and it allowed a daily take of 25 fish of all protected species combined, which included black bass, walleyes, and northern pike. In 1935, additional daily limits were established for trout and bluegills (25 each daily), ring perch (yellow perch) or bullheads (50 each daily), and 15 daily for all other species combined. The combination limit (walleyes, black bass, and northern pike) was lowered to eight fish daily in 1947 (Figure 19.2). The limit was again lowered to six fish daily in 1950, and the six fish could be a combination of walleyes and northern pike.

By the 1950s, fisheries management had added these simple human management tools to the ongoing programs of fish stocking, rough fish removal, and construction of water diversions to keep lakes full. The management philosophy was moving from the idea that there was no limit in the amount of fish that could be taken to the realization that protection

Figure 19.2 Surveying anglers and setting harvest regulations has always been a fundamental aspect of fisheries management. (Photo courtesy of C. Schmalz)

was required. At that time, the only fisheries data that existed were qualitative in nature – fishing was slow or good, and fish populations were high or low.

Scientific approach

The first creel *surveys* in the 1950s were used to get accurate, *quantitative* data on angler use, harvest, and preference. These surveys, which were the first attempt to measure whether fisheries management worked, gave fisheries biologists data that enabled them to measure the amount of fishing, and how many fish, by species, anglers were catching.

Modern fisheries management in South Dakota began in earnest in 1950 when the SDGFP hired its first two fisheries biologists, Robert Gibbs and Marvin Allum. Allum later became a professor at South Dakota State College (now SDSU). Prior to that time, the SDGFP fisheries staff consisted of hatchery workers, game wardens who enforced fisheries statutes, and engineers who worked on stream diversions and dam construction.

The State's colleges had developed courses dealing with ichthyology, biology, and zoology (See Chapter 22). Information about fisheries and the aquatic habitat necessary to support good fisheries was becoming better understood and defined. Common thinking had moved from exploitation to conservation and the efficient use of fisheries resources. Many fish and wildlife biologists were aware of a new underlying philosophy to their practical management efforts.

That philosophy was "the land ethic" presented in a new book titled *A Sand County Almanac* (Leopold 1949). The author, Aldo Leopold, coined the term "the land ethic," which defined man's responsibility to wisely use natural resources so that future generations could also enjoy them. The land ethic remains an underlying philosophy of fisheries management as we know it today.

Leopold wrote of the cleavage between two groups in their views of conservation: "For group A, the basic commodities are sport and meat; the yardsticks of production are ciphers of the take in pheasants and trout. Group B, on the other hand, worries about a whole series of biotic side-issues." One of those side issues that worried Leopold was the stocking of several species of trout in the same water and the subsequent hybridization that would lead to less fertile hybrid trout (Leopold 1918).

Taxing anglers improves fishing

The year 1950 will be remembered as a significant year in the history of fisheries management nationwide because it was the year the Sport Fish Restoration Act, commonly called the Dingell-Johnson (D-J) program, was enacted. This act provided for an excise tax on the manufacture of fishing equipment to be collected by the U.S. government and then be prorated to each state based on the acres of public fishing water and the number of fishing licenses sold. D-J money was to be cost-shared with nonfederal money. South Dakota now had funds designated for fisheries projects.

The D-J program funded objective-driven projects, surveys, and research. D-J funding was also used to develop fishing access to lakes and streams and to prepare lake maps. In 1985, the program was expanded with the Wallop-Breaux Amendment that added a portion of the motorboat fuel tax to fund fisheries development. South Dakota receives about $3.4 million annually from this program.

The D-J Program also requires objective-driven management strategies. Fisheries management was now based on science. Within the Department, a division of fisheries investigations was established in 1952 with Bill Clothier in charge. Early studies were about fish food habits, angler catches, fish assemblages, and fish populations. These data enabled biologists to determine the status of fish populations, how much harvest a given lake could support, species interactions, predator-prey relationships, the relationship of a fish population to its habitat, fish diseases, and the relationship between clean water and healthy fish populations.

The collective thought of biologists had evolved another step from the early 1900s to the 1950s—from the basic conservation of fishes to the management of fish populations

by scientific principles. An early scientific theory was management for *maximum* sustainable yield, which allowed a
maximum harvest by numbers and pounds while not impact-
ing fish population reproductive potential. This theory gave
way to *optimum* sustainable yield, where harvest limits are set
to maintain species densities, certain population size structure,
or specific catch rates. The gradual change from striving for
maximum to striving for optimum yield was a move across the
cleavage of Leopold's A group thinking to B group thinking.

Examples of fisheries research studies in 1952 were: creel
survey of Angostura Reservoir, identifying species composition of Lake Hendricks, determining reproductive success of
bass and bluegills of Eureka Lake, population assessment in
Bowdle-Hosmer Reservoir, population survey of Wall Lake,
winterkill of Lake Sinai, Black Hills stream survey, and population analysis and fish tagging study of Lake Kampeska.
Conclusions suggested that underutilized fish populations
were causing game fish populations to become stunted. This
conclusion helped justify the decision to remove rough fishes,
and to move and stock other fishes.

By 1960, fishing had become well established as an outdoor form of recreation and women were now required to
purchase a license. Fishing license sales had doubled since
World War II. Fisheries rules were changed to apply to individual species and were applied statewide. Population dynamics
were better understood; age, growth, and mortality rates of
fish populations were calculated before managers made recommendations for the new fishing regulations.

The 1960s and 1970s marked continued infusion of scientific studies into fisheries work in the state. SDGFP increased
its fisheries research and management staff, the Department
of Wildlife and Fisheries Sciences was established at SDSU,
and cooperative research between SDGFP and SDSU began.
Cooperative research has expanded in recent years to include
the University of South Dakota, Black Hills State University,
and the South Dakota School of Mines and Technology. Since
1970, many graduate students have been trained in fisheries
science and their work as published in theses and dissertations

has provided sound scientific information that has been used to manage the state's fisheries (see Chapter 21).

Business practices introduced

Strategic planning has become a way of doing business, and budget requests are supported by operational strategies that address specific objectives. The fisheries biologists adopted methods of planning and accountability that had made private companies successful. Both SDGFP and the Wildlife Division have mission statements based on authorities granted by statutes and the South Dakota constitution. Strategic plans support these missions.

The mix of changing issues, staff priorities, leadership philosophies, and management outcomes make strategic planning dynamic and sometimes difficult and challenging. However, planning enables fisheries managers to focus on management decisions and to develop specific strategies for a specific water body. Fisheries management has become more objective-driven as studies provide feedback on the success of management practices. Data are the basis for the rules now in place, not opinion.

By the end of the 20th century, the science of fisheries management was well established, but there was general uncertainty about one big piece of the fisheries management equation—the human dimension.

Human dimensions research

Human dimensions became a part of the fisheries vocabulary in South Dakota in 1993 when Dr. Larry Gigliotti was hired to lead studies of the human dimensions of managing a fishery. The first extensive human dimensions study evaluated anglers' preferences for species, type of water to fish, and style of fishing (Gigliotti 1996). A basic principle of human dimensions is that anglers fish for many different reasons and seek many different benefits from fishing.

The responses to his first ever angler survey enabled him to categorize South Dakota anglers into human dimension groups called utilitarian shore anglers, natural shore anglers, total experience anglers, utilitarian boat anglers, and nature/ trophy anglers. Each group had different opinions and expectations of fishing South Dakota waters. However, they all rated a fishing spot very highly if there was good water quality and accessibility—catching fish was rated lower on the lists of each group. It seems the public was in Leopold's B Group (worried about side issues).

That first-ever 1993 angler survey and the subsequent human dimensions research are basic to management of the state's fisheries today. Creel surveys now contain several questions related to angler opinions in addition to questions related to the number of anglers and fish caught. For example, 65% of resident anglers were satisfied with their fishing experience in 2003; more nonresidents (70%) were satisfied. In 2003, anglers said that they went fishing because it was a social activity in the out-of-doors that was relaxing. Only 8% went fishing for food. That year anglers fished about 3 million angler days of effort, which was a 0.4 million increase over 1999 fishing. Anglers harvested 1.8 million walleye, 1.8 million yellow perch, and about 200,000 each of largemouth bass, northern pike and trout. Maintaining that amount of harvest demands a lot of effort from fish managers.

Training in "customer service" has been a consequence of the awareness of the human dimension in fishing. Fisheries managers now receive training to help them develop and implement rules that will be most accepted by anglers, yet meet management objectives.

When rule changes are developed, public involvement and preference are an integral part of the data considered. Involving user groups in the decision-making process has benefited fisheries management in South Dakota. Information gained has helped define issues that require attention, opportunities that must be expanded, or problems that need fixing. Angler preferences may be accommodated when several options exist for solving a particular problem.

Public involvement has allowed user groups to develop ownership, which in turn generates support for the management approach that the data suggest. This process has evolved to the point that anglers are surveyed on specific issues *prior to* the staff or SDGFP Commission making recommendations or decisions.

An open Commission

The SDGFP Commission is an eight-person group given the responsibility by the legislature to promulgate game and fish rules as well as approve Wildlife Division budgets. This nonpartisan group, comprised of half landowners and half non-landowners and representing the relative population split east and west of the Missouri River, are appointed by the governor to a four-year term subject to a single reappointment or until replaced.

The commission is quite accessible to the public, meeting 10 times annually and at several locations around the state. All rules proposed for consideration are given public notice and a formal hearing, so anyone interested in commenting on the rule proposal has an opportunity to do so in writing or in person. This process ensures that all interested parties have access to due process, and sometimes the Commission meetings are quite lively while the public has its say.

In addition to this citizen involvement, SDGFP fisheries staff conduct public involvement activities appropriate to the issues at hand, which may involve person-to-person discussions, open houses, public information meetings, workshops, mailed questionnaires, telephone surveys, and mailed back question and answer information. For example, public relations became extremely important in the late 1980s when managers of the Lake Oahe fisheries began getting some bad news from their survey data.

Survey data warns managers

In 1990, significant changes in fisheries rules were implemented specifically for Lakes Oahe, Francis Case, and Sharpe. Study results showed walleye populations were being overfished and in general, only two year classes dominated the walleye population in Lake Francis Case. The average size of fish in the angler creel had declined to about 13 inches.

If natural reproduction failed one year, then only one year class would dominate the population. Such a situation could lead to a crash. Although catch rates remained acceptable, average fish size continued to decline. Anglers wanted larger fish, preferred catch rates to remain attractive, and favored unlimited opportunities to fish. The statewide daily limit for walleyes was six with no size restrictions.

The fisheries staff was convinced walleye populations could not sustain such harvest pressure. Most remaining walleyes were below a size preferred by anglers. The fisheries staff conducted a public involvement effort, carefully presented all existing data and approached the Commission with a recommendation of five walleye daily, 10 in possession, and a 14-inch minimum size restriction during April, May, and June; and a maximum size restriction of one fish equal to or greater than 22 inches.

Following the standard 30-day comment period and formal rule hearing, the Commission modified the proposal and passed the new rule limiting anglers to four walleye daily, eight in possession statewide, with the 14-inch minimum size restriction during April, May, and June for Lakes Sharpe and Francis Case. The commission did not accept the maximum size restriction, however. This first major break from statewide rules was a radical change, but was supported with data and for the most part was also supported by the public.

The new regulations have been effective for the Missouri River, where catch rates have remained high, the walleye population has expanded, and several mature-year groups have become well established in the populations of each reservoir. Missouri River walleye harvest and catch prior to the regulation change were approximately 500,000 annually during the late 1980s. By 1995, total catch was over one million fish and harvest was 700,000 (many fish are caught and released). Angler use continued to increase and angler satisfaction remained high.

Today, the angler pressure on fish populations is greater than ever as anglers have become quite mobile with new boats and fishing gear and are highly sophisticated and efficient at finding and harvesting fish. Fishing regulations have been devised to allow unlimited fishing opportunity, while restricting harvest to levels that do not harm the fisheries or the fish size distribution. It is a delicate balance that is only possible through diverse regulations that are lake and species specific and are based on management objectives and science.

Habitat management

Territorial laws established in 1885 required construction of fishways on all dams on the Big Sioux, Cheyenne and Dakota (Jim) rivers. This law was amended in 1887 to include all flowing water in the territory. Beginning in 1919, industrial mills on eastern South Dakota streams were required to have screens on their millraces to protect fish from being trapped. The number of mills may have prompted laws regarding fish passageways to protect fish. Legislative approval of such laws suggests folks understood the importance of free movement of fish up and down rivers and streams in the territory.

The first water pollution law, passed in 1899, prohibited sawdust from being placed in streams containing fish. This may have been the first formal recognition that water quality was important to the well-being of fish (See Chapter 9).

Building water

While fisheries managers believe they have control over fisheries, availability of water is what really determines the quality of fisheries. Trends in fishing license sales show the relationship between water and license sales - license sales tend to be higher during times of high water levels in lakes and streams than during extended droughts.

The Department of Game and Fish sponsored construction of 20 new lakes in 1927 because of the drought conditions. Water conditions were poor and people felt the state should help by funding construction of these waters. The Big Sioux River to Lake Poinsett diversion was constructed to store floodwater in Lake Poinsett, as well as to maintain lake levels. Annual reports discuss water diversions as a way to keep water levels high in receiving waters, to help manage flood runoff, and to provide fishable waters. Another 14 new lakes were constructed in 1932.

Many small dams were constructed statewide during the governmental work projects of the Civilian Conservation Corps (CCC) and Works Progress Administration (WPA). By 1958, these groups had built 90,000 stock dams. Many of these were built by the successor to the WPA, the Soil Conservation Service. The CCC worked in the Black Hills and the WPA laborers worked mainly in the rest of the state. Many of these small CCC and WPA impoundments continue to provide fisheries today and the state has established a plan to maintain the small dams currently in public ownership.

The State's water building programs were greatly expanded by the Bureau of Reclamation that built large reservoirs in the Black Hills, and by the U. S. Army Corps of Engineers that constructed the great reservoirs on the Missouri River. The involvement of several agencies requires cooperation in management.

Cooperation with the neighbors

Border waters are managed in cooperation with neighboring states, but in the past, rules for border waters often times were different than statewide rules. Annual "border water" meetings are held with fisheries managers in Iowa, Minnesota, and Nebraska to standardize regulations.

Another challenge to fisheries managers is the joint management of certain waters with other governmental agencies, such as tribal governments, U.S. Forest Service, U.S. Fish and Wildlife Service, U.S. Army Corps of Engineers, Bureau of Reclamation, and in some instances, local governments. At times, state fisheries managers do not have adequate control to implement the best practices in these situations.

A case in point is the cooperative management with the U.S. Army Corps of Engineers on the Missouri River and its reservoirs. Fisheries managers have no direct control over water management; however, they do have an opportunity to provide annual fisheries management recommendations designed to enhance or benefit Missouri River fisheries.

These *interjurisdictional waters* are challenging to manage. Water levels in the spring during spawning are critical to successful year-class establishment of both forage and game fish. State managers can only make recommendations and sound fisheries management may not occur because of other priorities or political pressure beyond heir control.

Habitat improvement

Habitat improvement projects are accomplished annually by the SDGFP. Examples are stream improvements in the Black Hills to improve riparian habitat, in-stream structure, in-lake habitat (Figure 19.3), overhead stream cover, or water quality. Many small CCC or WPA dams have been repaired or cleaned out to regain the original features of the lake.

A new trend is for fisheries managers to become involved with terrestrial habitat because it is an important issue mostly from a water quality perspective. The Conservation Reserve Program (CRP) provision to the federal farm bill program has had measurable benefits to water quality. While this program is not managed by fisheries biologists, CRP improves the state's fisheries by reducing erosion. Water quality is a challenging habitat issue facing fisheries managers today; so is water quantity.

High water conditions in northeastern South Dakota have created 50,000 acres of fishable water not previously available during modern memory. Bitter Lake, once a few thousand acres of prime waterfowl marshland, is now a 10,000-acre lake. Waubay Lake, which is currently one

Figure 19.3 Angler groups often help with habitat management. Here, an open-water trough has been cut in lake ice so that trees (weighted) can be submerged for fish spawning habitat and cover. (Photo by C. Berry)

large 17,500-acre lake, was formerly composed of five smaller lake basins. It currently supports a world-class fishery. These lakes, examples of what has happened in numerous lake basins in the northeast, have stimulated fish reproduction and growth and subsequent fishing activities never before seen in that area. Management challenges also come with the new water.

Management of the fisheries in the new lakes is objective-driven *adaptive management* that considers the fishes, the habitat, and the angler. Adaptive management is a management approach that gives fisheries managers the flexibility to manage in concert with users' demands while "adapting" to existing conditions. Surveys are conducted to monitor adult fish populations, young-of-year fish abundance, population statistics, forage abundance, and angler use, preference, and attitude.

Managers combine this with habitat information to evaluate whether fisheries are meeting their stated objectives. Where fisheries do not meet management expectations, managers provide recommendations for improvement in the form of new fisheries regulations, habitat improvements, fish stocking, or some combination of these.

Change through the years

Fisheries management has evolved in South Dakota for over 100 years, from the days of learning what fish were in the Territory to the scientific, applied research studies and surveys done today. The contemporary management challenge is to maintain fisheries in the face of increasing angler demand, competition for the same aquatic resources, and continued habitat degradation.

The quality of current fisheries could not be maintained without the help of fisheries staff. The old adage, "let Mother Nature take its course," will not provide a fisheries capable of meeting public demand today. This is especially true where natural reproduction is inadequate, habitat quality is a limiting factor, or current angler demand cannot be met.

Fishing has changed from the days of "take all you can" to these days of a regulated harvest. Biologists now manage harvest with a growing group of sometimes sophisticated fishing regulations. Commonly used today in South Dakota are minimum length limits, protected slot limits, maximum size limits, reduced creel limits, bait restrictions, and catch-and-release restrictions. As a result, fishing may be as good today as it ever has been. This is a true management success story.

The evolution of fisheries management has been long and interesting. From the first reported observations of fish by Lewis and Clark in 1804, to the first transportation of fish by horse and buggy, to territorial fishing laws, to statehood, to the formation of a state game and fish agency, to the Sport Fishing Restoration Act, to the hiring of the first fisheries biologists, to establishment of a fisheries program at SDSU and to biologically supported management by objective, South Dakota fisheries have been well cared for by programs that benefit citizens and nonresidents alike. The solid principles of today's fisheries management will be the basis for the long-term care of this important public trust resource well into the future.   ❖

Chapter 20

Fishing in South Dakota: Reservations and Tribal Lands

Kent C. Jensen

> **"** Lack of respect for growing, living things soon leads to lack of respect for humans too. **"**
>
> — Lakota Chief Standing Bear

Fish and fishing have played an important part in the lives and livelihood of tribes that called the northern Great Plains home. Originally, fresh and dried fish were an important supplement to buffalo meat and served as a source of fresh protein during times when the buffalo herds were absent (Calvin Jumping Bull, personal communication). Plains tribes used bone fish hooks, and fish traps constructed from young willow, cottonwood, or other readily available materials (see Chapter 2).

One Tribe, the Miniconjou, appeared to have a unique relationship with the prairie rivers and their associated fish resources. The name Miniconjou is translated as meaning "planters beside the water," and they traditionally placed a small fish in the soil along with the seeds they planted.

Today the tribes in South Dakota manage rivers, lakes, and ponds for recreational fishing. About 9% of South Dakota is in reservation lands that include substantial portions of Lake Oahe, the Missouri River and some of its tributaries, and some Northeastern lakes. The Cheyenne River Sioux Tribe Reservation has 370 miles of riverine habitat and 380 ponds, lakes and reservoirs.

Tribal fisheries biologists and game wardens may belong to the Native American Fish and Wildlife Society. The Society was formed in 1983 to foster a communication network between self-determined tribal fish and wildlife managers throughout the U. S. A key person in the early development of the Society was Ron Skates, a Cheyenne River Sioux Tribe member. Ron was a game warden for the Tribe and later had a long career in the U. S. Fish and Wildlife Service. He coordinated Service activities with those of many tribes in the west, and was a member of the Board of Directors of the Society. The Society grew into an organization of professional biologists, natural resource managers, technicians, and conservation law enforcement officers.

Each tribe in South Dakota sells fishing licenses and has a cadre of enforcement officers, water quality specialists and biologists who do surveys and cooperate with Federal agencies in other investigations. The purpose of this chapter is to review the history of the tribes in South Dakota, and give details about their fish management programs.

1868 Fort Laramie Treaty

Tribes of the Northern Great Plains included the Lakota, Dakota, and Nakota that occupied much of the present day state of South Dakota and neighboring states. The Dakota and Nakota tribes occupied what is now much of eastern North Dakota, eastern South Dakota, northeastern Nebraska, and western Minnesota. The 1868 Fort Laramie Treaty placed the Lakota on one large Great Sioux Reservation west of the

Missouri River. The terms Lakota, Dakota, and Nakota represent both the language and the people known as the Sioux. Originally there were several nations of people who all spoke one language with mutually understandable dialects.

The Lakota, Dakota, and Nakota peoples moved from just west of the Great Lakes into the Dakotas in pursuit of buffalo. While they were often at war with other tribes, active resistance against settlers started only when the settlers began decimating buffalo herds during the mid-1800s. This touched off the Plains Wars, which led to intervention by the Federal government, and, subsequently, to the Fort Laramie Treaty, which established the boundaries of the Great Sioux Nation Reservation and guaranteed its people freedom from outside intervention.

Under the Fort Laramie Treaty, the Lakota Sioux agreed to a territory encompassed by the western slopes of the Black Hills, the Niobrara River on the south, the Missouri River on the east, and the Cannonball River to the north. Additionally, the 1868 Treaty provided that the Missouri River was a part of the Great Sioux Reservation across its width to the eastern

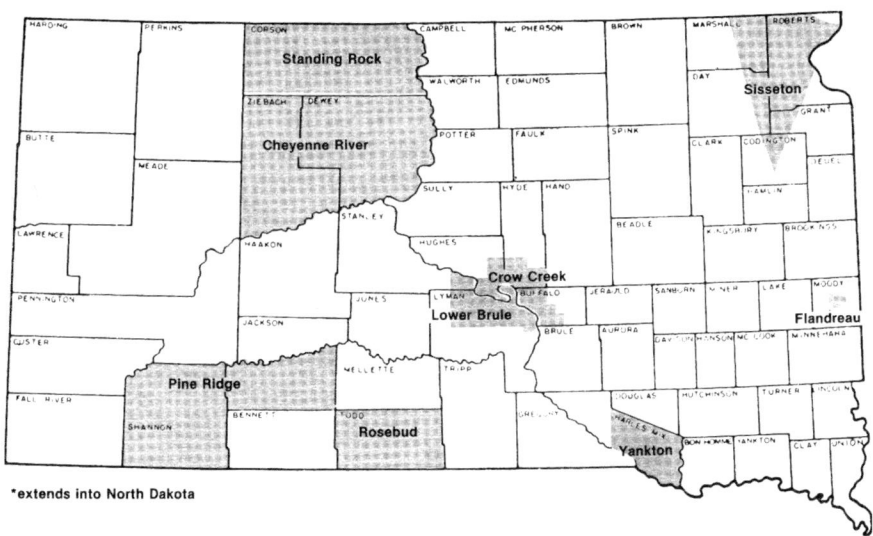

Figure 20.1 **Map showing location of Native American reservations in South Dakota.**

shore of the river. In South Dakota today, the Missouri River and several of its major western tributaries, the Cheyenne, White, Grand, and Moreau rivers, are bordering or entirely within the five west river reservations (Figure 20.1).

Tribal "Nations"

An 1889 act of Congress dissolved the large reservations of the 1868 treaty and established separate reservations for the tribes. Up until this point in time, the tribes were primarily nomadic within the general area of the northern Great Plains. However, the decimation of the wild bison populations essentially forced the tribes to a dependency on government food, and onto the relatively small reservations that exist today.

It is imperative to understand the separate and independent nature of tribal nations and their associated reservations in order to understand their present-day relationship to land and water resources, including the fisheries resources within the various reservations.

The Lakota (sometimes called Tetons –"prairie dwellers") have seven bands:
Oglala ("they scatter their own" or "dust scatterers")
Sicangu (or Brule –"burnt thighs")
Hunkpapa ("end of the circle")
Miniconjou ("planters beside the water")
Sihasapa (or Blackfeet, not to be confused with the separate Blackfeet Tribe)
Itazipacola (or Sans Arcs –"without bows")
Oohenupa ("two boilings" or "two kettles")

The Dakota Sioux (also called Santee Sioux) have four bands living in South Dakota, as well as in Minnesota, Nebraska, and North Dakota. They are the Mdewakantonwon, Wahpeton, Wahpekute, and Sisseton. The Nakota, or Yankton Sioux, have three bands living in South Dakota, North Dakota, and Montana: the Yankton, Upper Yanktonai, and Lower Yanktonai.

Fishing on Tribal Lands

Since the inception of the reservation system, fish have been a source of food for tribal people, especially those in villages adjacent to rivers and streams. In villages like Red Shirt on the Pine Ridge Reservation, fish from the Cheyenne River were an important food and in many instances were the only fresh meat available. In recent years, however, fishing in the Red Shirt community, and in many reservation towns that border prairie rivers like the Cheyenne, has dwindled or ceased because of concerns over polluted waters and the health of the fish (Mike Catches Enemy, personal communication).

Tribes issue licenses to maintain control of who fishes on their land. The first Indian reservation with its own licenses was the Rosebud Reservation of South Dakota in the late 1950s. The Crow Creek Sioux Tribe of South Dakota began issuing licenses in 1961. Regulations vary by reservation (Tables 20.1, 20.2). The Cheyenne River Sioux Reservation

Table 20.1 Species limits for harvest on waters of the Cheyenne River Sioux Tribe (CRST) Reservation, 2004.

FISH SPECIES	CRST IMPOUNDMENTS	LAKE OAHE, CHEYENNE AND MOREAU RIVERS
Largemouth Bass	6	6
Smallmouth Bass	5	5
White Bass / Hybrid Striped Bass	N/A	Unlimited
Northern Pike	Unlimited	6
Muskellunge and/or Tiger Musky	Unlimited	1
Walleye, Sauger, and/or Saugeye	5	10
Yellow Perch	25	Unlimited
Rock Bass	N/A	Unlimited
Bluegill	25	Unlimited
Crappie	Unlimited	25
Other Sunfish, bullheads and rough fish	Unlimited	Unlimited
Trout and/or Salmon (15-inch minimum)	5	5
Catfish (Channel, Flathead, Blue)	6	Unlimited
Paddlefish	N/A	1
Shovelnose and/or Lake Sturgeon	N/A	1
Rainbow Smelt	N/A	15 gallons
Pallid Sturgeon	CLOSED	CLOSED

Table 20.2 **Fish species regulations and Master Angler fish weights for the Rosebud Reservation, 2004.**

SPECIES	DAILY LIMIT	SIZE RESTRICTIONS	MASTER ANGLER QUALIFYING WEIGHT
Largemouth Bass	10 total bass	12-inch minimum	5 lbs
Smallmouth Bass	10 total bass	12-inch minimum	2.5 lbs
Northern Pike	6		15 lbs
Walleye	6	12-inch minimum	8 lbs
Saugeye	6	12-inch minimum	2.5 lbs
Muskellunge	1	35-inch minimum	20 lbs
Paddlefish	0	CLOSED	
Brook Trout	6 total trout		1.5 lbs
Brown Trout	6 total trout		5 lbs
Rainbow Trout	6 total trout		4 lbs
Crappie	25		2 lbs
Sunfish	100		1.25 lbs
Yellow Perch	50		1.5 lbs
Catfish	10		15 lbs
Bullhead	100		2 lbs
Carp	No Limit		20 lbs

directly adjacent to the Standing Rock Reservation does not charge tribal members for licenses, whereas the Sanding Rock Tribe does.

Most Indian reservations that issue licenses have a multi-tiered price structure. Although some licenses have a face value printed on them, most do not. This enables the issuing agent to charge the appropriate fee to the purchaser. For example, if you want to fish on the Standing Rock Reservation in North Dakota, you are issued a license and receive a stamp. The fee varies. An enrolled tribal member from the Standing Rock Reservation is charged $5. A member of another Indian reservation is charged $10. Non-Indians are charged $30. Tribal members over 60 years of age are not charged.

Violators of fish and game laws on reservations have committed a civil offense and must appear in the reservation court system. Failure to appear sends the case to Federal Court because fish and game violations on reservations are also covered by the Lacey Act. The Lacey Act covers activities on Indian Trust Lands and is enforced by the Department of

Interior, usually the Fish and Wildlife Service or Bureau of Indian Affairs.

Interpretation of where tribal jurisdiction ends in relation to the Missouri River has created some problems over the years. On the Cheyenne River Reservation tribal authorities were issuing citations to nonmembers who were fishing from a boat on the west bank of Lake Oahe (on the Reservation side) without tribal fishing licenses. When the cases ended up in court, it was determined that the Tribe only had jurisdiction to the shoreline on the west bank of the Missouri. The Tribe maintains it should have jurisdiction to the east bank of the Missouri River (now under Lake Oahe) as stipulated in the 1868 Fort Laramie Treaty. They also believed they should have *some* jurisdiction because most anglers on Lake Oahe launch boats and fish along the west bank (Joanna Murray, personal communication).

Another contentious issue on the Missouri River was over the "take" lands. The Flood Control Act of 1944 authorized seizure of treaty lands to construct a dam system in the Missouri River basin with the promise that any land found not to be necessary for the system's operation would be returned to the Lakota people. In 1999, 200,000 acres along the banks of the Missouri River were returned to the State of South Dakota, and to the Cheyenne River Sioux and Lower Brule Sioux tribes (other tribes did not choose to participate in the deal).

The legislation, known as the Wildlife Mitigation Act, includes funding for fish and wildlife enhancement projects. Returning the take lands has reduced the number of legal jurisdictional battles and provided a stable source of funding for Tribal fisheries programs along the river. These funds have allowed the tribes to implement some needed habitat management activities. For example, the Lower Brule Tribe is constructing a dike system that will stop shoreline erosion and restore fish and wildlife habitat.

Tribal History and Fishing Regulations

Pine Ridge Reservation

The Oglala Sioux are descended from the bands of Tetons who moved into the Dakotas from the area just west of the Great Lakes. In late December 1890, troops from the United States Calvary engaged a group of Sioux under Chief Big Foot at Wounded Knee Creek, which resulted in the slaughter of Indian men, women, and children. The Oglala Sioux Tribe, organized under the Indian Reorganization Act of 1934, operates under a constitution and bylaws approved in 1936. They now occupy the 2,778,000-acre Pine Ridge reservation in southwest South Dakota.

When the Great Sioux Reservation was subsequently divided, the Oglala were assigned to the Pine Ridge Reservation, the largest of the Lakota reservations. Current Tribal enrollment is 39,700, and trust acreage of the reservation is 1,783,741 acres. The reservation has a diversified agricultural economy with thousands of acres devoted to winter wheat, alfalfa, millet, and safflower; commercial interests are in timberland, pastureland, and rangeland.

Two reservoirs, White Clay and Oglala, offer hunting, fishing, and boating facilities. Fishing licenses are required for Tribal members and nonmembers alike, and can be purchased from the Tribal Parks and Wildlife Headquarters in Kyle, SD.

Rosebud Reservation

The Rosebud Reservation in southern South Dakota is adjacent to the Pine Ridge Reservation on the west and touches the Nebraska border on the south. Trust acreage is 954,571 acres; the reservation was established in 1889. Towns on the reservation include Mission, Rosebud, Parmelee, St. Francis, and Okreek.

The Rosebud Reservation was the first Indian reservation to issue hunting or fishing stamps. Current (2007) license fees are: resident annual, $11; nonresident annual, $27; nonresident five-day, $17. Senior or handicapped tribal members receive a free fishing card.

The Rosebud Tribal Game, Fish and Parks Department sponsors a Master Angler Award for trophy fish. A 5-pound largemouth or a 2.5-pound smallmouth bass make the angler "a Master" (Table 20.2).

Cheyenne River Reservation

The Cheyenne River Reservation, a rolling prairie containing washes, buttes, streams, and rivers, is located in north-central South Dakota. To the immediate north is the Standing Rock Sioux Reservation. Lake Oahe on the Missouri River forms the eastern boundary. The original reservation, consisting of 2,700,000 acres, was established in 1889. In 1909 and 1910, unallocated and unsold land was opened for homesteading to non-Indians, which resulted in their ownership of almost half of the original reservation. Additional acres were inundated by Lake Oahe. Currently, the Cheyenne River Reservation consists of 1,419,499 acres, of which 47% of the original reservation land area is owned by nonmembers of the Cheyenne River Sioux Tribe.

Tribal enrollment is 14,200 and consists of bands of Miniconjou, Sihasapa, Oohenupa, and Itazipacola. The reservation has 19 communities, 13 of which are Indian. Tribal headquarters are located in Eagle Butte, the largest town on the reservation.

Fishing regulations on the reservation vary according to tribal member status and water body. The fishing season is open year-round and harvest of many species is unlimited. Tribal members and nonmember spouses of tribal members receive free licenses. License fees for nontribal members or nonresidents are $10 per year.

Of the 380 ponds, lakes and reservoirs, about 75 are actively managed by stocking with an annual mix of about 50,000 largemouth bass, smallmouth bass, black crappie, bluegill, yellow perch, and walleye supplied by Federal fish hatcheries. The Fisheries Management Plan has a goal of supplying 20,000 angler hours of fishing annually as (1) basic yield fishing from rivers and (2) put and take fishing from ponds. Survey data help Tribal biologists determine stocking needs. Riparian zone protection projects (e.g., willow planting) on selected water bodies help curb erosion.

Standing Rock Reservation

The Standing Rock Reservation straddles the border between North and South Dakota. Its largest land mass is in South Dakota, and topography is typical prairie, with buttes rising as high as 2,000 feet above the surrounding prairie. The Cannon Ball River constitutes the northern border; Lake Oahe constitutes the eastern border; Perkins County, SD, and Adams County, ND, forms the western border; and the Cheyenne River Reservation boundary line constitutes the southern border.

The Standing Rock Sioux Tribe is part of the Great Sioux Nation and it includes the Hunkpapa and Sihasapa bands of the Lakota Nation, and the Hunkpatinas and Cuthead bands of the Yanktonias of the Dakota Nation. Tribal enrollment is 12,700 and trust acreage of the reservation is 846,291 acres.

The Standing Rock Sioux Tribe's economy centers on cattle ranching, farming, and grazing permit leases to private cattle interests. Fishing and boating are popular in the Lake Oahe area (Figure 20.2).

Lower Brule Reservation

The Lower Brule Reservation consists of prairie land in central South Dakota. Topography ranges

Figure 20.2　Sitting Bull sculpture overlooks the upper Oahe Reservoir. The large granite bust was done by sculptor Korzcak Ziolkowski, who started the Crazy Horse Monument in the Black Hills. (Photo by C. Berry)

from steep, rough terrain near the Missouri River to rolling hills elsewhere. The main town on the reservation is Lower Brule, which is approximately 75 miles southeast of the state capital, Pierre.

Various treaties and executive orders have reduced the original reservation to its current size of 135,002 trust acres. Tribal enrollment is about 2,355. Though semiarid, the land supports moderate farming ventures, and the Tribe owns several farms that produce navy beans, potatoes, and the third-largest crop of popcorn in the nation. Historically, ranching has been the Tribe's most successful enterprise.

Tribal members have established a guide service for hunting, fishing, and water-sports on 150-mile-long Lake Sharpe. A tribally owned RV park and campground are located on the lake.

The Tribe is trying to stop shoreline erosion that affects uplands and cultural sites (e.g., burial grounds) and is filling the reservoir. Cooperating with the Corps of Engineers, the Tribal engineers and biologists have designed an experimental dike some 900 feet off shore. Behind the dike are restored wetlands, side channels and other aquatic features that were part of the historic landscape of the Missouri River.

Crow Creek Reservation

The members of the Crow Creek Reservation are largely made up of peoples known as the Wiciyela who speak the Nakota dialect. When the Great Sioux Reservation was reduced to about one-tenth its size in 1889, the Crow Creek Reservation was one of three smaller reservations created at that time. Many Nakota and Lakota families were randomly assigned to live on the Crow Creek Reservation, and many extended families were split up in the process.

The Crow Creek Reservation consists of 122,531 acres located on the Missouri River in central South Dakota. The Crow Creek Sioux Tribe has a constitution and bylaws that were approved in 1949. The Crow Creek Sioux was the first Tribe to issue pictorial hunting and fishing license stamps. There is no formal fish management program but fishing along the Tribe's eastern shoreline of the 80-mile long Lake Sharp is said to be "very good." The Tribe's wildlife department offers guided fishing and hunting trips.

Lake Traverse Indian Reservation

The people presently living on the Lake Traverse Reservation are descendants of the Santee Sioux (Dakota) Tribe, which occupied much of southern Minnesota and northern Iowa. They are known as the Sisseton-Wahpeton Sioux Tribe. Sisseton means "Marsh Village" and Wahpeton means "Village Among the Leaves."

The 106,932-acre reservation is located in northeast South Dakota with about 2,600 acres in North Dakota. Of this area, one third is owned by the Tribe and two thirds by tribal members. The Sisseton-Wahpeton Sioux Tribe enforces fish, wildlife, and environmental laws within the exterior boundaries of the reservation (both trust and private lands). Tribal biologists manage several lakes for walleye fishing by stocking fry and monitoring the walleye population. The Tribe harvests baitfish (e.g., fathead minnows) from wetlands for sale at bait shops.

Yankton Reservation

The high prairie of the Yankton Reservation is located along the Missouri River just north of the Nebraska border. Ft. Randall Dam and Lake Francis Case are located adjacent to the western boundary of the reservation. This reservoir is used for water recreation, including fishing, swimming, and boating. The reservation maintains numerous camping and docking facilities and is the site of 19th-Century archeological sites and burial grounds.

The Nakota bands claimed approximately 2 million acres of land when white settlers began arriving on the Great Plains. Unlike other Indian tribes, the Yankton Sioux never took up arms against the U.S. government; however, like the other tribes, they ceded the vast majority of their territory in return for a reservation of 435,000 acres between the Des Moines and Missouri rivers. The reservation now has 36,561 trust acres.

Flandreau Reservation

The Flandreau Reservation, located in southeastern South Dakota near the western border of Minnesota, is approximately 45 miles north Sioux Falls. The reservation was established by an act of Congress in 1934; trust acreage today is 2,183 acres.

Under terms of the Fort Laramie Treaty of 1868, the Santee Sioux Indians were allowed to homestead if they renounced their tribal membership. Seventy-five families homesteaded near Flandreau, SD, between 1868 and 1873. They had no federal aid except a school, established in 1870. The Flandreau Santee Sioux gained federal recognition in 1934 by the Indian Recognition Act. Current tribal enrollment is 680.

A casino, built in 1990, drives the economy and employs approximately 75% non-Indians. Corn and soybeans are grown on 1,726 acres of reservation land, and cattle and buffalo graze on 844 acres of pasture. Fishing opportunities on the Flandreau Reservation are limited to the Big Sioux River, which flows through the town of Flandreau. State laws and licenses are honored by the Tribe.

Mitakuye Oyasin "All My Relatives"

Mitakuye Oyasin is the traditional "amen" that many Dakota, Nakota, and Lakota often say at the end of an important statement or prayer. The translation means "All My Relatives" and is the way of affirming that whatever was said should cover all relatives. "All relatives" is a very broad group, which is best described by the famous saying attributed to Chief Seattle (1854), "This we know…the earth does not belong to man, man belongs to the earth. All things are connected, like blood that connects one family. Whatever befalls the earth befalls the children of the earth. Man did not weave the web of life—he is merely a strand in it. Whatever he does to the web, he does to himself."

Tribal governments are sovereign and all tribes have a branch of government that manages and conserves natural resources on reservations (Box 20.1). Programs for farming, ranching and hunting are larger than those for fishing, but reservations in South Dakota have some important fishing and boating opportunities. All tribes are working on improving fishing, fish management programs, water quality programs, and natural resource education on tribal lands. ❖

20.1

List of Tribal Fish and Wildlife Headquarters in South Dakota

For an overview of out-of-doors activities available at each reservation, see *http://www.state.sd.us/oia/tribes.asp*

- Cheyenne River Sioux Tribe, Game, Fish and Parks Department, P.O. Box 590, Eagle Butte, SD 57625.
- Crow Creek Sioux Tribe, Department of Natural Resources, P.O. Box 50, Fort Thompson, SD 57339.
- Flandreau Santee Sioux Tribe, Department of Natural Resources, P.O. Box 283, Flandreau, SD 57028.
- Lower Brule Sioux Tribe, Game and Fish Department, P.O. Box 187, Lower Brule, SD 57548.
- Oglala Sioux Tribe (Pine Ridge Reservation), Department of Parks and Recreation, P.O. Box 570, Kyle, SD 57752.
- Sisseton-Wahpeton Sioux Tribe, P.O. Box 509, Agency Village, SD 57262.
- Standing Rock Sioux Tribe, Game and Fish Department, P.O. Box D, Fort Yates, ND 58538.
- Rosebud Sioux Tribe, Game, Fish and Parks Department, P.O. Box 300, 1165 Circle Drive, Rosebud, SD 57570
- Yankton Sioux Tribe, Fish and Wildlife Department, P.O. Box 248, Marty, SD 57361.

Chapter 21

Fisheries Research and Monitoring

Michelle A. Bouchard and Kenneth F. Higgins

" In the spring of 1884, various pieces of apparatus were purchased, including a valued compound microscope… it was the first such instrument in Dakota Territory. There was no previous need for exact magnification, buffaloes and Indians being quite large. "

— Dr. Edward Churchill

" I have a *strong belief* that there are too few walleye in the lake, and that we ought to stock more walleyes. The boys at the bait shop agree and they ought to know because they talk to anglers every day. Therefore, I recommend stocking more walleyes." This is a statement made by a game and fish commissioner at a fictitious commission meeting. Should the other commissioners agree to direct the agency to stock more walleyes because of the strong belief of an important person? Or, should the commission ask agency biologists to study the walleye population, evaluate the survey data, and then make a recommendation?

Science should trump *belief* when managing a public resource and spending public funds. Fisheries management programs are anchored on a research-based information system. However, the public is often unsure about the process of scientific research and sometimes even skeptical of its value (Miller 2007). However, for every dollar spent on agriculture research at South Dakota State University, about six dollars are returned in economic value (AES 2005). Similar data are not available for fisheries research but anglers pump over $200 million annually into the South Dakota economy. New information from fisheries research surely contributes to this and other parts of the $5 billion recreation industry in South Dakota.

One purpose of this chapter is to review the process of fisheries research that is taught in University classrooms (for example see *http://wfs.sdstate.edu*) and advocated by the American Fisheries Society (e.g., Waters and Erman 1990, Brown and Austen 1996). The second purpose is to review the history of fisheries research in South Dakota and analyze trends in research topics and products. We built a chronology of fisheries investigations (Box 21.1) using data from an exhaustive collection of published articles about fisheries resources in South Dakota from 1861 to 2004 (Bouchard et al. 2006).

The Scientific Process

The methods for conducting scientific research begin with making observations, reading other research reports, and stating a working hypothesis. A working hypothesis is a conjectural statement of our best understanding of the truth. An example hypothesis is "more fish will be found in areas where habitat is complex compared to areas where habitat is simple." The next steps in the research process are designing an experiment, collecting data, interpreting results, and synthesizing

21.1
Chronology of Fisheries Research Emphasis from 1861 to 2004

1861-1925   Fish fossils, initial observations on species present

1926-1945   Fish food habits and fish anatomy
First water pollution (cyanide) investigations

1946-1965   Missouri River reservoir studies
Fish inventories continue, lakes and
    streams classified statewide
First creel survey investigations
First common carp control investigations

1966-1980   Commercial fishing and recreational fishing
    assessments
Studies of wetland fishes
Big Stone Power Plant cooling ponds used
    for aquaculture
Lake renovation with fish poisons, non-native
    fish studies
Fish habitat studies with watershed perspective
Sonar used to locate and count fish schools

1981-1995   Fish culture and bait fish studies
Larval fish research
New methods of measuring fish growth and
    populations
New fishing regulations and human dimensions
    studied

1996-2004   Warmwater river and stream fish inventories,
    temporal comparisons
Fish predator-prey relationships, oxytetracycline
    used in mass fish marking
Fish passage through hydropower dams
Bioenergetics studies, diet analysis, caloric
    content of prey species
Fish-waterfowl diet competition studies

the results with those of other research studies to accept, reject or modify the hypothesis.

Much of our knowledge of fishes and fisheries is based on descriptive studies. Descriptive studies were the most frequent type of study in the early years of fisheries investigations, and they are still valuable. Descriptive studies tell us about the location of fishes, such as finding the northern redbelly dace for the first time in the Grand River Basin (Morey and Berry 2004). Another example of knowledge obtained through a descriptive-type study would be the results from a study of the channel catfish diet in Lake Oahe (Hill et al. 1995). This study is descriptive because it does not ask the "why" question, *i.e.,* why do channel catfish in Lake Oahe mostly eat aquatic insects? The data agree with that from an earlier study on Lake Francis Case, so the researchers have the beginnings of a working hypothesis about channel catfish diets in Missouri River reservoirs.

As science matured and fisheries research methods improved, researchers began to make conclusions derived from associations and correlations. A common type of correlative study is to collect fish and record the kind of habitat where they were and were not collected. After completing the steps in the research process, the researchers might conclude (for example) "largemouth bass density will increase with increasing water clarity and aquatic vegetation coverage" (Guy and Willis 1991). Such knowledge is *correlative*, but does not necessarily show cause and effect (Ratti and Garton 1994).

Fisheries experiments that determine cause and effect and ask the "why" question are possible when an experimental design enables the researcher to control the experiment. Control of variables is usually possible in fish hatcheries or laboratories where a number of tanks can be used. For example, to study the effects of a pesticide on flathead chub, a native species in South Dakota, five levels of a toxin were added to 10 exposure chambers (two chambers for each toxin level) with seven chubs per chamber where water quality was the same in all chambers (Fisher et al. 1999). This type of experiment isolated the toxin as the cause of the fish health problems and deaths, and more importantly, the experiment could be repeated by other researchers. Fisheries researchers may also conduct controlled experiments by using replicate cages in a natural setting (e.g., Roell et al. 1986).

As another example, walleye mortality from South Dakota fishing tournaments could be distinguished from handling mortality by a controlled experiment (Graeb et al. 2005). The researchers marked the left fin of fish collected by electro-fishing (the research "control" fish) and the right fin of fish collected by tournament anglers, and held the fish together in tanks with lake water. The mortality of tournament fish pro-gressively increased at tournaments held later in the season while mortality of control fish remained zero.

In chemistry and physics, there is a greater ability to con-trol and modify variables associated with an experiment than in fisheries science. This ability to control variables results in experiment repeatability under the same conditions to con-firm previous results. In fisheries, or any biological science involving field work, it is difficult to control and modify variables or exactly repeat a study. For example, the popu-lation dynamics of brown and rainbow trout in Gary Creek was determined in 1988 (Milewski and Willis 1989). The researchers found about 37 brown trout and 43 rainbow trout per 100 yards, but the study is not repeatable (either at a different location or even at the same location) because of differences in factors such as mortality rates, growth char-acteristics, or hatching success resulting from variations in weather conditions, habitat modifications, or any number of other changeable factors.

Inventories and Monitoring

Inventories are conducted to determine the location of fishes and the status of their populations. Most inventories have been done by graduate students who can focus their studies on a particular water body for a short time (e.g., 2–4 years). Inventories can be single events, but the most powerful information comes when an inventory can be repeated over time using the same methods. For example, about 93% of the fishes in the James, Vermillion and Big Sioux rivers have persisted since earlier (1970s) studies (Shearer and Berry 2003).

Monitoring is simply repeating the inventories over time. Most fish monitoring in South Dakota has been conducted by state fisheries biologists who work on lakes (e.g., Erickson et al. 2001), Missouri River reservoirs (e.g., Lott et al. 2004) and Black Hills streams (e.g., James and Erickson 2006). Less fish monitoring has been done on warm water rivers although inventories of river fishes have been thorough (See Chapter 3).

Applied and Basic Research

Research can be applied or basic but both use the scientific process. Applied research generally implies that study results can be implemented directly in fisheries management or angling regulations. For example, a study of road culverts alerted the Department of Transportation about certain culverts that probably blocked fish migration (Wall and Berry 2004).

Basic research results fill information gaps and address theoretical questions. For example, a study of muskellunge metabolism indicated that metabolism changed seasonally and as the fish grew from juvenile to adult (Chipps et al. 2000). The metabolic information (also called bioenergetics) is basic knowledge about the anatomy and physiology of muskellunge. Both basic and applied research are necessary to maintain a sustainable fisheries resource in South Dakota. Applied research seems never ending because the variety of aquatic habitats, fish populations, and anglers seems to continually change. Basic research helps us understand the changes, and the fundamental principles about how aquatic habitats and fish species function. Understanding fundamentals helps us make predictions and form hypotheses about unstudied habitats and species (Johnson 2002).

The state of our current knowledge of fisheries is incomplete. Because of this, biologists are often faced with the dilemma of making decisions based on insufficient information. This dilemma leads to questions such as—"Is water release from a dam the best way to stop under-ice fish kills?" The answer to such a question is not "yes" or "no." The answer should be expressed thusly, "based on our current knowledge this is the best practice that can be implemented."

As our knowledge base increases we often learn that the best practice at some previous time was really not the most appropriate solution. Although modifying solutions has been called "waffling" and "flip-flopping" by some, and the terms are used to diminish scientific credibility, the approach is unavoidable in science, as well as in other practices where the objective is to find truth.

Under the banner of a relatively new approach called "adaptive management," fisheries managers should monitor the benefits of a solution and then change management strategy when appropriate. Adaptive management can enhance scientific inquiry and has been proposed as part of the solution to managing the Missouri River (CMRES 2002). Under most circumstances, all a biologist can do is use the best information available and then strive to obtain better information using the scientific method.

Where's the Scientific Proof?

Statements about theories such as "there is no scientific proof," and the converse, "it has been proven scientifically" are commonly heard. Such statements indicate a lack of understanding about the scientific method. The misconception that science deals with certainties is far from the truth—science deals with probabilities because scientists usually study samples, not entire populations. Statistics (the mathematics of probability) enable scientists to quantify the level of confidence they have in their predictions about the population based on data from the sample.

The major drawback to the scientific method is that it can never "prove" that theories about fisheries management (also, evolution, gravity, relativity, global warming or wildlife management) are correct. However, we can continue to test hypotheses associated with these theories and from the results, strengthen or weaken the theories.

The primary strength of the scientific method is that when studies are correctly designed, implemented, and interpreted, human biases and beliefs are eliminated from the equation. There is a great difference between human belief and human knowledge; the scientific method provides knowledge-based information. Theories or hypotheses that cannot be scientifically tested are not really theories or hypotheses—they are beliefs.

Early Fisheries Inventories

Early research in South Dakota was done mostly on the Missouri River by exploration and description. Lewis and Clark named their camp of July 24, 1804 "White Catfish Camp 10 miles above Platte" because one of the party caught a white catfish, "its eyes small and tail much like that of a dolphin." The fish was probably a channel catfish (Moring 1996). Although there is little documentation of fish from their travels through South Dakota, Lewis and Clark's river maps and journals are invaluable to naturalists.

Bailey and Allum (1962) summarized the early fish inventories in South Dakota. The first fishes cataloged in South Dakota were collected in 1853 by a railroad survey party commissioned to map a route through the Great Plains to the Pacific Ocean. Charles Girard, a naturalist from the Smithsonian, accompanied the survey and collected fish while the party traversed the Missouri River. His primary collection from South Dakota came from the Fort Pierre area where he reported finding spotted suckers, plains minnows, creek chubs, highfin suckers, channel catfish, and paddlefish. Girard later listed 23 fish species for the Missouri River. A decade later, Edward Cope described eight fish species collected from Battle Creek (now believed to be LeBeau Creek) north of the Moreau River (Cope 1891).

Federal Studies

The U.S. Fish Commission funded early studies in and near South Dakota. The Commission's goals were to suggest solutions for declining food fish populations. The Commission hired Seth Meek in 1889 to report on the fishes of Iowa. His report lists 33 fish species found below the falls in Sioux Falls and at the mouth of the Big Sioux River near Sioux City (Meek 1895). The report includes descriptions of the river bed and the abundance of fish collected. Albert Woolmann was funded to find a hatchery site in Minnesota. He studied the Big Stone and Traverse lakes in 1892 and documented fish species, fish abundance, water quality, vegetation, and food availability.

Barton Evermann and Ulysses Cox were commissioned to find suitable locations for trout hatcheries in 1892 (Evermann and Cox 1896). They studied rivers and lakes in southern South Dakota and in the Black Hills (Evermann 1893). They recommended prairie sites for hatcheries for pond and river species, and Black Hills sites for trout hatcheries. This conclusion seems to be common sense to us today, but the early biologists knew little about fish habitat needs and less about water conditions in South Dakota, where they reported 49 fish species.

One of the largest fisheries research programs in South Dakota was the Department of Interior's Reservoir Research Program that operated from stations in Yankton, Pierre and Mobridge. In the 1950s, the Nation had hundreds of new reservoirs and South Dakota had some of the largest. But, the science of reservoir management was new. There was even a myth (hypothesis?) that reservoirs were "biological deserts."

The Mobridge station studied methods for commercial fishing in reservoirs using the Research Vessel *Hiodon*, which may be the largest research vessel ever to sail on Lake Oahe. The 45-ft vessel had twin diesel engines, hydraulic winches for pulling nets, the newest echo sounder technology, and a fish-hold capacity of 20 tons (Greenwood and Boussu 1967).

The new reservoirs were also seen as the best way to meet the increasing demand for recreational fishing (Gottschalk 1967). Federal researchers were part of the North Central Reservoir Investigations team that operated from 1962 to 1975. The team produced many reports, the most valuable were produced toward the end of the project when fisheries trends could be summarized over the years of reservoir filling and later operation of the reservoirs (e.g., June 1976, 1977; Walburg 1976, 1977). For example, in Lewis and Clark Lake, the abundance of fish exploded and then decreased about 66% while the number of species declined 20%. The researchers saw early signs of shoreline erosion and silt deposition that foreshadowed problems we face today with the growing delta filling the upper end of the reservoir.

It is ironic that today, some 30 years after the reservoir program closed, the Department of Interior is again increasing its research and monitoring of the Missouri River. However, today, the research is on the remnant 100-mile portion of the *free flowing* river instead of on the reservoirs. The goal of the new project, which began in 2005, is to monitor the status and recovery of endangered pallid sturgeon and other native riverine fishes. Work in South Dakota is linked to similar efforts in other regions of the river (Drobish 2005 a, b), and is planned as a long-term monitoring and research effort. Research topics have included evaluating stocking programs of the Gavins Point Fish Hatchery and investigating the life history of the paddlefish.

South Dakota Department of Game and Fish and Parks

The South Dakota Department of Game and Fish was created in 1909. Most of the early work in fisheries was devoted to raising and stocking fish, removing rough fish, constructing dams and fishways, and collecting license fees. During the drought of the 1930s, most fisheries work was fish rescue from the low-water lakes and maintaining fisheries in Black Hills streams (Hipschman 1959).

Although a fisheries branch had been in place since 1909, it was not until 1950 that a full-time fisheries biologist was hired. Another biologist was hired in 1952 when the Division of Fisheries Investigations was created in response to the 1950 Dingell-Johnson (D-J) Act. This Act, which generates money by taxing sales of fishing equipment, is used for improving and restoring fishery resources. It was a huge boost for fisheries research in all states. Each state receives a share of the revenues based on its fishing license sales and land and water area. The first statewide survey done with D-J funds was completed in 1952. To date, the D–J Act has funded over 600 fisheries projects in South Dakota (Table 21.1).

University Research

A new era in biological research began in 1920 when Dr. Edward Churchill was hired as a Zoology professor at the University of South Dakota (USD) (Churchill 1962). Originally from Iowa, Dr. Churchill had been working in Virginia with crabs and oysters for the U.S. Bureau of Fisheries. He taught at USD for a few years until the Biology Department split into the Departments of Zoology and Botany. The newly developed Department of Zoology was placed under the direction of Dr. Churchill.

University of South Dakota

The formation of the Zoology Department led to development of a graduate program. In 1926, Dr. Churchill mentored South Dakota's first fisheries graduate student, Miss Louella Cable. Her thesis was titled "The food habits of the black bull-

Table 21.1 Number and percent of reports by subject funded by the Federal Aid to Sport Fishing Program.

STUDY TYPE	1950s	1960s	DECADE 1970s	1980s	1990s	2000s
Limnological	1 (1%)	8 (10%)	4 (3%)	4 (4%)	0	1(2%)
General surveys/habitat assessment	8 (14%)	0	49 (37%)	22 (23%)	23 (14%)	4 (7%)
Surveys —Missouri River and reservoirs	11 (19%)	12 (15%)	13 (10%)	15 (15%)	9 (5%)	4 (7%)
Surveys—Black Hills	9 (15%)	7 (9%)	6 (4%)	0	2 (1%)	0
Surveys— lakes	13 (22%)	11 (14%)	8 (6%)	0	24 (15%)	12 (21%)
Surveys—rivers	0	3 (4%)	0	2 (2)	4 (2%)	1 (2%)
Fish stocking/spawning assessment	1 (1%)	13 (17%)	21 (16%)	32 (33%)	43 (26%)	24 (42%)
Creel/angler surveys	5 (9%)	9 (12%)	14 (11%)	8 (8%)	31 (19%)	8 (14%)
Habitat improvement	11(19%)	10 (13%)	11 (8%)	6 (6%)	3 (2%)	0
Baitfish/commercial harvest	0	5 (6%)	7 (5%)	3 (3%)	11 (7%)	3 (5%)
Age and growth methods	0	0	0	4 (4%)	9 (6%)	0
Food habits	0	0	0	2 (2%)	4 (2%)	0
Tournament	0	0	0	0	1 (1%)	0

head of eastern South Dakota." Churchill believed that the black bullhead was the most economically important fish in South Dakota. He wrote, "While distained by the sportsmen, bullheads are everybody's fishes; anyone from Grandma down to the baby can catch them."

During his time at USD, Dr. Churchill mentored seven graduate students in fisheries. It was said that if you were unsure what you would like to study, Dr. Churchill would have you studying fish. Most of his early studies dealt with the diets of fish and his later studies were mostly about fish anatomy. Miss Cable went on to be the first woman fisher-

21.2
Dr. Louella E. Cable

Louella Cable, born in 1900, was raised southwest of Chamberlain. She was educated in a one-room schoolhouse, attended Dakota Wesleyan University and then the University of South Dakota (USD) and pursued her dream of becoming a zoologist. She was elected to Phi Beta Kappa, received a B.A. degree and then an M.A. Degree in 1927. Her research topic was bullhead diets. She was the first graduate student in South Dakota to study fisheries.

Louella joined the U.S. Bureau of Fisheries (Beaufort, N.C., 1927) as an illustrator, but the next year she became the first woman Aquatic Biologist hired by the Bureau. She discovered and named a new species of goby, *Eleotrica cableae*, commonly known as Cable's goby, illustrated many articles and books, and invented a new fish tag.

In 1950, she moved to the Great Lakes Fishery Investigation Center in Ann Arbor, Michigan, and began studying ciscoes and lake trout of the Great Lakes. She received her Ph.D. degree in 1959 from the University of Michigan. She retired from the U.S. Fish and Wildlife Service in 1970 at the age of 70. She continued her interest in research and art until her death in 1986. Her estate funds an annual scholarship for a USD biology student.

ies biologist hired by the U.S. Bureau of Fisheries and later received her Ph.D. degree (Box 21.2).

Dr. Churchill also received State funding to gather as much information as possible about the fishes of South Dakota. For three summers he inventoried lakes, rivers, and creeks for fish composition, abundance, and food habits. He found 81 species of fish in the state including one new species of minnow that was later named for Dr. Churchill, *Hybognathus churchilli* (although the fish name has since changed). His studies led to the 1933 book *The Fishes of South Dakota*, the first comprehensive publication of fishes in the state (Churchill and Over 1933).

Dr. James Schmulbach joined USD in 1959 and added a fisheries management course. During his years at the university, Dr. Schmulbach published papers on fish in South Dakota and became known for his work with sturgeon and paddlefish on the Missouri River. He supervised the work of 31 fisheries graduate students. The first fisheries Ph.D. student in the state, Arthur McDonald, graduated under Dr. Schmulbach in 1966 with a dissertation on the "Modification of agonistic behavior in fish." He studied the interactions between predator (largemouth bass) and prey (green sunfish) in laboratory tanks. Most of the early research done under Dr. Schmulbach was on food habits, age and growth, and life history studies. Later students also studied fish distributions, population abundance, and ecology.

Recent research by other faculty at USD has involved anatomy, physiological effects of stress, and evolution. It is ironic that evolution was a contentious topic when Dr. Churchill was first hired at the University. He asked the University administrators, "how do you want your world taught—round or flat?" The concern about teaching evolution quieted and the University continued to offer the course, although under another name.

South Dakota State University

The first graduate level classes in fisheries at South Dakota State University (SDSU) were offered in 1957. As with USD, black bullheads were also the subject of the first thesis done in 1958—this time on the diet of black bullheads in Lake Poinsett. In 1963, the Department of Wildlife Management was formed, and later changed its name to the Department of Wildlife and Fisheries Sciences. In 1993, the department moved into the Northern Plains Biostress Laboratory. To date, the department has had over 140 graduate students in fisheries (Figure 21.1).

Early studies at SDSU primarily involved fish age and growth rates, life histories, and food habits. Later research included stocking strategies, population and distribution studies, bioenergetics, and assessment of management strategies. Walleye has been the species most researched over the last three decades, accounting for about 19% of all studies (Table 21.2). Techniques to gather data have improved through the years. The department now uses telemetry, sonar, underwater cameras, Global Positioning System technology, and portable water quality test kits for field work. In the laboratories, several types of microscopes are used to study cellular and sub-cellular anatomy, and chemistry laboratories detect micro quantaties of toxins and enzymes.

Figure 21.1 Fisheries graduate students of the early 1970s shown holding a clipboard would eventually use hand-held computers by the end of their careers. (Coop Unit photo)

The long tenures of Churchill and Schmulbach at USD is matched by Dr. Charles Scalet at SDSU. Scalet joined the faculty in 1973 and has been Head of the Department of Wildlife and Fisheries Sciences since 1976. His research has primarily concerned pond fisheries ecology, but recently he has studied and published on University administration. He

Table 21.2   Number of studies on 19 fish species by researchers at South Dakota State University.

FISH SPECIES	1950s	1960s	1970s	1980s	1990s	2000s	TOTALS
		NUMBER OF STUDIES BY DECADE					
Black bullhead	1		2	2			5
Bigmouth buffalo		3					3
Emerald shiner	1				1	1	3
Walleye			1	9	8	6	24
Black crappie			3		2	2	7
White crappie			2		1	1	4
Yellow perch			2	1	2	2	7
Paddlefish			2	1			3
Fathead minnow			1		2		3
Channel catfish			2	1	1	2	6
Rainbow trout			2	8	2		12
Bluegill			2	8	1	1	12
Largemouth bass			2	8	5	1	16
Common carp				1	2		3
Northern pike				1	3		4
Muskellunge				4			4
Grass carp				3			3
Rainbow smelt					3	1	4
Saugeye					1	3	4

co-authored a textbook for use in introductory fish and wildlife classes nationwide (see Chapter 22).

South Dakota Cooperative Unit

In 1960, the federal government created the Cooperative Unit Program Act. This act established the Cooperative Research Units that were jointly supported by the federal government (U.S. Fish and Wildlife Service), land grant universities, and the state. The intent was to stimulate education and increase the number of people with advanced degrees in fisheries and wildlife science. The stimulus worked well at South Dakota State College (now University) because the Department of Wildlife and Fisheries Sciences was formed to accept the new Federal wildlife and fisheries researchers. Initially, both USD and SDSU wanted the Fisheries Cooperative Research Unit. After consultations with senators, congressmen, and the state government, the Cooperative Fishery Research Unit joined the already-established Cooperative Wildlife Research Unit at SDSU in 1965.

The Cooperative Unit Program was almost closed in the 1980s during government cutbacks in environmental research and again in the 1990s when the program was moved to a new Federal Agency named the National Biological Survey. Politicians disagreed with or did not understand the new agency, and closed it within a few years (Wagner 1999). The Cooperative Unit Program is now a part of the biology discipline of the U.S. Geological Survey.

During the last 40 years, Cooperative Unit students have produced about 75 theses or dissertations and published more than 90 papers on South Dakota fishes.

Changes in Research Funding and Capabilities

Funding for fisheries research has increased greatly over the years due to federal legislation (Figure 21.2). In 1960, the National Environmental Policy Act (NEPA) was passed to prevent or eliminate damage to the environment and was fol-

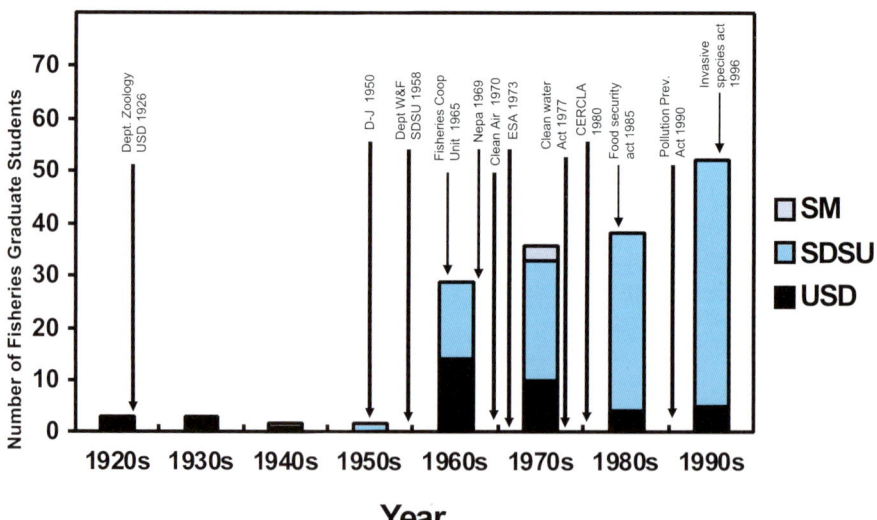

Figure 21.2   Number of fisheries graduate students in South Dakota and a timeline of increasing funding opportunities. SM = School of Mines, SDSU = South Dakota State University, USD = University of South Dakota.

lowed by the Clean Air and Endangered Species Acts. In the 1980s, Congress passed the Clean Water Act, Comprehensive Environmental Response, Compensation and Liability Act (also know as Superfund), and the Pollution Prevention Act. These three acts dealt with reducing pollutants, setting standards, increasing efficiency in water use and natural resources, and cleaning up hazardous waste.

The Food Security Act of 1985 had provisions designed to protect wetlands. Out of this act came the Conservation Reserve Program and the Wildlife Incentive Program. The National Invasive Species Act was passed to fund preventative management, record keeping, and sampling techniques for aquatic invasive (often exotic) species. In 2001, Congress created the State Wildlife Grants program. It is the Nation's core program for conservation actions on non-game fishes to prevent them from becoming endangered.

Early fisheries research in South Dakota was limited to inventories of what fishes were present and where, and to studies of fish size, age, growth rate, and food habits. Researchers were well equipped if they had nets to capture fish, and a measuring board and scales for measuring fish length and weight. With the advent of "scientific fisheries management" in the 1950s came new research techniques that the public often thought of as "silly." In the 1959 Annual Report of the Department of Game, Fish and Parks is this sentence, "…the department may tag fish or put radio transmitters on deer or any of what will appear to the layman unconnected or even silly things."

Today's advanced technology has enabled research in new areas of aquatic science, such as genetics, energetics, biomonitoring, nutrient recycling, predator-prey interactions, food web structure effects, energy dynamics, and trophic ecology. New technology has produced a pin-head sized fish tag that carries information that can be read by a special tag detector and decoder. Electronic advances have resulted in bio-telemetry tags that send information about the fish's location and heart rate. A boat-mounted hydroacoustic machine can assess fish size, distribution, and abundance in Missouri River

Reservoirs, and even estimate the number of rainbow smelt in Lake Oahe. Recent advances in remote sensing technology have enabled better mapping. Global Position Systems (GPS) enable determining the exact sample sites or animal locations, whereas Geographic Information Systems (GIS) technology enables landscape-scale views of layers of data (e.g. water, geology, roads, vegetation, fish locations).

All of these modern techniques, and many more, are being used today in South Dakota by agency and University fisheries researchers. In concert with advances in fish research technology, the development of larger and faster computer technology and statistical capabilities has enabled fisheries biologists to manipulate and integrate much larger and more complex data sets than in past decades.

Research on fishes, fish habitat, and anglers has progressed from descriptive surveys of the past to the advanced research studies of today. These studies have been done by agency biologists, university professors, and graduate research assistants. Each of these research studies has been done to meet an information need, and they have followed scientific methods of hypothesis testing, sample design, data management, and statistical analysis. All of these research studies have added to our knowledge of the state's aquatic resources.

The future of fisheries research in the state is bright because of South Dakota's 2010 Initiative. The Initiative has several goals concerning an increased role of research in economic development, specifically in the agriculture and natural resources sectors. Fishing generates more that $50 million in retail sales in South Dakota. Research on game fish, water quality and quantity, and angler opinions will ensure that fishing remains a big business in South Dakota. Research on non-game species will ensure that the state's natural heritage is conserved, while basic research will produce a more accurate understanding of how our aquatic environment might respond to future changes in climate or land use. Enhancing outdoor opportunities and quality of life will depend on past and future research that will show all of us how to be responsible stewards of our aquatic resources, of which fish are a primary component. ❖

Chapter 22

Education and Outreach

Charles G. Scalet

“ Kind and amount of schooling is of great importance to a professional career…but less so than the personal aptitudes of the student. A pre-existing enthusiasm for wildlife and its conservation is the first essential. ”

— Dr. Aldo Leopold, 1948

Fisheries education and outreach activities in South Dakota have steadily increased during the last century. Numerous state and federal entities and professional societies have always had a variety of fisheries education and outreach programs (Box 22.1). The primary centers of fisheries education and outreach, however, have been two universities, South Dakota State University (SDSU) and the University of South Dakota (USD). In 1962, South Dakota College became South Dakota State University.

Universities generally have multiple missions and each contributes differently to education and outreach. A three-mission approach (teaching, research, and outreach) is used at land grant universities such as SDSU (Dunkle 2003). The first mission is teaching (Figure 22.1). Teaching involves coursework and advising at both the undergraduate (bachelor's degree) and graduate levels (master's and doctoral degrees). Most undergraduate students take four to five years to complete a degree. Most master's students take two to three years beyond their bachelor's degree to complete their academic preparation. Most Ph.D. students require four to five years beyond their master's degree to complete their doctoral work

The second mission is research, which involves conducting scientific investigations to obtain data and knowledge. University research usually includes graduate students, although there is a growing tendency to incorporate undergraduates into this process.

The third mission is outreach and one facet of outreach involves the dissemination of information to state citizens in a variety of non-classroom settings. These activities can

Figure 22.1 Fisheries education and research goes on in lecture halls (a), laboratories (b), on lakes and streams (c), and in floating classrooms on the Missouri River (d).

22.1

Non-University Fisheries Education

Government Agencies: The Department of Game, Fish and Parks and the U. S. Fish and Wildlife Service are two government agencies that provide loads of fisheries information by phone, e-mail, letters, brochures, pamphlets, booklets, maps, and public meetings.

Professional Societies: The Dakota Chapter (American Fisheries Society) distributes a poster of Dakota fishes and "fish cards" (like baseball cards) that young students like. The Chapter also provides continuing education classes to fisheries professionals.

Citizen Groups: The South Dakota Wildlife Federation, Isaac Walton League, Sierra Club, and other citizen groups help conserve South Dakota's fisheries. The Ikes' Save Our Streams (SOS) Program is a winner.

include anything from personal consultations to producing published information for use by citizens interested in fisheries (Box 22.2). The level of effort that a university program can put into its missions in any specific field, such as fisheries, is primarily determined by the number of faculty members who have expertise in a particular area.

The purpose of this chapter is to provide an historical reference to fisheries education and outreach primarily at universities in South Dakota. The information is presented in chronological order and focuses on efforts at SDSU and USD.

Early University Programs

University educational efforts directed at South Dakota fisheries began with ichthyology courses taught in the zoology programs at USD and SDSU. In other parts of the country subjects such as fish culture, commercial fisheries, and marine

22.2

Recent Books About Aquatic Resources by South Dakota Educators

Bandas, S. J. and K. F. Higgins. 2004. A field guide to South Dakota turtles. SDCES EC 919. Brookings: South Dakota State University. 36 pp.

Brown, M. L. and C. S. Guy, editors. (2007). Analysis and interpretation of freshwater fisheries data. American Fisheries Society, Bethesda, Maryland.

Fischer, T. D., D. C. Backlund, K. F. Higgins, and D. E. Naugle. 1999. A field guide to South Dakota amphibians. SDAES Bulletin 733. Brookings: South Dakota State University. 62 pp.

Murphy, B. R. and D. W. Willis, editors. 1996. Fisheries techniques, 2nd edition. American Fisheries Society, Bethesda, Maryland. 732 pp.

Neumann, R. M. and D. W. Willis. 1994. Guide to the common fishes of South Dakota. SDCES EC 899. Brookings: South Dakota State University. 60 pp.

Scalet, C. G., L. D. Flake, and D. W. Willis. 1996. Introduction to wildlife and fisheries: an integrated approach. W. H. Freeman and Company, New York. 512 pp.

Willis, D. W., M. D. Beem, and R. L. Hanten. 1990. Managing South Dakota ponds for fish and wildlife. South Dakota Game, Fish and Parks, Pierre. 70 pp.

fisheries were offered in academic settings, but these topics were not taught in South Dakota.

A major event in South Dakota fisheries education and outreach occurred in 1939 when an undergraduate wildlife and fisheries educational curriculum was initiated at SDSU; the program was a part of the Entomology-Zoology Department. The impetus for starting this curriculum was provided by the growing awareness of and the need for university programs that educated students in the management and conservation of wildlife and fisheries resources. The shift of academic attention from just ichthyological and fish cultural studies to a more management and conservation approach occurred at many U.S. universities during the 1930s and 1940s. This was also a time when federal and state agencies involved with fisheries and wildlife resources were growing in size and responsibilities.

The early wildlife and fisheries curriculum at SDSU bore limited resemblance to the course requirements for students of today. It was agriculturally oriented and had few courses that specialized in fisheries; the curriculum at SDSU was more directed towards wildlife in these early years. SDSU undergraduate students in the 1940s were required to take courses in zoology, mathematics, social science, composition, speech, chemistry, botany, fishes, birds, mammals, and wildlife management, but they also had to take coursework such as surveying, dairy science, turkey production, crop production, farm forestry, military science, accounting, and engineering drawing.

The 1950s

By 1950, the wildlife and fisheries curriculum at SDSU had divided into two curricula; one was agriculturally oriented and the other was more conservation oriented. Both, however, still contained strong agricultural components. During this period, SDSU undergraduate enrollment in wildlife and fisheries increased from 40 students in 1950 to 121 in 1959.

The 1950s also brought an increased emphasis on research. U.S. universities had always had research as one of their missions, but that emphasis was now increasing. For research to increase in scope, universities had to expand graduate programs. USD had offered graduate work in zoology at the master's degree level since 1927 (Churchill 1962). SDSU initiated master's degree graduate work specifically in fisheries in 1957. The first completed master's degree at USD that involved fishes was in 1927; the first completed master's degree in fisheries at SDSU was in 1958.

In the 1950s, James C. Schmulbach joined the Zoology Department at USD in the fisheries area and would be at that institution into the 1990s teaching fisheries courses and conducting research with graduate students. Marvin O. Allum was the fisheries person at SDSU, but that program was still skewed toward wildlife.

Fisheries courses have been taught at both universities for many years. In addition, both schools also have taught a variety of other courses that deal with aquatic systems. While some of these aquatic courses are not directed specifically at fishes, their contributions to fisheries education are significant and will be included in this chapter.

The 1960s

During this period numerous events occurred that would change the face of fisheries teaching, research, and outreach in the state. A major event at USD was the construction of the Churchill-Haines Laboratories, which housed the Zoology Department. Improved facilities increased the type and scope of research that could be conducted. James C. Schmulbach, who remained the sole fisheries-oriented faculty member at USD, usually had two to three fisheries graduate students a year and taught courses such as ichthyology and fisheries management.

At SDSU a number of significant changes occurred during this period. In 1963, the wildlife and fisheries curriculum, which had been in the Entomology-Zoology Department, was moved to a newly formed department. The Department of Wildlife Management was established and a couple years later the name was changed to the Department of Wildlife and Fisheries Sciences. In 1965, the South Dakota Cooperative Fisheries Research Unit was formed; a Cooperative Wildlife Research Unit had been formed in 1963. Both units were attached to the Department of Wildlife and Fisheries Sciences at SDSU. The units represented a cooperative effort among SDSU; South Dakota Game, Fish and Parks; the U.S. Bureau of Sport Fisheries and Wildlife; and the Wildlife Management Institute (Goforth 2006). The Cooperative Fisheries Unit brought two federal fisheries scientists to the SDSU campus and their efforts were directed at graduate education, research, and outreach.

Formation of the Department of Wildlife and Fisheries Sciences and the Cooperative Fisheries Unit resulted in SDSU becoming the center of fisheries academic teaching, research, and outreach activities in the state. In the 1960s, SDSU generally had three faculty members working in the fisheries area at any one time. University fisheries faculty members during that period were Marvin O. Allum, followed by Norman D. Schoenthal, and then John G. Nickum. Cooperative Fisheries Unit faculty members were Alfred C. Fox and Richard A. Tubb, followed by Richard L. Applegate. In addition, Raymond L. Linder, who was with the Cooperative Wildlife Unit, pioneered research on South Dakota wetlands.

Graduate work in the Department of Wildlife and Fisheries Sciences at SDSU expanded rapidly in the 1960s. At the start of the decade, graduate student numbers averaged under 10 per year; at the end of the decade, there were over 20 per year and approximately half were in fisheries. Undergraduate student numbers fluctuated during this period and reached a high of 206 in 1969.

The fisheries curriculum at SDSU in the 1960s included undergraduate courses, but also had fisheries and aquatic graduate-level classes in fisheries science, limnology, wetland management, and aquatic ecology. In 1965, the undergraduate curriculum moved away from its agricultural underpinnings and only had a single program that stressed conservation. This was also a period when new areas, such as statistics, were added to requirements and an increased emphasis was placed on basic biological and physical sciences. The 1960s was also a period of rapid expansion in job opportunities for fisheries graduates with federal and state agency hiring increasing dramatically.

The 1960s saw another important event at USD. In 1961, a Ph.D. program was initiated in the Zoology Department. The Ph.D. program provided USD with an opportunity to expand its research effort since such programs allow for a depth of research not possible at the master's degree level. USD lost its doctoral program in biology in the late 1960s. The Zoology Department at USD was merged with the Biology Department in 1968, but the direction of the program remained essentially the same (Cummins et al. 1982).

The 1970s

During this period USD still had James C. Schmulbach as its sole fisheries faculty member, while SDSU generally had four fisheries faculty members at any one time. Richard L. Applegate continued in the Cooperative Fisheries Unit and Donald C. Hales and Robert S. Benda were also associated with the unit in the 1970s. University fisheries faculty during this period were John G. Nickum, followed by Charles G. Scalet. A new fisheries position was added at SDSU in the late 1970s and was filled by Timothy C. Modde.

The 1970s also saw an expansion of fisheries coursework at SDSU. Fisheries management had always been a component in the undergraduate wildlife management course and in 1975, fisheries management was separated into its own course.

This period was a time of increased graduate work at SDSU. The decade started with the department having approximately 20 graduate students a year and ended with an average of 30 per year with half in fisheries. Undergraduate student numbers in the Wildlife and Fisheries Sciences program at SDSU fluctuated in the 1970s. The early part of the decade saw large numbers that were stimulated by increased environmental awareness and the first Earth Day in 1970. Undergraduate enrollment reached an all-time high of 218 students in 1972, but dropped to half that by 1979.

The 1970s also was a time when the Department of Wildlife and Fisheries Sciences at SDSU received its first South Dakota Cooperative Extension Service position to cover both wildlife and fisheries outreach. The Cooperative Extension Service is located at SDSU because it is the land grant university in the state. The new position provided basic wildlife and fisheries information to state citizens. Three people filled the position during the 1970s including John L. Schmidt, Victor Van Ballenbergh, and W. Alan Wentz. The position was terminated in 1980 when the Cooperative Extension Service made numerous reductions in natural resource programming.

The 1980s

This period saw changes in the fisheries faculty at SDSU. On the university faculty Timothy C. Modde resigned and was replaced by David W. Willis. Charles G. Scalet remained on the faculty. Richard L. Applegate and Robert S. Benda left the Cooperative Fisheries Unit and were replaced by Charles R. Berry and Walter G. Duffy. In 1984, the two units were combined into one, the South Dakota Cooperative Fish and Wildlife Research Unit. In 1989, Daniel E. Hubbard joined the SDSU faculty; he was aquatic oriented because of his work on wetlands.

In the middle of the 1980s a blue ribbon citizens' panel studied the South Dakota Cooperative Extension Service. One of their recommendations was for the service to hire an extension aquaculture specialist to provide increased outreach to state citizens in the expanding area of aquaculture. Marley D. Beem was hired in 1986 to fill that position, but the position was eliminated in 1988 as a part of a series of changes in the Cooperative Extension Service.

Wildlife and Fisheries Sciences enrollments at SDSU were stable during the 1980s. Undergraduate enrollment fluctuated between 118 and 89; graduate enrollment remained at about 30, with half being fisheries students.

Coursework in the fisheries and aquatic areas at SDSU expanded in both the undergraduate and graduate areas. A freshman Introduction to Wildlife and Fisheries Management course and graduate-level courses in Aquaculture, Fish Structure and Function, and Advanced Fisheries Management were valuable additions. Use of computers was also rapidly expanding, a course in computer science was added to the undergraduate requirements, and computers were integrated into other coursework and research. The South Dakota School of Mines was a leader in applying computer technology to

22.3

Engineering and Fisheries

The South Dakota School of Mines (SDSMT) has geology and engineering programs with an aquatic emphasis. Courses include: surface and groundwater hydrology, hydraulics, surface water investigation techniques, and watershed modeling.

Dr. Larry Stetler models flows in Spearfish Creek to predict fish habitat. The seasonal flows and water chemistry produced a "cement rind" on the streambed that hurts fish spawning.

Dr. Scott Kenner studies water quality monitoring, habitat statistics, lake models, and "ecohydraulics" (hydraulic models simulate fish habitat). He has studied urbanization effects on rainbow and brown trout in Rapid Creek.

The SDSMT program compliments the biology programs at USD and SDSU. For example, a Ph.D. dissertation by Jack Erickson was titled "Stress responses of brown trout (*Salmo trutta*) within an urbanized reach of stream in the Black Hills of South Dakota." The trout were not stressed by suspended sediment or stream temperature during stormwater runoff events. Erickson is a Biologist bringing SDSMT engineering into the State fisheries program.

civil engineering issues related to water quality and to fish habitat in lakes and streams (Box 22.3).

The 1990s

Changes occurred in the 1990s that were significant to fisheries education, research, and outreach in South Dakota. In 1990 a Ph.D. program in biology was started at SDSU and USD as a joint effort between the two universities and included, among others, the Biology Department at USD and the Department of Wildlife and Fisheries Sciences at SDSU. With the addition of this program, research efforts were expanded. The first fisheries Ph.D. student at SDSU graduated from the program in 1993 (Christopher S. Guy). Researchers in the new field of genomics (fish genetics) at USD and at Black Hills State University helped us better understand how to conserve South Dakota fishes (Box 22.4).

The Northern Plains Biostress Laboratory, completed at SDSU in 1993, provided new and expanded space for the Department of Wildlife and Fisheries Sciences. Prior to this time, the department was in cramped, outdated facilities. The new structure provided a major impetus for improved and

22.4

Fish Genetics Education and Research

Livestock breeds, strains, hybrids, and other genetic types are part of South Dakota agriculture—and aquaculture. Fish culturists and managers study fish genotypes to conserve existing genetic resources by using new genetic tools, techniques and expertise.

"Old fashioned" genetic research was done in the 1980s and 1990s when researchers used muscle proteins to identify strains of fish, migratory populations of walleye, and walleye-sauger hybrids (e.g., Ford 1978, Flammang and Willis 1993, Brown and Flammang 1995, Ward and Berry 1995).

Modern fish geneticists study DNA. The new field of *genomics* is the study of heriditary information to advance medicine, industry, and fisheries management.

At the University of South Dakota, Dr. Paula Mabee studies evolution using fish as an experimental animal (Mabee 2006); Dr. Hugh Britten works with population genetics of many animals, including fish (Britten et al. 1997).

At the Center for the Conservation of Biological Resources at Black Hills State University, Dr. Shane Sarver's genomics team is using DNA techniques to help manage trout, walleye and the endangered Topeka shiner.

expanded teaching, research, and outreach activities in the fisheries and aquatic areas.

Also at SDSU, the number of university fisheries faculty was expanded to three with the addition of a new position. Charles G. Scalet and David W. Willis were joined by Michael L. Brown. The Cooperative Fisheries Unit continued to have two faculty members. Charles R. Berry remained; Walter G. Duffy was replaced by Steven R. Chipps. Thus, the Department of Wildlife and Fisheries Sciences at SDSU now had five fisheries faculty.

Fisheries and aquatic classes at SDSU also expanded. Undergraduate coursework was expanded with the addition of Human Dimensions in Wildlife and Fisheries, Wildlife and Fisheries Techniques, Limnology, and Integrated Natural Resource Management. Graduate work at SDSU expanded with course additions in Wetland Ecology and Management, Fish Culture, Ecology of Aquatic Invertebrates, Aquatic Trophic Ecology, and Stream Ecology and Management.

Wildlife and Fisheries Sciences enrollment at SDSU fluctuated during this period. In 1990 undergraduate enrollment was 131 students, in 1999 it was 132. In 1995 the highest enrollment ever in the undergraduate program was reached with 225. Graduate enrollment at SDSU increased from 36 graduate students in 1990 to the mid-40s by the end of this period. The Ph.D. program also changed the composition of graduate students with approximately one quarter now Ph.D. students.

In 1993, James C. Schmulbach retired at USD and was replaced by Bruce A. Barton. During this period courses in Ichthyology, Aquaculture and Fish Health, Fisheries Management, and Limnology were periodically taught at USD as combined undergraduate/graduate-level courses. Generally there were one or two fisheries graduate students a year in the USD program.

In 1999, USD also started the Missouri River Institute, which was created to develop and promote scholarly research, education, and public awareness related to the natural and cultural resources of the Missouri River Basin.

The Early 2000s

SDSU added Brian D. Graeb to the fish faculty. Dr. Graeb, a Ph.D. product of SDSU, brought expertise in large lake and reservoir ecology, having studied Lake Michigan and Missouri River reservoirs and their fish assemblages. This brought the fishing faculty number at SDSU to six. At USD, The Missouri River Institute expanded and construction of a National Missouri River Research and Education Center at Ponca State Park, Nebraska, was begun in 2002. In 2003, Bruce A. Barton left the Biology program at USD and was replaced in 2004 by Daniel A. Soluk, an aquatic ecologist.

Undergraduate student numbers at SDSU increased throughout the period with 254 students in 2006. Graduate student numbers at SDSU increased to 59 in 2006, with approximately half being fisheries and aquatic students. A new undergraduate course, Fisheries and Wildlife Biometrics, was added. New graduate courses in Quantitative Fisheries Science and Natural Resources Policy and Administration were added. In addition, in 2006 the Department was granted its own Ph.D. program; it was no longer a part of the umbrella Biological Sciences Ph.D.

The undergraduate curriculum for Wildlife and Fisheries Sciences majors at SDSU encompassed a wide variety of coursework. Included were three courses in writing, two courses in speaking, two social sciences courses, two humanities courses, two mathematics courses, one statistics course, one computer science course, three chemistry courses, a physics course, five general biology courses, two botany courses, and 10 vertebrate biology and wildlife and fisheries sciences courses. Graduate student course requirements remained as they had always been with requirements tailored to the individual student.

Since inception of a wildlife and fisheries curriculum at SDSU, approximately 1,350 undergraduate Wildlife and Fisheries Sciences majors have completed their degrees. Approximately 200 M.S. and 15 Ph.D. fisheries students also have graduated.

The Future

The future of fisheries education, research, and outreach in South Dakota should remain strong. A constantly increasing public awareness of the societal values provided by fisheries and aquatic resources should continue to fuel growth in this area. The economic, ecological, educational, sociocultural, aesthetic, and recreational values of South Dakota's fisheries are important to the state, its citizens, and its visitors.   ❖

Chapter 23

South Dakota Fisheries: Predictions for the Next 20 Years

C. Berry, K. Higgins, D. Willis and S. Chipps

" A people without a vision will perish.**"**

— Proverbs 29:8

Any successful enterprise must have a vision of the future. Managing the fisheries of South Dakota is similar to managing a business—make predictions, have a plan, and carry out the plan. Game, Fish and Parks Division Director Doug Hansen has said that his vision of the future is encapsulated in what a South Dakotan might say to a visitor. Hansen would like to truthfully say, "You will like South Dakota. It has wonderful wildlife populations and a wide variety of opportunities to enjoy them." Predictions for the next 20 years will be needed so that we can achieve this vision. The purpose of this chapter is to summarize the predictions made by the contributors to this book, and by other knowledgeable people who play various roles in the recreational fishing industry.

Before looking forward, let's take a quick look back at the tremendous developments in recreational fishing. In the mid-1950s, a new theory called structure fishing caused anglers to visualize what lay beneath the surface. Closely following this concept was a little black box called a Fish Lo-K-Tor invented by Carl Lowrance (the person who also proposed using South Dakota salamanders as bait, see Chapter 15). Anglers abandoned wooden rental boats for the new aluminum boats.

The Nation undertook new programs in the 1960s to protect water quality. The policy guiding the new water conservation programs was embodied in the National Environmental Policy Act, which stated that harmony between man and the environment would be a national policy. Cleaning up water pollution has undoubtedly helped improve fishing and enjoyment of South Dakota's lakes and rivers.

The advent of tournament fishing boosted the entire fishing industry. A new breed of professional angler advertised new lures and fishing techniques. For South Dakota walleye anglers, the Lindy Rig complete with instructions was a hot new idea. Graphite rods, electric trolling motors, bottom bouncers, planer boards and other new ideas took off in the 1980s. Navigation aids and super lines—in every year there seemed to be an improvement in fishing equipment. And, the number of anglers increased.

New reservoirs and stock dams opened up tremendous fishing opportunities. There were salmon on the Dakota prairies! The number of Game, Fish and Parks staff increased and was better educated. Fisheries scientists developed new tools and passed the information to managers to apply to the fishes, the aquatic habitat, and angling regulations. There were new fisheries (e.g., musky) and new game fish (e.g., saugeye). Communication improved among and between anglers, scientists, and managers. In 1985 there were two computers in the fish and wildlife faculty offices at South Dakota State University. Today, many students have two computers each!

Today, fishing in South Dakota is in great shape. Most South Dakotans rate healthy wildlife populations and clean environment as important to them (Gigliotti 1999). There is more water then ever before. Anglers have a number of species to catch—salmon, catfish, trout, walleye, crappie, northern pike, yellow perch and many other less sought after species. River fishing especially is likely to produce a variety of caught fish (Doorenbos et al. 1996, Wickstrom and Schuckman 2006).

Future Driving Factors

Some important forces will shape the future of fishing in South Dakota. The forces are demographics, economic development, and fishing technology. The numbers of anglers will increase by 25% nationwide in the next 20 years. Angling promoters say the rate of increase is less than that of the population increase, but there will be more anglers fishing the same amount of water. Demographic data indicate that many more people will be fishing because the "baby boomer" generation will have ample money and time to fish and many will try (or return to) fishing as entertainment. Whether more young people will fish in the future is not clear, but certainly more young people in future generations will fish than will hunt. Some groups will want to market fishing and tourism, while others will begin to wonder how more anglers will improve their own fishing experience.

Economic development will probably trump conserving and preserving open spaces for wildlife and water for fish. The public will hear "there is no conflict between economic growth and environmental protection," however, toward the end of the next 20 years, the public will begin to wonder if "sustainable" economic growth is possible, and whether sustainable growth is compatible with sustainable fish populations (Czeck et al. 2004, Reed and Czeck 2005). Fisheries sustainability has eluded us to date because of an increasing human population and the associated demands on natural resources (Knudsen and MacDonald 2000). People interested in other water sports and in developing waterfront property will be more vocal and numerous than anglers at public meetings addressing the future of a lake or river.

Predicting the future of fishing equipment is easily done by looking at the pages of any outdoors magazine. Twenty-five years ago, the first issue of *In-Fisherman* proposed a formula for fishing success: Fish + Location + Presentation = Success (Zernov 2000). Anglers now have a better understanding of the fish's anatomy and physiology, and they know how to find fish by knowing their habitat needs and sensory abilities. Hence, the "Location" and "Presentation" variables are becoming easier to solve, leaving "Fish" as the primary constraint to fishing success.

One reads about voice-activated electric motors, trolling motors that automatically read depth while steering and adjusting for the wind, underwater cameras, three-dimensional sonar, and other electronics. Global Geographic Positioning Systems return anglers to a spot and sonar memory pictures fine-tune the position as downriggers raise and lower lures automatically over bottom structure. High-tech engines will match larger, high-tech boats needed to travel to distant fishing locations. Locating fish has become a science, not a guessing game. Former "gadgets" are now common. Communication about a hot bite is instant because of cell phones and computers. One thing is sure—in this Internet age it should be easier to buy a license.

Contributors' Predictions

Most of the contributors to this book made some predictions about the future of South Dakota fish and fishing. These predictions are more sharply focused on South Dakota than the mega-trends mentioned above, but the influences of demographics, economic development, and fishing technology are real. We have partitioned predictions into lists about anglers (Table 23.1), about the fishes (Table 23.2), and about the aquatic habitat (Table 23.3). Most of the predictions were mentioned in several chapters so there is agreement among contributors.

Predictions about anglers (Table 23.1) focuses not on the increased numbers of anglers and their fishing skills, but on their need to engage in preserving their sport. Future anglers will have to be more involved in protecting their sport than ever before. This will require fewer trips to the sporting goods store and more trips to county commission meetings. When the discussion turns to water, anglers definitely "have a dog in the fight" and are really some of the most knowledgeable people about local conditions because of their time spent outdoors.

Anglers will also have to be better conservationists. Self-imposed standards will evolve among anglers toward more

Table 23.1   Contributors' list of future fisheries issues relating to anglers.

- Anglers more involved and educated about fish management and water issues
- Anglers increase awareness of local governments and citizens about the value of fishing
- Anglers will be more numerous and better informed
- Anglers will have better equipment that will increase harvest
- Anglers will need to accept more restrictive regulations
- Fishing guides will have to help fish managers educate anglers
- Consumption advisories for mercury (perhaps other contaminants)

responsible conduct in exercising the privileges and freedoms conveyed by a fishing license (Hannon 2000). Anglers will stock fish, but it will be the fish they *just caught and released* instead of a hatchery-reared fish. Stocking hatchery fish may become more important, so expect the need for more hatcheries or expanded hatchery production. One contributor wrote, "Climate and water fluctuations won't allow natural reproduction to produce all of the fish needed by anglers." Trout fisheries will likely remain a put-and-take system in the Black Hills reservoirs.

Access to fishing opportunities is an issue that looms in the future although it is not as hot as access to hunting opportunities. Indeed, the motto of the South Dakota Wildlife Federation is "preserving the hunting and fishing heritage in

South Dakota," and this group joins other advocacy groups to support preserving, and even increasing, access to fishing opportunities. Anglers seek more access to state waters whereas private landowners sometimes feel that their land is being used unfairly or without compensation. Access issues change when water levels rise and fall in lakes and wetlands, and when road right-of-ways are debated or blocked. Tribal lands with access to fisheries will likely be a plus to anglers, especially nonresidents. Demand for guide services might increase.

Several predictions about recreational fish populations (Table 23.2) focus on the impacts of non-native species and rough fish populations that might change native fish assemblages to the detriment of recreational fishes. Anglers already know that Asian carp are here (probably to stay!), but few anglers (and biologists) can see the changes that the exotics are causing in native fish populations. Measuring these changes is the job of the fishery biologists. Their data on fish diets, growth rates and population sizes will be the first evidence of a changing fish fauna in South Dakota. Anglers can help stop the spread of many exotics, but anglers should be prepared to understand that a paddlefish population decline might be because of competition with Asian carp and not because of management, over-fishing, or dams.

Table 23.2 Contributors' list of future fisheries issues relating to fishes.

- Fish assemblages will change because of exotics and habitat alteration
- Missouri River reservoirs will have more bass and rough fish
- Walleye fishing will be maintained by stocking
- Endangered species will be contentious; specialized fish hatcheries built for rare fish
- Studies of stocked fish performance after being stocked
- Nuisance fish populations may increase; GFP subsidizes commercial fishermen
- Climate and water conditions will hamper fish reproduction to support angling
- A fish museum will be used for research, education, and economic development
- Fish hatcheries will be upgraded; new aquaculture techniques developed

Several predictions address the non-game fish species and the growing role of Game, Fish and Parks (GFP) in protecting declining populations of fishes and preserving aquatic biodiversity. Increasing the funding and biological expertise in non-game programs and habitat management will probably happen, but GFP will need support from recreational anglers for more money for conservation projects. Protecting non-game fishes will undoubtedly benefit game fish populations as well.

The definition of a "recovered species" will change. Conserving endangered species will still mean conserving fish in natural habitats by fostering natural reproduction, but for some species, conservation in managed areas or even in hatcheries will be enough to remove them from the endangered species list (Coble and Scott 2006). The word "recovered" will be used less and the word "stabilized" will be used more when referring to endangered species. We repeat the earlier quotation "Fisheries sustainability has eluded us to date because of increasing human population and associated demands on natural resources (Knudsen and MacDonald 2000).

Almost all predictions about aquatic habitat (Table 23.3) are about the possible decline in quality and quantity of water in rivers, lakes and reservoirs. In South Dakota, citizens are the first conservationists because they must be concerned

Table 23.3 Contributors' list of future fisheries issues relating to aquatic habitat.

- Missouri River reservoirs will be managed as one system; deltas become an issue
- Missouri River reservoirs lose coldwater fishery during droughts
- Watershed degradation will affect fish production in small impoundments
- "Big" agriculture from field to factory may change the land and waters
- The changing eastern lakes will require regular fish data and management
- River management will require regular fish data and management
- Black Hills development and livestock grazing will stress Black Hills streams
- Stream restoration is the only way to develop more cold-water stream miles
- Water conservation important because most sources have been developed
- Need for water for humans will trump fish needs
- Watershed-scale planning and education for complex problems

enough to request agency intervention when a problem is perceived, and work with agencies on watershed management—a bottom up approach to conservation rather than "command-and-control." Unlike the past half-century when many new aquatic habitats were developed, the next 20 years will be more about managing the existing waters and perhaps restoring some degraded waters.

Two aquatic habitats are "extra special" in South Dakota—the Black Hills streams and the Missouri River reservoirs. In the Black Hills, few stream miles remain to be improved (58 miles were rehabilitated since 1976) and no new reservoirs are anticipated. On the Missouri River, water rights and natural flow regime issues will remain as fish and wildlife support, recreation, navigation, flood control, agriculture and domestic uses compete for river services. Every reservoir in South Dakota is filling with sediment, and in the next 20 years there will be more discussion about what to do about the growth of "deltas." The deltas offer a new type of habitat for fish and wildlife, which is good, but they also block water flow, impede navigation, and provide new environs for exotic species, which is bad.

The definition of riverine habitat will continue to change and expand to include the **watersheds**. Historically, the main rivers were acknowledged by the public, but there was little thought about headwaters, connected wetlands, or the uplands that yield surface and groundwater to the main river channel. In the next 20 years, the terms watershed and watershed management will be used more often and educational programs may contain more holistic, multidisciplinary views of river conservation than are available today. The public will hear about a National Fish Habitat Initiative that will protect, restore and enhance aquatic habitats (SFBPC 2004).

Game Fish and Parks Staff's Predictions

Doug Hansen, Wildlife Division Director penned his vision of the future in 1994, and asked State fisheries biologists to predict the major issues facing the Division in the future. Although somewhat dated now, their issues closely match those identified today. The biologists listed agency image, angler ethics and behavior, meeting the desires of anglers for access and adequate fish numbers, providing adequate information on opportunities, landowner issues, conserving aquatic biodiversity with license money, endangered species, habitat degradation, and cultural (particularly Native American) issues.

When asked about underlying trends, the biologists listed increased urbanization, decreased opportunities for young people to learn fishing, changing demographics (e.g., aging) and economic issues pitted against environmental issues. It seems that the South Dakota fisheries biologists' "crystal ball" was showing a future similar to that described by many professional anglers and our chapter contributors.

Top fisheries issues in other states (Mather et al. 1995) match the predictions made by the South Dakota biologists. The top issues being addressed by other state agencies are: assessing the effectiveness of harvest regulations, evaluating fishing pressure, fish stocking, lack of recruitment, conflict between policy and human dimensions, habitat deterioration, water quantity (e.g., keeping instream flows), water quality (e.g., contaminants) and introduced species.

Other Predictions

A few outliers are on the lists of future predictions by contributors to this book. First, there may be a movement back to the fishing basics. Although the old fashioned "backlash" in an open-faced reel is uncommon now, a backlash of sorts may occur in fishing. Some anglers will be asking "Is there an imaginary line, beyond which "fair chase" no longer registers, and man reigns supreme over fish?" (Zernov 2000). "Back to basics" may be a rod, a hook, some bait, the fish and the angler. And, the fish may be a carp, bullhead or other species instead of the glamour fish of the day (Stange 2000).

The future of fishing will include debates about the wisdom of genetic manipulation of species. Growth genes from carp might be inserted into rainbow trout. New strains of fish will be developed and exotic species that are tolerant to pollution or that grow large may be planned for the waters of

North America. Such national issues are important to South Dakota because fish migrate, and we are connected with 29 other states in the Mississippi River drainage.

Tourism will increase in South Dakota because fish and game are plentiful and the pace of life is slower that in other states. More versatile and more mobile anglers will continue to visit South Dakota even though regulations may become more restrictive. Studies show that reducing harvest doesn't reduce tourism and fishing interest. The tourists may bring hitch-hiking pests such as exotic aquatic species, plants, or fish pathogens (e.g., whirling disease).

Another challenge of fishery management over the next 20 years will be to understand and deal with a more diverse clientele. New immigrants bring different values and may be very aggressive anglers. The ethnic composition will change and with it the attitudes and concerns of anglers as well as their languages (Quinn 2000).

Certain angling groups (e.g., bass anglers, walleye anglers) will continue to be strong but other fishing interests will grow (e.g., trophy catfish angling). Non-consumptive users must be recognized as constituents (Quinn 2000). Family fishing will grow. Boat manufacturers are producing new boat styles to accommodate various kinds of water sports, and the new boats will be as comfortable as an automobile. At the same time, increases in the number of single parent and two-wage earner families may decrease opportunities for young people to learn about fishing.

The subject of global warming is often mentioned as a driving force for the future. That may be, but in the next 20 years, the climate changes will be small and the resultant changes in fisheries and aquatic habitats will only be seen in the research data (Poff et al. 2002). Increased temperatures will bring longer growing seasons, milder winters, hotter summers and perhaps longer drought episodes (reviewed by Johnson et al. 2005). The most vulnerable fisheries may be those in west-to-east oriented rivers that are already warm, and in the lower elevations of coldwater trout streams. Today, the summer maximum temperatures in the White River approach levels that are stressful to channel catfish (Fryda 2001). A little temperature increase will begin stressing the channel catfish population. A one-degree rise in mean July air temperature is predicted to decrease trout habitat by 16% in Wyoming (Keleher and Rahel 1996), and these predictions may be meaningful for Black Hills streams also (Simpson 2007).

Smallmouth bass may be the fish of the future if waters become cleaner, warmer, and clearer. These conditions combined to foster populations of smallmouth bass, as was recently found in the Cheyenne River below Angostura Dam. However, smallmouths can be easily fished out, so anglers may have to accept length limits or even catch-and-release rules. The smallmouth bass could be the angler's "canary in the mine"—its appearance indicates that water quality is improving. Also, watch for a northern pike explosion when higher water returns to the Missouri River reservoirs and re-floods new vegetation on the mud flats.

Finally, one demographic about fishery biologists and professors is of concern. The data show that "mirroring society as a whole, fewer biologists will fish regularly." There will be almost a complete turnover in staffs of the Federal and State agencies as the baby boomers retire in the first part of the next 20 years. This demographic change in the regional agency offices, and in the general angling public, will have yet unknown consequences because "predictions are hard to make, especially about the future" (Yogi Berra).

The last lines in the poem in the Preface of this book are "God give thee strength, O gentle trout, to pull the rascal in." The rascal—the angler—won't be pulled into the water as the poet suggests, but might be pulled into the sport of fishing as have thousands of South Dakotans. Protecting our fishing heritage depends on educated citizens, biologists and anglers. We hope that the information in this book helps protect that heritage in some small way.

To go fishing is the chance to wash one's soul with pure air, with the rush of the brook, or with the shimmer of sun on blue water…it is a discipline in the equality of men—for all men are equal before fish. – Ernest Hemingway

References

Abel, A. 1921. Trudeau's description of the upper Missouri. The Mississippi Valley Historical Review 8(1/2):149-179.

AES (Agriculture Experiment Station). 2005. Return on your investment. AESadmin205, 2-page brochure available at http://sdaes.sdstate.edu/ (accessed 7/07).

Alex, L. 1973. An analysis of fish utilization at four initial middle Missouri sites. Master's Thesis, University of Wisconsin, Madison.

Amundson, F. 2002. Estimated use of water in South Dakota, 2000. Open-File Report 02-440. U.S. Geological Survey Information Services, Building 810, Box 25286, Federal Center, Denver, CO, 80225.

Anderson, M. 1980. Water quantity and quality of three adjacent Black Hills watersheds. Master's Thesis. South Dakota School of Mines and Technology.

Antonides, B. 1993. What it means to be a Conservation Officer. South Dakota Conservation Digest 60(July/August):10-11.

Audubon, M. 1897. Audubon and his journals. The Missouri River journals. Pages 447-537 *in* Vol. I and Pages 1-95 *in* Vol. II. Charles Scribner and Sons, Publisher, New York.

Backlund, D. 1995. The blacknose dace. South Dakota Conservation Digest 62 (March/April):18-19.

Bailey, R., and M. Allum. 1962. Fishes of South Dakota. Miscellaneous Publication No. 119, Museum of Zoology, University of Michigan, Ann Arbor.

Bakker, K., D. Naugle, and K. Higgins. 2002. Incorporating landscape attributes into models for migratory grassland bird conservation. Conservation Biology 16:1638-1646.

Bandas, S., and K. Higgins. 2004. Turtles of South Dakota. South Dakota State University, Box 2140B, Brookings.

Barton, B., and B. Taylor. 1996. Oxygen requirements of fishes in northern Alberta rivers with a general review of the adverse effects of low dissolved oxygen. Water Quality Research Journal of Canada 31:361-409.

Baxter, G., and M. Stone. 1995. Fishes of Wyoming. Wyoming Game and Fish Department, Cheyenne.

Benson, N. 1968. Review of fishery studies on Missouri River main stem reservoirs. U.S. Fish and Wildlife Service Research Report 71, Washington, D.C.

Bergman, R. 1947. Trout. Alfred Knopf Publisher, New York, NY.

Berry, C. 2003. The Missouri River ecosystem: exploring the prospects for recovery. Book Review in: Wetlands 23:208-211.

Berry, C. 2000. Great Plains rivers and fishes of South Dakota. South Dakota Conservation Digest 67(May/June):14-19.

Berry, C., and B. Young. 2004. Fishes of the Missouri National Recreational River, South Dakota and Nebraska. Great Plains Research 14:89-114.

Berry, C., W. Duffy, R. Walsh, S. Kubeny, D. Schumacher, and G. Van Eeckhout. 1993. The James River of the Dakotas. Pages 70-86 *in* L. Hesse, C. Stalnaker, N. Benson, and J. Zuboy. Proceedings of the symposium on restoration planning for the rivers of the Mississippi River ecosystem. Biological Report 19, U.S. Department of the Interior, National Biological Survey, Washington, D.C.

Berry, C., M. Wildhaber, and D. Galat. 2004. Fish distribution and abundance. Volume 3 of Population structure and habitat use of benthic fishes along the Missouri and lower Yellowstone rivers. Available from https://www.nwo.usace.army.mil//html/pd-e/benthic_fish/Benthic_Fishes_Volume_3.pdf (accessed 7/07).

Bich, J., and C. Scalet. 1977. Fishes of the Little Missouri River, South Dakota. Proceedings of the South Dakota Academy of Science 56:163-177.

Bishop, G. 1981. Occurrence and fossilization of the *Dakoticancer* assemblage, upper Cretaceous Pierre shale, South Dakota. Pages.383-413, *in* J. Gray and others, Editors. Communities of the Past. Hutchinson Ross Publishing Company, Stroudsburg, PA.

Black, P. 1991. Watershed hydrology. Prentice-Hall, Inc., Englewood Cliffs, NJ.

Black, P. 1996. Watershed hydrology. Ann Arbor Press, Inc., MI.

Blackwell, B. 2001. A comparison of adult and sub-adult walleye movements and distribution in lakes having simple and complex morphometry. Doctoral dissertation, South Dakota State University, Brookings.

Boettcher, S., W. Johnson, S. Kronberg, and F. Gartner. 1998. The Mortenson Ranch: cattle and trees at home on the range: A restoration guidebook. South Dakota State University, Agriculture Communications, Brookings.

Bouchard, M., K. Higgins, D. Gardner, C. Berry, and S. Chipps. 2006. An annotated bibliography: published articles (1861-2004) about the fish resources of South Dakota. SDSU, Brookings, Bulletin No. 751 (http://agbiopubs.sdstate.edu/articles/B751.pdf) or available from SDSU, Box 2140B, Brookings.

Bowen, J. 1970. A history of fish culture as related to the development of fishery programs. Pages 71-94 *in* N. G. Benson, Editor. A century of fisheries in North America. Special Publication 7, American Fisheries Society, Bethesda, MD.

Braaten, P. 1993. The influence of habitat structure and environmental variability on habitat use by fish in the Vermillion River, South Dakota. Master's Thesis. South Dakota State University, Brookings.

Braaten, P., and C. Berry. 1997. Fish associations with four habitat types in a South Dakota prairie stream. Journal of Freshwater Ecology 12:477-489.

Bradstad, R. 1930. Sioux Falls sewage treatment works. Sewage Works Journal 2(1):65.

Britten, H., B. Riddle, P. Brussard, R. Marlow, and T. Lee. 1997. Genetic delineation of management units for the desert tortoise in northeastern Mojave Desert. Copeia 1997:523-530.

Broughton, J., and K. Potter. 2002. South Dakota baitfish harvest 2001 summary. South Dakota Department of Game, Fish and Parks Report 03-03, Pierre.

Brown, B. 1993. State game wardens, a century of wildlife law enforcement begins. South Dakota Conservation Digest 60(July/August):2-3.

Brown, M., and D. Austen. 1996. Data management and statistical techniques. Pages 17-62 *in* B. Murphy and D. Willis, Editors. Fisheries Techniques, Second Edition. American Fisheries Society, Bethesda, MD.

Brown, M., and M. Flammang. 1995. Walleye genetics. South Dakota Conservation Digest 62 (March/April):5-7.

Carl, C. 1941. South Dakota sewage works short course. Sewage Works Journal 13(4):824-827.

Carlson, B., and C. Berry. 1990. Population size and economic value of aquatic bait species in palustrine wetlands of eastern South Dakota. The Prairie Naturalist 22:119-128.

Carrels, P. 1987. The most useless river there is. Pages 124-134 *in* Western water made simple. Island Press, Washington, D.C.

Carrels, P. 1999. Uphill against water: the great Dakota water war. The University of Nebraska Press, Lincoln.

Carson, R. 1962. Silent spring. Houghton-Mifflin Company, Boston, MA.

Carter, J., D. Driscoll, and J. Williamson. 2002. The Black Hills hydrology study. U.S. Department of the Interior, U. S. Geological Survey, Fact Sheet FS-046-02.

Carter, J., D. Driscoll, J. Williamson, and V. Lindquist. 2002. Atlas of water resources in the Black Hills area, South Dakota. U.S. Department of the Interior, U.S. Geological Survey. 120 pp. http://pubs.usgs.gov/ha/ha747/, accessed 7/07.

Chaisson, R., and W. Radke. 1991. Laboratory anatomy of the perch. 4th Edition. Wm. C. Brown Publishers, Dubuque, IA.

Chaney, E., W. Elmore, and W. Platts. 1990. Livestock grazing on western riparian areas. U.S. Environmental Protection Agency, Eagle, ID.

Chase, A. 2001. Music discrimination by carp (*Cyprinus carpio*). Animal Learning and Behavior 29:336-353.

Chipps, S., D. Clapp, and D. Wahl. 2000. Variation in routine metabolism of juvenile muskellunge: evidence for seasonal metabolic compensation in fishes. Journal of Fish Biology 56:311-318.

Christensen, C., and J. Stephens. 1967. Geology and hydrology of Clay County, South Dakota. South Dakota Geological Survey Bulletin 19, Pierre.

Churchill, E., and W. Over. 1933. Fishes of South Dakota. S. D. Department of Game and Fish, Pierre.

Churchill, E., and W. Over. 1938. Fishes of South Dakota. South Dakota Department of Game, Fish and Parks. Pierre.

Churchill, E. 1944. The effect of the effluent from Sioux Falls sewage treatment plant on the fauna of the Big Sioux River. Proceedings of the South Dakota Academy of Science 24:43-48.

Churchill, E. 1962. Three thousand coyotes and I. State University of South Dakota. Vermillion.

Clark, S., W. Willis, and C. Berry. 1991. Indexing of common carp populations in large palustrine wetlands of the northern plains. Wetlands 11:163-172.

Clayton, L. 1967. Stagnant-glacier features of the Missouri Coteau in North Dakota. Pages 25-46 *in* L. Clayton and T. F. Freers, Editors. Glacial geology of the Missouri Couteau and adjacent areas. North Dakota Geological Survey, Miscellaneous Series 30, Grand Forks.

CMRES (Committee on Missouri River Ecosystem Science. Water Science and Technology Board, Divisions of Earth and Life Studies, National Research Council). 2002. The Missouri River ecosystem, exploring the prospects for recovery. National Academy Press, Washington, D.C. (Read online free at http://www.nap.edu, accessed 7/07)

Coble, D., and J. Scott. 2006. Recovery management agreements offer alternative to continuing ESA listings. Fisheries 31(1):35-36.

Coker, R., and J. Southall. 1915. Mussel resources in tributaries of the upper Missouri River. Bureau of Fisheries Document No. 812, Appendix IV to the Report of the U. S. Commissioner of Fisheries for 1914. Government Printing Office, Washington, D. C.

Cope, E. D. 1891. On some fishes from South Dakota. American Naturalist 1891:25:654-658.

Coughlin, P., and K. Higgins. 1995. Flight counts as an index of wood duck abundance along prairie streams in South Dakota. Proceedings of the South Dakota Academy of Science 74:65-70.

Courtenay, W., and J. Stauffer. 1984. Distribution, biology, and management of exotic fishes. The Johns Hopkins Press. Baltimore, MD.

Crawford, C. 1907. Inaugural Address to the 10th Legislative Session. State Printing Office, Pierre.

Cross, F., and R. Moss. 1987. Historical changes in fish communities and aquatic habitats in plains streams of Kansas. Pages 155-165 *in* W. Matthews and C. Heins, Editors. Community and evolutionary ecology of North American stream fishes. University of Oklahoma Press, Norman.

Cross, F., R. Mayden, and J. Stewart. 1986. Fishes in the western Mississippi basin (Missouri, Arkansas, and Red rivers). Pages 363-412 *in* C. Hocutt and E. Wiley, Editors. The zoogeography of North American freshwater fishes, John Wiley and Sons, Inc., NY.

Cross, F. 1970. Fishes as indicators of pleistocene and recent environments in the Central Plains. Pages 241-257 *in* W. Dort, Jr. and J. Jones, Jr., Editors. Pleistocene and Recent Environments of the Central Great Plains. The University Press of Kansas, Lawrence.

Cross, F., and D. Huggins. 1975. Skipjack herring, *Alosa chrysochloris*, in the Missouri River basin. Copeia 1975:382-385.

Cummins, C., R. Hilderbrand, and S. Ward. 1982. The College of Arts and Sciences 1882-1982: A History. College of Arts and Sciences, University of South Dakota.

Cunningham, G., R. Olson, and S. Hickey. 1995. Fish surveys of the streams and rivers in south-central South Dakota west of the Missouri River. Proceedings of the South Dakota Academy of Science 74:55-64.

Cvancara, A., L. Clayton, W. Bickley, Jr., A. Jacob, A. Ashworth, J. Brophy, and others. 1971. Paleolimnology of late Quaternary deposits: Siebold Site, North Dakota. Science 171:172-174.

Czeck, B., P. Angermeier, H. Daly, P. Pister, and R. Hughes. 2004. Fish conservation, sustainable fisheries, and economic growth: no more fish stories. Fisheries 29(8):36-37.

Dahl, T. E. 1990. Wetlands losses in the United States-1780s to 1980s. U.S. Department of the Interior, Fish and Wildlife Service, Washington, D.C.

DENR (Department of Environment and Natural Resources). 2004. The 2004 South Dakota integrated report for surface water quality assessment. South Dakota Department of Environment and Natural Resources, Pierre.

DENR (Department of Environment and Natural Resources). 1993. State water planning process. South Dakota Department of Environment and Natural Resources, Pierre.

DENR (Department of Environment and Natural Resources). 2001. Report on Coteau Lakes—Deuel County ordinary high water mark investigation. South Dakota Department of Environment and Natural Resources, Pierre.

DENR (Department of Environment and Natural Resources). 1994. South Dakota water quality, water years 1998-2003 (streams) and water years 1993-2003 (lakes). DENR, Pierre, SD. 220 pp.

DeVoto, B., Editor. 1953. The journals of Lewis and Clark. Houghton-Mifflin Company, Boston, MA.

DeWitt, E., J. Redden, D. Buscher, and A. Wilson. 1989. Geologic map of the Black Hills area, South Dakota and Wyoming. USGS I-1910, Misc. Investigation Series Map, USGS, Reston, VA.

Dieter, C., and T. McCabe. 1989. Factors influencing beaver lodge-site selection on a prairie river. American Midland Naturalist 122:408-411.

Dieterman, D., and C. Berry. 1994. Fishes in seven streams of the Minnesota River drainage in northwestern South Dakota. Proceedings of the South Dakota Academy of Science 73:23-30.

Dieterman, D., and C. Berry. 1998. Fish community and water quality changes in the Big Sioux River. The Prairie Naturalist 30:199-224.

Dieterman, D., and D. Galat. 2004. Large-scale factors associated with sicklefin chub distribution in the Missouri and lower Yellowstone rivers. Transactions of the American Fisheries Society 133:577-587.

Doorenbos, R. 1998. Fishes and habitat of the Belle Fourche River, South Dakota. Master's Thesis, South Dakota State University, Brookings.

Doorenbos, R., C. Berry, and G. Wickstrom. 1999. Ictalurids of South Dakota. Pages 377-389 in E. Erwin and others, Editors. Catfish 2000, Symposium 24, American Fisheries Society, Bethesda, MD.

Doorenbos, R., D. Dieterman, and C. Berry. 1996. Recreational use of the Big Sioux River, Iowa and South Dakota. South Dakota Department of Game, Fish and Parks, Special Report 96-14, Pierre.

Dornbush, J. 1971. The impact of water quality criteria standards on the development of the Big Sioux River near Sioux Falls, South Dakota. Report for OWRR Project No B-005-SDAK, Water Resources Institute, South Dakota State University, Brookings.

Dowd, E., and L. Flake. 1985. Great blue heron nesting biology on the James River in South Dakota. The Prairie Naturalist 16:159-166.

Driscoll, D. 1994. Black Hills hydrology study: U. S. Geological Survey Open-File Report 4-344, U. S. Geological Survey, Rapid City, SD.

Drobish, M., Editor. 2005a. Pallid sturgeon population assessment program. U.S. Army Corps of Engineers, Omaha District, Yankton, SD.

Drobish, M., Editor. 2005b. Missouri River standard operating procedures for sampling and data collection. U.S. Army Corps of Engineers, Omaha District, Yankton, SD.

Dryer, M., and A. Sandovol. 1993. Recovery plan for the pallid sturgeon (*Scaphirhynchus albus*). Available from U. S. Fish and Wildlife Service, Bismarck, ND. (also http://www.fws.gov/moriver/Pallid%20Sturgeon%20Activities.htm, accessed 7/07)

DuBois, R., T. Margenav, R. Stewart, P. Cunningham, and P. Rasmussen. 1994. Hooking mortality of northern pike angled through the ice. North American Journal of Fisheries Management 14:769-775.

Duchossois, G. 1993. Geology of Hughes County, South Dakota. South Dakota Geological Survey Bulletin 36, Pierre.

Duehr, J. 2004. Fish and habitat relations at multiple spatial scales in Cheyenne River Basin, South Dakota. Master's Thesis. South Dakota State University, Brookings.

Duffy, W. 1996. Population dynamics, production and prey consumption of fathead minnows in prairie lakes and wetlands. South Dakota Department of Game, Fish and Parks, Report 96-6, Pierre.

Dunkle, A. 2003. The College on the hill: a sense of South Dakota State University history. South Dakota State University Alumni Association, Brookings.

Earle, S. 1937. Fish culture is big business in the United States. The Progressive Fish-Culturist 31:1-29.

Eddy, S. 1969. The freshwater fishes. Wm. C. Brown Publishers, Dubuque, IA.

Eddy, S., and J. Underhill. 1974. Northern fishes, with special reference to the upper Mississippi Valley (third edition, revised and expanded). University of Minnesota Press, Minneapolis.

Eddy, S., R. Tasker, and J. Underhill. Fishes of the Red River, Rainy River, and Lake of the Woods, Minnesota, with comments on the distribution of species in the Nelson River drainage. Occasional Paper 11, Bell Museum of Natural History, University of Minnesota, Minneapolis.

Ellis, M. 1937. Detection and measurement of stream pollution. Bulletin of the Bureau of Fisheries 48(1937):365-437.

Erickson, J., and R. Koth. 2002. Black Hills of South Dakota fishing guide. South Dakota Department of Game, Fish and Parks, Pierre.

Erickson, J., L. Ferber, G. Galinat, G. Simpson, and E. Unkenholz. 2001. Statewide Fisheries Surveys, 2000. Surveys of Public Waters. Annual Report No. 02-11. South Dakota Department of Game, Fish and Parks, Pierre.

Evermann, B., and U. Cox. 1896. Report upon the fishes of the Missouri River Basin. Pages 325-429 *in* Report of the U. S. Commissioner of Fish and Fishes for 1894. U. S. Government Printing Office, Washington, D. C.

Evermann, B. 1893. The ichthyologic features of the Black Hills region. Proceedings: Indiana Academy of Science 1893:73-78.

Evetts, M. 1979. Upper Cretaceous sharks from the Black Hills region, Wyoming and South Dakota. The Mountain Geologist 16:59-66.

FEMA (Federal Emergency Management Agency). 1993. Interagency hazard mitigation team request, in response to May-July 1993 flood disaster State of South Dakota. FEMA-999-DR-SD, Federal Emergency Management Agency, Denver, CO.

Fischer, T., D. Backlund, K. Higgins, and D. Naugle. 1999. Field guide to South Dakota amphibians. SDAES Bulletin 733. South Dakota State University, Brookings.

Fisher, C. 1996. Population characteristics and habitat selection of walleye in the Big Sioux River, South Dakota. Master's Thesis, South Dakota State University, Brookings.

Fisher, R., M. Jenkins, and W. Fisher. 1987. Fire and the prairie-forest mosaic of Devil's Tower National Monument. American Midland Naturalist 17:250-257.

Fisher, S., G. Galinat, and M. Brown. 1999. Acute toxicity of carbofuran to adult and juvenile flathead chubs. Bulletin of Environmental Contamination and Toxicology 63:385-391.

Flammang, M., and D. Willis. 1993. Comparison of electrophorograms and external characteristics for distinguishing juvenile walleyes and saugeyes. The Prairie Naturalist 25:255-260.

Flint, R. 1955. Pleistocene geology of eastern South Dakota. U.S. Geological Survey Professional Paper 262, U.S. Government Printing Office, Washington, D.C.

Ford, R. 1978. Evaluation of four strains in rainbow trout fingerling stockings in Black Hills reservoirs. South Dakota Department of Game, Fish and Parks, Report No. 83-1, Pierre.

Ford, R. 1970. The water spoilers. South Dakota Conservation Digest 37(September/October):10-11.

Ford, R. 1988. Black Hills stream inventory and classification, 1984 and 1985. Report 88-1, South Dakota Department of Game, Fish and Parks, Pierre.

Forman, S., R. Oglesby, and R. Webb. 2001. Temporal and spatial patterns of Holocene dune activity on the Great Plains of North America: megadroughts and climate links. Global and Planetary Change 29:1-29.

Frissell, C., W. Liss, C. Warren, and M. Hurley. 1986. A hierarchical framework for stream habitat classification: viewing streams in a watershed context. Environmental Management 10:199-214.

Fryda, D. 2001. A survey of the fishes and habitat of the White River, South Dakota. Master's Thesis, South Dakota State University, Brookings.

Fuller, P., L. Nico, and J. Williams. 1999. Nonindigenous fishes introduced into inland waters of the United States. American Fisheries Society, Bethesda, MD.

Galat, D., C. Berry, W. Gardner, J. Hendrickson, G. Mestl, G. Power, C. Stone, and M. Winston. 2005. Spatiotemporal patterns and changes in Missouri River fishes. American Fisheries Society Symposium 45:249-291.

Gallagher, W. 1990. Dinosaurs: creatures of time. New Jersey State Museum, Bulletin 14, Trenton.

Game, Fish and Parks. 1959. Looking back past 50 years, 1959 annual report. South Dakota Department of Game, Fish and Parks, Pierre.

Gierach, J. 1990. Sex, death and fly fishing. Simon & Schuster Publisher, NY.

Gigliotti, L. 2005. Report to survey participants: fishing in South Dakota—2003, fishing activity, harvest, and angler opinion survey. South Dakota Department of Game, Fish and Parks, Pierre.

Gigliotti, L. 1996. A good fishing spot. South Dakota Conservation Digest. 63(March-April):2-5.

Gigliotti, L. 1999. Environmental and wildlife attitudes of South Dakota residents. South Dakota Conservation Digest. 66(January/February):16-19.

Gigliotti. L. 2003. Fishing in South Dakota—2003: fishing activity, harvest and angler opinion Survey. http://www.sdgfp.info/Publications/AnglerOpinionSurvey03.pdf , accessed 7/07.

Glover, R. 1975. Black Hills watershed inventories, 1969-1974, South Dakota. DJ Project F-15-R-10, Study No. V, Job 14. South Dakota Department of Game, Fish and Parks, Pierre.

Goforth, R. 2006. The cooperative fish and wildlife research unit program: serving the nation since 1935. U.S. Geological Survey Special Publication, Reston, VA. www.coopunits. org, accessed 7/07.

Gordon, S. 1955. How to fish from top to bottom. Stackpole Books, Mechanicsburg, PA.

Gordon, T. 1965. American trout fishing. Alfred A. Knopf Publisher, NY.

Gottschalk, J. 1967. Scientific management of reservoir fisheries. Pages 1-6 *in* Reservoir Fishery Resources Symposium, Southern Division, American Fisheries Society, Bethesda, MD.

Gourneau, J., and R. Hanten. 1987. South Dakota baitfish harvest summary 1986. Report GFP 87-8, South Dakota Department of Game, Fish and Parks, Pierre.

Graeb, B., J. Shepherd, D. Willis, and J. Sorensen. 2005. Delayed mortality of tournament-caught walleye. North American Journal of Fisheries Management 25:251-255.

Graham, L., E. Hamilton, T. Russell, and C. Hicks. 1986. The culture of paddlefish—a review of methods. Pages 78-94 *in* J. Dillard, L. K. Graham, Editors. The paddlefish: status, management and propagation. American Fisheries Society, Special Publication Number 7, Bethesda, MD.

Greenwood, M., and B. Boussu. 1967. Efficient fishing systems—potential for increasing effectiveness of commercial fisheries in reservoir management and utilization. Pages 467-476 *in* Reservoir fishery resources symposium, Southern Division, American Fisheries Society, Bethesda, MD.

Guy, C., and D. Willis. 1991. Relationships between environmental variables and density of largemouth bass in South Dakota ponds. Proceedings of the South Dakota Academy of Science 70:109-117.

Hammer, D. 1989. Constructed wetlands for wastewater treatment—municipal, industrial, and agricultural. Lewis Publishers, Chelsea, MI.

Hampton, D., and C. Berry. 1997. Fishes of the mainstem Cheyenne River in South Dakota. Proceedings of the South Dakota Academy of Science 76:11-25.

Hannon, D. 2000. Largemouth bass. In-Fisherman 25(3):36-42.

Hansen, D. 1981. Angler and other recreational uses of the James River, South Dakota 1975-1979. South Dakota Department of Game, Fish and Parks Final Report 81-6, Pierre.

Harksen, J. C. 1969. The Cenozoic history of southwestern South Dakota. Pages 11-28 *in* J. C. Harksen and J. R. Macdonald. Guidebook to the major Cenozoic deposits of southwestern South Dakota. South Dakota Geological Survey Guidebook 2.

Harland, B. 2003. Survey of the fishes and habitat of western South Dakota streams. Master's Thesis. South Dakota State University, Brookings.

Harland, B., and C. Berry. 2004. Fishes and habitat characteristics of the Keya Paha River, South Dakota-Nebraska. Journal of Freshwater Ecology 19:169-177.

Haug, J., M. Fosha, J. Abbott, C. Hjort, and R. Mandel. 1994. Exploring the Bloom site, 1993-1994. South Dakota Archaeology 18:29-71.

Hayer, C-A., S. Wall, and C. Berry. 2006. Evaluation of aquatic gap analysis fish distribution models, with emphasis on rare fish species in South Dakota. Final report for State Wildlife Grant Number T-9, South Dakota Department of Game, Fish and Parks, Pierre.

Heakin, A., K. Neitzert, and J. Shearer. 2006. Summary of environmental monitoring and assessment program (EMAP) activities in South Dakota, 2000-2004. Scientific Investigations Report 2006-2007, U.S. Geological Survey, Washington, D.C.

Hesse L., G. Mestl, and J. Robinson. 1993. Status of selected fishes in the Missouri River in Nebraska with recommendations for recovery. Pages 327-340 *in* L. W. Hesse and others, Editors. Proceedings of the symposium on restoration planning for the rivers of the Mississippi River ecosystem. National Biological Survey, Washington, D.C.

Hesse, L. 1987. Taming the wild Missouri River: what has it cost? Fisheries 12(2):2-9.

Hesse, L., J. Schmulbach, J. Carr, K. Keenlyne, D. Unkenholz, J. Robinson, and G. Mestl. 1989. Missouri River fishery resources in relation to past, present, and future stresses. Pages 352-371 *in* D. Dodge, Editor, Proceedings of the international large river symposium (LARS), Canadian Special Publication of Fisheries and Aquatic Sciences 106, Department of Fisheries and Oceans, Ottawa.

Higgins, K., E. Dowd-Stukel, J. Goulet, and D. Backlund. 2000. Wild mammals of South Dakota. South Dakota Department of Game, Fish and Parks, Pierre.

Hill, T., W. Duffy, and M. Thompson. 1995. Food habits of channel catfish in Lake Oahe, South Dakota. Journal of Freshwater Ecology 10:319-323.

Hipschman, D. 1959. Department history 1909-1959. Pages 13-73 *in* 1959 annual report of the Department of Game, Fish and Parks, looking back over the past 50 years. South Dakota Department of Game, Fish and Parks, Pierre.

Hoagstrom, C., C-A. Hayer, J. Kral, S. Wall, and C. Berry. 2007. Rare and declining fishes of South Dakota: a river drainage scale perspective. Western North American Naturalist 67:161-184.

Hoagstrom, C. 2006. Fish community assembly in the Missouri River basin. Doctoral dissertation, South Dakota State University, Brookings.

Hoagstrom, C., and C. Berry. 2006. Island biogeography of native fish faunas among Great Plains drainage basins: basin scale features influence composition. Pages 221-264 *in* R. M. Hughes, L. Wang, and P. Seelbach, Editors. Influences of landscapes on stream habitats and biological assemblages. American Fisheries Society, Symposium 48. Bethesda, MD.

Hodgins, R. 1971. Quality water must remain. South Dakota Conservation Digest March/April 1971:1.

Hogan, E. 1995. The geography of South Dakota. Pine Hill Press, Freeman, SD.

Holien, R. 2001. Lakes in peril. South Dakota Magazine 19(4):14.

Hughes, D. 2006. Trout from small streams. Stackpole Books, Mechanicsburg, PA.

Hunoff, B. 2004. Revival of Spearfish Falls. South Dakota Magazine 19(4):14.

Hunter, C. 1991. Better trout habitat: a guide to stream restoration and management. Island Press, Covelo, CA.

Isaak, D., W. Hubert, and C. Berry. 2003. Conservation assessment for lake chub, mountain sucker, and finescale dace in the Black Hills National Forest, South Dakota and Wyoming. U.S. Department of Agriculture, Forest Service, Custer, SD.

IWLA (Izaak Walton League of America). 1999. Spilling swill: a survey of factory farm water pollution in 1999. The Izaak Walton League Midwest Office, St. Paul, MN.

IWLA (Izaak Walton League of America). 2006. Guide to aquatic insects and crustaceans. Stackpole Books. Mechanicsburg, PA.

James, D., and J. Erickson. 2006. Statewide fisheries surveys, 2004. Surveys of public waters. Annual Report No. 06-06. South Dakota Department of Game, Fish and Parks, Pierre.

Johnson, B., J. Lott, W. Nelson-Stastny, and J. Riis. 1997. Annual fish population and sport fish harvest surveys on Lake Oahe, South Dakota 1996. South Dakota Department of Game, Fish and Parks, Annual Report No 97-15.

Johnson, C., B. Millett, T. Gilmanov, R. Voldseth, G. Guntenspergen, and D. Naugle. 2005. Vulnerability of northern prairie wetlands to climate change. BioScience 55:863-872.

Johnson, D. 2002. The role of hypothesis testing in wildlife science. Journal of Wildlife Management 66:272-276.

Johnson, R. 1997. The Vermillion River: managing the watershed to reduce flooding. Clay County Conservation District, Vermillion, SD.

Johnson, R., and K. Higgins. 1997. Wetland resources of eastern South Dakota. South Dakota State University, Brookings.

Johnson, R., K. Higgins, M. Kjellsen, and C. Elliott. 1997. Eastern South Dakota wetlands. South Dakota State University, Brookings.

Jordan, D., and B. Evermann. 1969. American food and game fishes. Dover Publications, Inc., New York, NY.

June, F. 1976. Changes in young-of-the-year fish stocks during and after filling of Lake Oahe, an upper Missouri River storage reservoir. U.S. Fish and Wildlife Service, Technical Paper 87, Washington, D. C.

June, F. 1977. Reproductive patterns of seventeen species of warmwater fish in the upper Missouri River. Environmental Biology of Fish 2:285-296.

Keenlyne, K., and L. Jenkins. 1993. Age at sexual maturity of the pallid sturgeon. Transactions of the American Fisheries Society 122:393-396.

Keleher, C., and F. Rahel. 1996. Thermal limits to salmonid distributions in the Rocky Mountain region and potential loss due to global warming: a geographic information system (GIS) approach. Transactions of the American Fisheries Society 125:1-13.

Kellert, S. 1999. The future of fishing in the United States, assessment of ways to increase sport fishing participation. Responsible Management National Office, Harrisonburg, VA.

Kelsch, S. 1994. Lotic fish-community structure following transition from severe drought to high discharge. Journal of Freshwater Ecology 9:331-341.

Keyser, E. 1993. Continuing the challenge. South Dakota Conservation Digest 60(July/August):15.

Killen, S., C. Suski, M. Morrissey, P. Dyment, M. Furimsky, and B. Tufts. 2003. Physiological responses of walleyes to live-release angling tournaments. North American Journal of Fisheries Management 23:1238-1246.

Kime, W. 1996. The Black Hills journals of Colonel Richard Irving Dodge. University of Oklahoma Press, Norman.

Kinnunen, R. 1996. Walleye fingerling culture in undrainable ponds. Pages 135-145 in R. C. Summerfelt, Editor. Walleye culture manual. NCRAC Culture Series 101. North Central Regional Aquaculture Center Publications Office, Iowa State University, Ames.

Kirby, D. 2001. An assessment of the channel catfish populations in the Big Sioux River, South Dakota. Master's Thesis, South Dakota State University, Brookings.

Kivett, M., and R. Jensen. 1976. Archeological investigations at the Crow Creek site (39-BF-11), Fort Randall Reservoir Area, South Dakota. Nebraska State Historical Society Publications in Anthropology 7: xvi+220 pages.

Knudsen, E., and D. MacDonald. 2000. Sustainable fisheries: are we up to the challenge. Fisheries 25(12):4, 43.

Kornfeld, M., and A. Osborn. 2003. Islands on the plains: ecological, social, and ritual use of landscapes. The University of Utah Press, Salt Lake City.

Kornfeld, M. 2003. Pull of the hills: technological structures around biogeographical islands. Pages 111-141 *in* M. Kornfeld and A. J. Osborn, Editors. Islands on the plains: ecological, social, and ritual use of landscapes. The University of Utah Press, Salt Lake City.

Kubeny, S. 1992. Population characteristics and habitat selection of channel catfish (*Ictalurus punctatus*) in the lower James River, South Dakota. Master's Thesis. South Dakota State University, Brookings.

Lange, C. 1998. Emergency on the James River. South Dakota Magazine 65(March/April):18-24

Larson, A. 2001. Optimal macroinvertebrate metrics for the assessment of a northern prairie stream. Master's Thesis, South Dakota State University, Brookings.

Larson, G. 1993. Aquatic and wetland vascular plants of the northern Great Plains. General Technical Report RM-238, Rocky Mountain Forest and Range Experiment Station, Fort Collins, CO.

Leap, R. 1974. Glacial geology and hydrology of Day County, SD. Doctoral dissertation, Pennsylvania State University, State College, PA.

Lee, D., C. Gilbert, C. Hocutt, R. Jenkins, D. McAllister, and J. Stauffer, Jr. 1980. Atlas of North American freshwater fishes. North Carolina State Museum of Natural History, Raleigh.

Leopold, A. 1918. Mixing trout in western waters. Transactions of the American Fisheries Society 37(3):101-103.

Leopold, A. 1948. Game management. Charles Scribner's Sons, New York, NY.

Leopold, A. 1949. A Sand County almanac. Oxford University Press.

Leopold, A. 1968. A Sand County almanac, and sketches here and there. Oxford University Press, New York, NY.

Leopold, L. 1994. A view of the river. Harvard University Press, Cambridge, MA.

Loomis, T., C. Berry, and J. Erickson. 1999. The fishes of the upper Moreau River basin. The Prairie Naturalist 31:193-214.

Loope, D., and J. Swinehart. 2000. Thinking like a dune field: geologic history in the Nebraska Sand Hills. Great Plains Research 10:5-35.

Loope, D., J. Swinehart, and J. Mason. 1995. Dune-dammed paleovalleys of the Nebraska Sand Hills: intrinsic versus climatic controls on the accumulation of lake and marsh sediments. Geological Society of America Bulletin 107:396-406.

Lott, J., R. Hanten, and K. Potter. 2004. Annual fish population and angler use, harvest and preference surveys on Lake Sharpe, South Dakota, 2003. South Dakota Department of Game, Fish and Parks, Report No. 04-15, Pierre.

Lovell, R., and L. Sackey. 1973. Absorption by channel catfish of earthy-musty flavor compounds synthesized by cultures of blue-green algae. Transactions of the American Fisheries Society 102:774-778.

Lucchesi, D., T. St. Sauver, B. Johnson, K. Hoffman, and D. Willis. 2004. Region III small impoundments strategic plan, 2004-2009. Special Report 04-13, South Dakota Department of Game, Fish and Parks, Pierre, SD.

Lundberg, J. 1975. The fossil catfishes of North America. University of Michigan Papers on Paleontology 11, iv+51 pages.

Mabee, P. 2006. Integrating evolution and development: the need for bioinformatics in evo-devo. BioScience 56:301-310.

Martin, J. 1987. Paleoenvironment of the Lange/Ferguson Clovis kill site in the Badlands of South Dakota. Pages 314-332 *in* R. Graham, H. Semken, Jr., and M. Graham, Editors. Late Quaternary mammalian biogeography and environments of the Great Plains and prairies. Illinois State Museum, Springfield.

Martin, J., B. Shumacher, D. Parris, and B. Grandstaff. 1998. Fossil vertebrates of the Niobrara Formation in South Dakota. Dakoterra 5:39-54.

Martin, J., R. Alex, L. Alex, J. Abbott, R. Benton, and L. Miller. 1993. The Beaver Creek Shelter (39Cu779): a Holocene succession in the Black Hills of South Dakota. Plains Anthropologist 38(145):17-36.

Martin, R. 1973. The Java local fauna, Pleistocene of South Dakota: a preliminary report. Bulletin of the New Jersey Academy of Science 18:48-56.

Mason, J., J. Swinehart, R. Goble, and D. Loope. 2004. Late-Holocene dune activity linked to hydrological drought, Nebraska Sand Hills, USA. Holocene 14:209-217.

Mather, M., D. Parrish, R. Stein, and R. Muth. 1995. Management issues and their relative priority within State Fisheries Agencies. Fisheries 20(10):14-21.

Mayden, R. 1987. Faunal exchange between the Niobrara and White river systems of the North American Great Plains. The Prairie Naturalist 19:173-176.

McClelland, M. 1989. Crankbaits: a guide to casting and trolling depths of 200 popular lures. Fishing Enterprises Press, Pierre, SD.

McClelland, M. 1991. Walleye in shallow water. Fishing Enterprises Press, Pierre, SD.

McClelland, M. 1996. Walleye trouble shooting. Fishing Enterprises Press, Pierre, SD.

McClelland, M. and L. McClelland (Illustrator). 1996. Bank fishing secrets. Fishing Enterprises Press, Pierre, SD.

McCoy, R., and D. Hales. 1974. A survey of eight streams in eastern South Dakota: physical and chemical characteristics, vascular plants, insects and fishes. Proceedings of the South Dakota Academy of Science 53:202-219.

McMahon, T., A. Zale, and D. Orth. 1996. Aquatic habitat measurements. Chapter 4. Pages 83-120 in B. Murphy and D. Willis. Fisheries techniques, 2nd edition. American Fisheries Society, Bethesda, MD.

MDNR (Minnesota Department of Natural Resources). 2005a. Lake information report for Big Stone Lake. Available: http://www.dnr.state.mn.us/lakefind/showreport.html?downum=06015200 (accessed 7/07).

MDNR (Minnesota Department of Natural Resources). 2005b. Lake information report for Lake Traverse. Available: http://www.dnr.state.mn.us/lakefind/showreport.html?downum=78002500 (accessed 7/07).

Meek, S. 1895. Notes on the fishes of western Iowa and eastern Nebraska. Bulletin of the U. S. Fish Commission for 1984, 14:133-138.

Meyer, E. 1975. Coe I. Crawford and the persuasion of the progressive movement in South Dakota. Doctoral dissertation, University of Minnesota, Minneapolis.

Milewski, C., and D. Willis. 1989. Reproduction, recruitment, and survival of brown and rainbow trout in a prairie coteau stream. The Prairie Naturalist 21:147-156.

Milewski, C. 2001. Local and systemic controls on fish and fish habitat in South Dakota rivers and streams: implications for management. Doctoral dissertation, South Dakota State University, Brookings.

Milewski, C., C. Berry, and J. Gilbertson. 1997. Watershed management workshop for the James, Vermillion, and Big Sioux rivers. Agriculture Experiment Station Bulletin 720, South Dakota State University, Brookings.

Milewski, C., C. Berry, and D. Dieterman. 2001. Use of the index of biological integrity in eastern South Dakota rivers. The Prairie Naturalist 33:135-151.

Miller, J. 2007. Scientific Literacy - - How do Americans stack up? Report presented at the Annual Meeting of the American Association for the Advancement of Science, (www.aaas.org, accessed 7/07).

Monroe, D. 1967. Commercial fishing industry survey South Dakota. South Dakota Department of Game, Fish and Parks, Report No. 67-5, Pierre.

Morey, N., and C. Berry. 2004. New distribution record of the northern redbelly dace in the northern Great Plains. The Prairie Naturalist 36(4):257-260.

Moring, J. 1996. Fish discoveries by Lewis and Clark and Red River Expeditions. Fisheries 21(7):6-12.

Moum, K. 1993. Wildlife law enforcement then and now. South Dakota Conservation Digest 60(July/August):4-9.

Moyle, P., H. Li, and B. Barton. 1986. The frankenstein effect: impact of introduced fishes on native fishes in North America. Pages 415-426 in R. Stroud, Editor, Fish culture in fisheries management. American Fisheries Society, Bethesda, MD.

Muhs, D., and V. Holliday. 1995. Evidence of active dune sand on the Great Plains in the 19th century from accounts of early explorers. Quaternary Research 43:198-208.

Muth, R., and J. Schmulbach. 1984. Downstream transport of fish larvae in a shallow prairie river. Transactions of the American Fisheries Society 113:224-230.

NAS (National Academy of Science). 1995. Science and the endangered species act. National Academy Press, Washington, D. C.

Nelson-Stastny, W. 2004. Battle of the Mighty Mo. South Dakota Conservation Digest 71(September/October):2-7.

Neumann, R., and D. Willis. 1994. Guide to the common fishes of South Dakota. EC899, South Dakota Cooperative Extension Service, SDSU, Brookings.

Newbrey, M., and A. Ashworth. 2004. A fossil record of colonization and response of lacustrine fish populations to climate change. Canadian Journal of Fisheries and Aquatic Sciences 61:1807-1816.

Newman, R., C. Berry, and W. Duffy. 1999. A biological assessment of four northern Black Hills streams. Proceedings of the South Dakota Academy of Science 78:185-197.

NFWFHF (National Fresh Water Fishing Hall of Fame). 2006. Official world and USA state fresh water angling records. National Freshwater Fishing Hall of Fame, Hayward, WI.

Nickum, J., and J. Sinning. 1971. Fishes of the Big Sioux River: an annotated list. Proceedings of the South Dakota Academy of Science 50:143-154.

Nielsen, L. 1993. History of inland fisheries management in North America. Pages 3-32 in C. Kohler and W. Hubert, Editors. Inland Fisheries Management in North America. American Fisheries Society, Bethesda, MD.

Norris, S. 2004. Only 30: a portrait of the Endangered Species Act as a young law. BioScience 54:288-294.

NRC (National Research Council). 2002. The Missouri River Ecosystem: Exploring the Prospects for Recovery. National Academy Press, Washington, D. C. 188 p. (Read the free electronic version at http://books.nap.edu, accessed 7/07).

Ossian, C. 1973. Fishes of a Pleistocene lake in South Dakota. Michigan State University Publications of the Museum; Paleontological Series 1(3):105-126.

Osthoff, R. 2006. Active nymphing: aggressive strategies for casting, rigging, and moving the nymph. Stackpole Books, Mechanicsburg, PA.

Page, L., and B. Burr. 1991. A field guide to freshwater fishes, North America north of Mexico. Houghton-Mifflin Company, Boston, MA.

Palmer, T. 1996. America by rivers. Island Press, Washington, D.C.

Parris, D., B. Grandstaff, and G. Bell, Jr. 2001. Reassessment of the affinities of the extinct genus Cylindracanthus (Osteichthyes). Proceedings of the South Dakota Academy of Science 80:161-172.

Payer, R., and C. Scalet. 1978. Population and production estimates of fathead minnows in a South Dakota prairie wetland. Progressive Fish Culturist 40:63-66.

Persons, S., and S. Hirsch. 1994. Hooking mortality of lake trout angled through the ice by jigging and set-lining. North American Journal of Fisheries Management 14:664-668.

Peterson, E. 1942. Warden activities during 1941. South Dakota Conservation Digest 9(February):7.

Péwé, T. 1983. The periglacial environment in North America during Wisconsin Time. Pages 157-189 in S. C. Porter, Editor. The late Pleistocene. Late-Quaternary environments of the United States. University of Minnesota Press, Minneapolis.

Pflieger, W. 1997. The fishes of Missouri. Missouri Dept. Conservation, Jefferson City, MO.

Pimentel, D., L. Lach, R. Zuniga, and D. Morrison. 2000. Environmental and economic costs of nonindigenous species in the United States. BioScience 50:53–65.

Pimentel, D., Editor. 2002. Biological invasions. Economic and environmental costs of alien plant, animal and microbe species. CRC Press. Boca Raton, FL.

Pinchot, G. 1947. Breaking new ground. Harcourt, Brace and Company, Inc. New York.

Poff, L., M. Brinson, and J. Day. 2002. Aquatic ecosystems and global climate change. Pew Center on Global Climate Change, Arlington, VA. www.pewclimate.org (accessed 7/07).

Poff, L., and J. Ward. 1989. Implications of stream flow variability and predictability for lotic community structure: a regional analysis of stream flow patterns. Canadian Journal of Fisheries and Aquatic Sciences 46:1805-1818.

Pope, K., and D. Willis. 1994. Changes in the brown trout population in Gary Creek, South Dakota, 1988-1993. Proceedings of the South Dakota Academy of Science 73:51-58.

Quaife, M., Editor. 1916. The journals of Captain Merewether Lewis and Sergeant John Ordway. The State Historical Society of Wisconsin. (1965 Edition).

Qualset, H. 1945. Lou's getting ready to go fishin. South Dakota Conservation Digest 12(6):6, 16.

Quinn, S. 2000. A future for fishery science. In Fisherman 25(3):98-100.

Ratti, J., and E. Garton. 1994. Research and experimental design. Pages 1-23 *in* T. Bookout, Editor. Research and management techniques for wildlife and habitat. 5th Edition. The Wildlife Society, Bethesda, MD.

Reebs, S. 2001. Fish behavior in the aquarium and in the wild. Comstock Publishing Associates, Cornell University Press, Ithaca, NY.

Reed, K., and B. Czeck. 2005. Causes of fish endangerment in the United States, or the Structure of the American Economy. Fisheries 30(7):36-38.

Reisner, M. 1993. Cadillac desert: the American west and its disappearing water. Penguin Books, NY.

Repetski, J. 1978. A fish from the Upper Cambrian of North America. Science 200:529-531.

Reynolds, B. 2004. Mastering pike on the fly: Strategies and techniques. Johnson Books, Boulder, CO.

Rieger, B. 2004. Demographics of western South Dakota wetlands and basins. Master's Thesis, South Dakota State University, Brookings.

Rieger, B., K. Higgins, J. Jenks, and M. Kjellsen. 2006. Demographics of western South Dakota wetlands and basins. Publication SDSU AES B748, Department of Wildlife and Fisheries Science, South Dakota State University, Brookings.

Roell, M., G. Schuler, and C. Scalet. 1986. Cage-rearing rainbow trout in dugout ponds in eastern South Dakota. The Progressive Fish-Culturist 48:273-278.

Rose, J. 2002. The neurobehavioral nature of fishes and the question of awareness and pain. Reviews in Fisheries Science 10:1-38.

Rose, J. 1999-2000. Do fish feel pain? In-Fisherman Magazine. December-January: 38-42.

Rounds, J. 2006. Basic fly fishing: all the skills and gear you need to get started. Stackpole Books, Mechanicsburg, PA.

Royce, W. 1972. Introduction to the fishery sciences. Academic Press, NY.

Ruelle, R., R. Koth, and C. Stone. 1993. Contaminants, fish, and hydrology of the Missouri River and western tributaries, South Dakota. U.S. Fish and Wildlife Service. Pages 449-490 *in* Hesse, L., C. Stalnaker, N. Benson, and J. Zuboy, Editors. Proceedings of the symposium on restoration planning for the rivers of the Mississippi River ecosystem. Biological Report 19, National Biological Survey, Washington, D.C.

Sando, S. 1991. Estimation and characterization of the natural stream flow of the White River near the Nebraska-South Dakota state line. Water-Resources Investigations Report 91-4096, U.S. Geological Survey, Huron, SD.

Sando, S. 1996. Thickness and volume of sediment in Pelican Lake, South Dakota, June 1994. U.S. Geological Survey, Water-Resources Investigations Report 96-4247, Rapid City, SD

Schaap, B., and S. Sando. 2002. Sediment accumulation and distribution in Lake Kampeska, Watertown, South Dakota. U.S. Geological Survey, Water-Resources Investigations Report 02-4171, Rapid City, SD.

Schmulbach, J., and P. Braaten. 1993. The Vermillion River: neither red nor dead. Pages 57-69 *in* L. Hesse, C. Stalnaker, N. Benson, and J. Zuboy. Proceedings of the symposium on restoration planning for the rivers of the Mississippi River ecosystem. Biological Report 19, U.S. Department of the Interior, National Biological Survey, Washington, D.C.

Schumacher, D. 1995. Aquatic macroinvertebrate production in predominant habitats of a warm-water stream: the James River, South Dakota. Master's Thesis, South Dakota State University, Brookings.

Schwalb, A., and W. Dean. 1998. Stable isotopes and sediments from Pickerel Lake, South Dakota, USA: a 12ky record of environmental changes. Journal of Paleolimnology 20:15-30.

SDDENR (South Dakota Department of Environment and Natural Resources). 1998. The 1998 South Dakota report to congress, 305(b) water quality assessment report, South Dakota Department of Environment and Natural Resources, Pierre, SD.

SDDGFP (South Dakota Department of Game, Fish and Parks). 1959. Looking back past 50 years. 1959 Annual Report. South Dakota Department of Game, Fish and Parks, Pierre.

SDDGFP (South Dakota Department of Game, Fish, and Parks). 2005a. Fishing in South Dakota. Available: http://www.sdgfp.info/Wildlife/fishing/Index.htm, accessed 7/07.

SDDGFP (South Dakota Department of Game, Fish, and Parks). 2005b. Northeast South Dakota 2004 fish sampling results. http://www.sdgfp.info/Wildlife/fishing/Index.htm, accessed 7/07.

SDDGFP (South Dakota Department of Game, Fish and Parks). 2005c. Southeast lakes survey summaries and net catches. http://www.sdgfp.info/Wildlife/fishing/Index.htm, accessed 7/07.

SDDGFP (South Dakota Department of Game, Fish and Parks). 2005d. Western South Dakota lake surveys and catches. http://www.sdgfp.info/Wildlife/fishing/Index.htm, accessed 7/07.

SDDGFP (South Dakota Department of Game, Fish and Parks). 2005e. Fishing handbook. Available: http://www.sdgfp.info/Publications/FishingHandbook.pdf, accessed 7/07.

SDGFP (South Dakota Department of Game, Fish and Parks). 1959. Looking back past 50 years. 1959 Annual report of the Department of Game, Fish and Parks, Pierre.

SDGFP (South Dakota Department of Game, Fish and Parks). 1994. Systematic approach to management: fisheries. South Dakota Department of Game, Fish and Parks, Wildlife Division, Pierre.

SDGFP (South Dakota Department of Game, Fish and Parks). 2002. Law enforcement program: program narratives and compiled program reports for calendar year 2002. South Dakota Department of Game, Fish and Parks, Pierre.

SDGFP (South Dakota Department of Game, Fish and Parks). 2006. South Dakota fishing handbook. South Dakota Department of Game, Fish and Parks, Pierre.

SDHS (South Dakota Historical Society). 1920. Nicollet and Fremont. South Dakota Historical Collections 10:69-129.

SFBPC (Sport Fishing and Boating Partnership Council). 2004. National Fish Habitat Initiative. Http://www.fws.gov/sfbpc, accessed 7/07.

Shearer, J. 2001. Temporal changes in fish communities and modification of the index of biotic integrity for the James River of the Dakotas. Master's Thesis, South Dakota State University, Brookings.

Shearer, J. 2003. Topeka shiner (*Notropis topeka*) management plan for the state of South Dakota. South Dakota Department of Game, Fish and Parks, Wildlife Division Report No 2003-10, Pierre.

Shearer, J., and C. Berry. 2002. Index of biological integrity utility for the fishery of the James River of the Dakotas. Journal of Freshwater Ecology 17:575-588.

Shearer, J., and C. Berry. 2003. Fish community persistence in eastern North and South Dakota rivers. Great Plains Research 13:139-159.

Shearer, J., and J. Erickson. 2005. Finescale dace: the reintroduction of a glacial relict. South Dakota Conservation Digest 72(May/June):10-11.

Sherrod, N. 1963. Late Pleistocene fish from lake sediments in Sheridan County, North Dakota. North Dakota Academy of Science 17:32-36.

Simon, J. 1946. Wyoming fishes. Bulletin No. 4, Wyoming Game and Fish Department, Cheyenne.

Simpson, G. 2007. Longitudinal water temperature changes in selected Black Hills streams during a period of drought. Completion Report No 07-01, South Dakota Department of Game, Fish and Parks, Pierre.

Smith, M. 1987. D. C. Booth, Fish Culture Pioneer. Fish and Wildlife News, June-July 1987: 16-20. See http://dcbooth.fws.gov/ for more information on the D. C. Booth Historic Fish Hatchery.

Smith, V., C. Kopplin, S. Wall, J. Jenks, and C. Berry. 2000. The South Dakota gap analysis project. South Dakota Conservation Digest 67(January/February):4-6.

Sneedon, L., V. Braithwaite, and M. Gentle. 2003. Do fish have nociceptors: evidence for the evolution of a vertebrate sensory system. Proceedings of the Royal Society 1270(1520): 1115-1121.

Soupir, C. 2003. South Dakota baitfish harvest summary, January 1—December 31, 2002. South Dakota Department of Game, Fish and Parks, Report 03-14, Pierre.

Soupir, C. 2004. South Dakota baitfish harvest, 2003 summary. South Dakota Department of Game, Fish and Parks, Report 04-20, Pierre.

Soupir, C., B. Blackwell, and M. Brown. 1997. Relative precision among calcified structures for white bass age and growth assessment. Journal of Freshwater Ecology 12:531-537.

Stange, D. 2000. Taking stock 2000. In Fisherman 25(3):12-14.

State Lakes Preservation Committee. 1977. A plan for the classification, preservation, restoration of lakes in northeastern South Dakota. State of South Dakota and the Old West Regional Commission Report, Pierre.

Stewart, R., and C. Thilenius. 1964. Black Hills lakes and streams. South Dakota lake and stream classification report for DJ Project F-1-R-13, Job Numbers 14 and 15, Department of Game, Fish and Parks, Pierre.

Stroud, R. 1986. Fish culture in fisheries management. American Fisheries Society, Bethesda, MD.

Stueven, E., and W. Stewart. 1996. 1995 South Dakota lakes assessment final report, South Dakota Department of Environment and Natural Resources, Pierre.

Stukel, S. 2003. Assessing the sustainability of fish communities in glacial lakes: habitat inventories and relationships between lake attributes and fish communities. Master's Thesis, South Dakota State University, Brookings.

Sundestrom, J. 1994. Pioneers and Custer State Park. J. Sundestrom, Custer, SD. Self published.

Sundstrom, L. 1996. A moveable feast: 10,000 years of food acquisition in the Black Hills. South Dakota Archaeology 19-20:16-48.

Thompson, J., O. Reichman, P. Morin, and others. 2001. Frontiers of ecology. BioScience 51:15-24.

Thwaites, R., Editor. 1904. Original journals of the Lewis and Clark Expedition. Volume 1. Arno Press (1969 Edition), New York, NY.

Tilberg, D. 1993. A full range of emotions. South Dakota Conservation Digest 60(July/August):12

Tomelleri, J., and M. Eberle. 1990. Fishes of the central United States. University Press of Kansas, Lawrence.

Towne, W. 1946. Depleted staff has delayed action in South Dakota. Water Works Engineering 99(18):1048-1049.

Towne, W. 1957. Raw sewage stabilization ponds in the Dakotas. Sewage and Industrial Wastes 29(4):377-395.

Tunison, A., S. Mullin, and O. Meehan. 1949. Survey of fish culture in the United States. The Progressive Fish-Culturist 11(1):31-69.

Underhill, J. 1957. The distribution of Minnesota minnows and darters in relation to Pleistocene glaciation. Occasional Papers: Number 7, Minnesota Museum of Natural History, University of Minnesota Press, Minneapolis.

Underhill, J. 1959. Fishes of the Vermillion River, South Dakota. Proceedings of the South Dakota Academy of Science 38:96-102.

Unkenholz, E., M. Brown, and K. Pope. 1997. Oxytetracycline marking efficacy for yellow perch fingerlings and temporal assays of tissue residues. The Progressive Fish-Culturist 59:280-284.

USGS (U.S. Geological Survey). 1987. Historical summary—South Dakota district. Pages 2-4 *in* Water-Resources Activities of the U.S. Geological Survey in South Dakota—fiscal years 1986-87. USGS Open File Report 87-383.

Valero-Garcés, B., K. Laird, S. Fritz, K. Kelts, E. Ito, and E. Grimm. 1997. Holocene climate in the northern Great Plains inferred from sediment stratigraphy, stable isotopes, carbonate geochemistry, diatoms, and pollen at Moon Lake, North Dakota. Quaternary Research 48:359-369.

VanZee, B., N. Billington, and D. Willis. 1996. Morphological and electrophoretic examination of *Stizostedion* samples from Lewis and Clark Lake, South Dakota. Journal of Freshwater Ecology 11(3):339-344.

Verrill, D., and C. Berry. 1995. Effectiveness of an electrical barrier and lake drawdown for reducing common carp and bigmouth buffalo abundances. North American Journal of Fisheries Management 15:137-141.

Wagner, F. 1999. Whatever happened to the National Biological Survey. BioScience 49:219-221.

Walburg, C. 1971. Loss of young fish in reservoir discharge and year-class survival, Lewis and Clark Lake, Missouri River. Pages 441-448 *in* G. Hall, Editor. Reservoir fisheries and limnology, Special Publication No. 8, American Fisheries Society, Bethesda, MD.

Walburg, C. 1976. Changes in the fish population of Lewis and Clark Lake, 1956—74, and their relation to water management and the environment. U. S. Fish and Wildlife Service, Research Report 79, Washington, D. C.

Walburg, C. 1977. Lake Francis Case, A Missouri River reservoir: changes in the fish population in 1954-75, and suggestions for management. U.S. Fish and Wildlife Service, Technical Paper 95, Washington, D. C.

Walden, H. 1947. Angler's choice: an anthology of American trout fishing. The MacMillan Company Publisher, NY.

Wall, S., and C. Berry. 2004. Road culverts across streams with the endangered Topeka shiner, *Notropis topeka*, in the James, Vermillion, and Big Sioux river basins. Proceedings of the South Dakota Academy of Science 83:125-135.

Wall, S., and C. Berry. 2004. Threatened fishes of the world: *Notropis topeka* Gilbert, 1884 (Cyprinidae). Environmental Biology of Fishes 70:246.

Wall, S., C. Berry, Jr., C. Blausey, J. Jenks, and C. Kopplin. 2004. Fish-habitat modeling for gap analysis to conserve the endangered Topeka shiner (*Notropis topeka*). Canadian Journal of Fisheries and Aquatic Sciences 61:954-973.

Walsh, R. 1992. Differences in fish abundance among habitat types in a warm-water stream; the James River, South Dakota. Master's Thesis. South Dakota State University, Brookings.

Ward, J. 1989. The four-dimensional nature of lotic ecosystems. Journal of the North American Benthological Society 8:2-8.

Ward, N., and C. Berry. 1995. Evaluation of skin pigmentation for identifying adult saugers and walleye-sauger F1 hybrids collected from Lake Sakakawea, North Dakota. The Progressive Fish-Culturist 57:302-304.

Warnick, D. 1970. Commercial fish industry survey (1968) South Dakota. South Dakota Department of Game, Fish and Parks, Report 70-5, Pierre.

Warnick, D. 1971. Commercial fish industry survey (1969) South Dakota. South Dakota Department of Game, Fish and Parks, Report 71-10, Pierre.

Warnick, D. 1971. Commercial fish industry survey (1970) South Dakota. South Dakota Department of Game, Fish and Parks, Report 71-13, Pierre.

Warnick, D. 1973. Commercial fish industry survey (1971) South Dakota. South Dakota Department of Game, Fish and Parks, Report 73-22, Pierre.

Warnick, D. 1977. Commercial fishing or rough fish control in South Dakota, some views and apparent values. South Dakota Department of Game, Fish and Parks, Bulletin 7, Pierre.

Waters, W., and D. Erman. 1990. Research methods: concept and design. Pages 1-34 *in* C. Schreck and P. Moyle, Editors, Methods for fish biology. American Fisheries Society, Bethesda, MD.

Watts, W., and H. Wright, Jr. 1966. Late-Wisconsin pollen and seed analysis from the Nebraska Sandhills. Ecology 47:203-210.

Wayne, W., J. Aber, S. Agard, and others. 1991. Quarternary geology of the northern Great Plains. Pages 441-476 *in* R. Morrison, Editor. Quaternary nonglacial geology; conterminous U.S.: the geology of North America, K-2. Geological Society of America, Inc., Boulder, CO.

Welker, T., and D. Scarnecchia. 2004. Habitat use and population structure of four native minnows (family Cyprinidae) in the upper Missouri and lower Yellowstone rivers, North Dakota. Ecology of Freshwater Fish 13:8-22.

Welsh, R. 1986. South Dakota baitfish harvest, summary 1985. South Dakota Department of Game, Fish and Parks, Report No. 86-12, Pierre.

Whalen, P. 1994. Source aquifers for Cascade Springs, Hot Springs, and Beaver Creek Springs in the southern Black Hills of South Dakota. Master's Thesis, South Dakota School of Mines, Rapid City.

Whitehead, J., D. Lipton, F. Lupi, and R. Southwick. 2005. Economic growth and environmental protection: a clarification about neoclassical economics. Fisheries 30(4):32-34.

Wickstrom, G. 2004. Annual fish population surveys of Lewis and Clark Lake, 2003. South Dakota Department of Game, Fish and Parks, Report 04-09, Pierre.

Wickstrom, G., and J. Schuckman. 2006. 2005 Angler use and harvest survey of the Missouri River in South Dakota and Nebraska from Fort Randall Dam to Gavins Point Dam. Game, Fish and Parks, Annual Report 06-16, Pierre.

Williams, J., C. Wood, and M. Dombeck. 1997. Watershed restoration principles and practices. American Fisheries Society, Bethesda, MD.

Williamson, J. 2000. Streamflow and water-quality data for Bear Butte Creek downstream of Sturgis, South Dakota, 1998-2000. Open-File Report 00-430. USGS Information Services, Box 25286, Denver, CO.

Willis, D., M. Beem, and R. Hanten. 1990. Managing South Dakota ponds for fish and wildlife. South Dakota Department of Game, Fish and Parks, Pierre.

Wilson, R., and T. Wilson. 1999. Bluegill fly fishing & flies. Frank Amato Publications, Portland, OR.

Wilson, S. 2002. Relation of habitat to fish community characteristics in small South Dakota impoundments. Master's Thesis, South Dakota State University, Brookings.

Wishart, D., Editor. 2004. Water. Pages 845-867 *in* The encyclopedia of the Great Plains. University of Nebraska, Lincoln.

Wood, E. 1953. A century of American fish culture, 1853-1953. The Progressive Fish-Culturist 15(4):147-162.

Xia, J., B. Haskell, D. Engstrom, and E. Ito. 1997. Holocene climate reconstructions from tandem trace-element and stable-isotope composition of ostracodes from Coldwater Lake, North Dakota, USA. Journal of Paleolimnology 17:85-100.

Young, R., and J. Hayes. 2004. Angling pressure and trout catchability: behavioral observations of brown trout in two New Zealand backcountry rivers. North American Journal of Fish Management 24:1203-1213.

Zernov, J. 2000. Electronics. The Future Never Rests. In Fisherman 25(3):122.

Zimmerman, L. 1985. Peoples of prehistoric South Dakota. University of Nebraska Press, Lincoln.

Index